1984

Fundamental Structures of Computer Science

William A. Wulf
Mary Shaw
Paul N. Hilfinger
Computer Science Department
Carnegie-Mellon University

Lawrence Flon
Computer Science Department
University of Southern California

ADDISON-WESLEY PUBLISHING COMPANY
Reading, Massachusetts • Menlo Park, California
London • Amsterdam • Don Mills, Ontario • Sydney

Reproduced from camera-ready copy provided by the authors. The copy was prepared at Carnegie-Mellon University on a photocomposer operated by the SCRIBE
text editing system.

Library of Congress Cataloging in Publication Data
Main entry under title:

Fundamental structures of computer science.

  Includes bibliographical references and index.
  1. Electronic digital computers—Programming.
2. Data structures (Computer science) I. Wulf,
William Allan.
QA76.7.F86   001.6'42   79-12374
ISBN 0-201-08725-1

ISBN 0-201-08725-1
   DEFGHIJK-DO-8987654

To the CMU Computer Science community:
this couldn't have happened elsewhere.

Preface

This textbook is designed to support an intermediate-level course in computer science. It is intended for students who have at least one or two semesters of programming and problem-solving experience and who wish to prepare for more advanced topics in computing. Ideally, the students will also have a semester's background in discrete mathematics.

The content and organization of the textbook was motivated by a single premise—that *programming is an engineering discipline*. Like other engineering desciplines, programming is concerned with building things, particularly *real* things of enormous practical importance. And, like other engineering, *good* programming is rooted in the application of science. Understanding that science is essential to being a good programmer.

This textbook is also an introduction to the science that underlies good programming. Many of the concepts involved are mathematical. Software concepts such as program organization, data structures, perform-ance analysis, and so on, have traditionally been taught only as programming techniques. In this textbook we support these programming techniques by relating programming ideas to the underlying mathematical concepts of automata, formal languages, data types, and so on. Through examples we illustrate the relation and utility of the mathematical concepts to practical contexts.

Much of the material presented in this textbook has traditionally been taught (in greater depth) in advanced undergraduate and graduate courses. We believe, however, that it is both necessary and appropriate to introduce this material at an early stage. Just as introductory calculus and physics are prerequisite to most engineering courses, the material presented here forms an appropriate background for courses in software engineering and advanced computer science. For the organization of this material, see the matrix on page 8 and the discussion immediately preceding it.

Using the Book

We originally designed this textbook for a second course in computer science, and since 1974 have used various generations of it for a one-semester sophomore course. One semester is not enough to cover all the material now contained in the textbook, but in a semester we can generally cover most of Parts One and Two and the theoretical material in Part Three. We also cover several of the example chapters in Part Four. Homework usually involves substantial programming exercises. Although we are aware that this is a large amount of material to cover in one semester, we have *not* observed that our students have any great difficulty with the *level* of the material.

Although we strongly believe that the material in the textbook is best presented early, that is in a second course in computer science, there are other ways that the book can be used. For example,

- As the basis of a two-semester course: This textbook contains ample material for a full-year course. In such a course, however, we would not only cover this material, but also emphasize advanced programming techniques and more advanced data structure representations.

- Merged with a software engineering course: Most of the material in this text is among the formal prerequisites for a rigorous treatment of software engineering. Half of a third or fourth year course in software engineering might profitably be spent on the chapters about formal models, specification, verification, and performance analysis.

- Incorporated into a theoretical computer science course: Much of the material in this course is predicated on a knowledge of discrete mathematics and formal logic and therefore constitutes a meaningful application of these formal mathematical ideas.

- As in-house training material for practicing programmers: Most of today's practicing programmers were educated when computing was little more than a collection of tricks. That's no longer the case. In order to compete with the recent graduates, the experienced professional will need to master the material we cover—especially the more formal material.

The language used throughout this book is essentially PASCAL. The material we cover, however, applies to programming in general, rather than to any particular language, and so we have felt free to deviate from PASCAL from time to time in presenting some material. Appendix C summarizes the major deviations. The programs in the example chapters have all been executed with a PASCAL compiler for the DEC PDP-10 originally developed at the University of Hamburg, Germany. We have, however, tried to avoid including in our examples any statements peculiar to that compiler.

Computer Science has made enormous advances in the past 10 to 15 years. It is no longer simply an art or an ad hoc collection of programming tricks. There is now a genuine body of science and mathematics that the programmer of the future must know and use. This book is hardly the last word that will be written on the subject, but it is a good beginning for new students and a sourcebook for the experienced professional.

Acknowledgments

The authors gratefully acknowledge the support and assistance of

- All those students over the past six years who have suffered through drafts of our material,
- Anita Andler, Sharon Carmack, and Dorothy Josephson, who (re)*typed much of the material, adapting to a series of dynamically developing computerized document production systems,
- The graduate students who helped teach sections of the course and who contributed material and comments: Sten Andler, Lee Cooprider, David Notkin, Gary Feldman, Lanny Forgy, Brian Reid, and Elaine Rich.
- Brian Reid, who developed and maintains SCRIBE, the document production system that allowed the manuscript of this textbook to be almost automatically produced,
- Bob Morgan, who gave some of us strong personal support,
- Jon Bentley, whose experiences with using the textbook in an industrial setting gave us new insights into ways to teach the material,
- A host of readers and reviewers, both known to us and anonymous, whose comments were invaluable, notably Terry Baker, Jon Bentley, Ron Brender, Mike Horowitz, David Jefferson, John Kender, David Lamb, David Levine, Tony Ralston, Brian Reid, Peggy Ross, Len Shustek, John Sowa, Bob Sproull, and Judy Zinnikas,

- The National Science Foundation and the Defense Advanced Research Projects Agency, which provided support for the research environment in which we learned much of what we teach here.

Pittsburgh, Pennsylvania   W.A.W.
April 1980   M.S.
                                                            P.N.H.
                                                            L.F.

Contents

ix

x   **Contents**

Contents   xi

Part Two/Fundamental Structures of Data

Introduction

Programming as an Engineering Activity

Programming is an engineering activity. Hence, good programming, like other good engineering, is rooted in the careful application of science to practical problems. In order to understand the rationale for the specific content and organization of the book, however, we first need to examine the ways programming is similar to, and different from, other engineering.

A computer is, of course, just a tool. But it is a "general purpose" tool, for it is not designed to solve just one problem. Instead, a computer is specialized to solve a particular problem by a program—a list of instructions that the computer obeys. Programming is the act of writing a program; it is, in effect, the act of building a specialized tool. It is this emphasis on "building something" that leads to our premise.

There is, however, an important difference between software engineering and other kinds of engineering. Most engineers must deal with physical limitations of the materials they use—weight and strength, for example. Achieving an acceptable design within these limitations is often *the* central problem faced by the engineer. Although there *are* a few limitations of computers such as the amount of memory and processor speed, these limitations do not matter for a large collection of programs. Instead, the limitations encountered in programming are most often related to our own, very human capabilities:

> *The principal limitations on the programs we write are often imposed by our inability to comprehend the design— not by the physical capabilities of computers.*

1

Most students using this book will have written programs of, say, 100 lines and will have a feeling for the effort involved. Now consider a 5000 line program—such a program would be considered moderately sized, but quite small in relation to some that exist. The 5000 line program is only 50 times as large, so if the 100 line program took a week to complete, you might be led to suspect that the larger one will take 50 weeks—that is, about a year.

Unfortunately, it usually takes much longer. The time required to write a program is related to its complexity, and only indirectly related to its size. The problem is that we need to produce, not 50 separate 100-line programs, but rather something closer to 50 interdependent 100-line programs. These interdependencies, if not properly controlled, can result in a virtual explosion of complexity, increasing the time necessary to produce our 5000-line program from many weeks to many years.

It is this nonlinear increase in complexity and, more to the point, our human limitations in dealing with that complexity, that are at the heart of the software engineering problem. Among the most important tools and techniques employed by the software engineer are those that control complexity. It should not be surprising that the material in this book is the scientific foundation of those tools and techniques.

Complexity and Abstraction

Programs are not the only, or even the most complex systems that humans must deal with. The national economy, for example, is far more complex; even the motion of the molecules in a simple physical object is many orders of magnitude more complex than the most complex program. However, no human *completely* understands the national economy or molecular structure in all their fine detail. In order to deal with complex situations we use a powerful technique—abstraction. We choose to ignore the details of complex systems; instead we develop *models* which reflect only certain important, macroscopic behavioral properties. Newton's laws of motion, for example, are a model of physical reality. This model abstracts from the enormous complexity of the motion of an object's component particles and describes gross properties of their aggregate behavior.

For nearly all purposes the precise motions of the molecules in an object are irrelevant. The only relevant information is expressed by summarizing the individual motions in some way. For example, we speak of the velocity of the object (which is actually the average velocity of all its molecules), its temperature (a measure of the kinetic energy of the molecular motion), and so on. When we summarize many details in a single property (velocity or temperature, for instance) we are *abstracting* from the details.

For some purposes these gross abstractions may be inadequate. For example, the structure of a crystal depends on the properties of its constituent molecules. The abstraction used to describe crystalline properties must contain more detail than is used to describe the temperature of an object that happens to be a crystal. If we wish to understand chemical reactions we must consider an even more detailed model: the atomic structure of our materials. Notice something very important. In each of these cases we merely use another *model*, another *abstraction* of reality. Each model contains just enough detail to explicate the phenomena under study. To analyze the motion or temperature of an object, we can totally ignore its molecular structure. To analyze its simple chemical properties, we can ignore, for example, the wave-like behavior of subatomic particles. Only if we were to study nuclear reactions would we need a still more finely grained model, namely quantum mechanics.

It is clear, then, that a deep philosophical assumption underlying modern science is that the complexity of reality can be understood by understanding a collection of models—some that describe macroscopic behavior by ignoring detail, others that successively explain increasingly microscopic behavior. Whether this simplifying assumption is entirely valid can be debated, but our limited intellectual capacity *forces* us to make it; without this assumption, we could not cope with the complexity surrounding us.

Let's return to programming. Here the notions of models and abstraction also play a central role, but the situation is also somewhat different. For the chemist and physicist, reality *is*; they propose models to describe and predict reality. For them, the degree to which a model is successful can be tested by performing experiments and comparing the results to predictions from the model. The programmer, on the other hand, is building an artificial reality. He can also use abstractions to control complexity, but he has the ability, in principle, to make these models exact. That is, he can *make* the program perform exactly as predicted by the model under all circumstances. Thus, in a sense, the notions of modeling and abstraction are even more powerful in programming than in the physical sciences. Still, in practice the abstract descriptions used in program specification are almost always, to some extent, incomplete, just as they are in the physical sciences; for example, we seldom specify the exact running times of our programs. Further, as the physical systems of chemistry and physics illustrate a hierarchy of models, each more detailed than the last, so is a similar hierarchical structure applicable to programming. An entire program is usually too complex to comprehend, so we'll describe it in terms of an abstract model.

That model is defined in terms that, in turn, must be implemented with fairly complex programs. Generally we need to explain each of these subprograms in terms of still more detailed models. Thus, in a large program you should expect to find many levels of abstraction. Only in this way we can hope to avoid an explosion in complexity.

Abstraction and Structure

Having determined that abstractions and models are essential, we now need to consider what constitutes a good abstraction or model. In the physical sciences a principle referred to as "Occam's Razor" has long been accepted as a criterion for choosing between competing models. In effect, Occam's Razor says "Pick the simplest model that adequately describes a phenomenon." The emphasis on simplicity is important; remember, the whole rationale for models is to accommodate our human limitations in dealing with complexity. The criterion for good programming abstractions is also simplicity. You should choose abstractions that are both simple (easy to understand) and yet adequate to describe the desired program behavior.

The classic Bohr atom is an excellent example of a simple model. There are only two kinds of parts: the nucleus and its orbiting electrons. Both parts can be characterized in terms of their weight and charge, and the relation between them is characterized in terms of the simple laws governing orbital motion.

In the programming domain, the "parts" we deal with are data objects and portions of programs. The relations between them are such things as how control flows from one part of a program to another and what kinds of assumptions each portion of a program makes about the variables it uses. As scientists, engineers, and programmers, we strive to make both the individual parts and the relationships between them as simple—as good—as possible.

An Approach to Good Programming

There are no absolute rules for building a good program. There are only the general properties that any good engineering design must have: it must perform its intended function well, and it must be cost-effective. Nevertheless, there is a set of general guidelines that *usually* lead to a well-engineered program. These guidelines are often referred to as (1) structured programming, (2) programming methodology, or (3) software engineering discipline. Much of the material in this textbook is directed at providing a good foundation for them, as is the following brief overview.

The difficulty in writing (and understanding) a large program can be characterized in terms of the conceptual distance between the language of the problem and the programming language. An inventory control problem, for example, speaks of part numbers, reorder points, sales, deliveries, warehouse space, and so on. None of these concepts is present in the programming languages we use to write the eventual program; instead we speak of integers and real variables, arrays and records, subroutines and loops. The programming task is to use the programming language facilities in a way that externally—to the user of the program— mimicks the problem's concepts. Because there is a large distance between the two, this mimicry may take a large amount of code.

As in many other areas, the resolution of a programming difficulty is implicit in a good statement of it. If the difficulty is the conceptual distance between the problem's language and programming languages, then we must reduce that distance. We can't hope to design a single programming language that contains the right concepts for all problems, but we can outline an approach to programming that has the same effect.

1. Analyze the problem, writing precise specifications of what constitutes a solution.

2. Specify abstractions suitable for expressing the solution of the problem, and write an abstract program using them.

3. Analyze the proposed abstract program; specifically,

   - Verify that the abstract program satisfies the problem specification.

   - Verify that the space and time consumption of the abstract program will be acceptable to its hypothetical users.

4. Implement the lower-level abstractions of step 2.

To accomplish step 4, we may invoke all of the steps 1–4 over again. Thus, the approach introduces several hierarchical levels of abstraction— each of which is well structured.

This is, of course, a statement of the ideal, since in practice the steps may not always be carried out in this order and there will inevitably be some iteration between them. However, we try to make the abstract program as clear and simple as possible, constructing it from a small number of simple parts with simple relationships between parts. Since this abstract program is written in the same terms as the problem is stated, the size of the program should be related to the intrinsic complexity of the task and *not* to the quirks of contemporary programming languages and computers. Generally this program will be surprisingly small; and, given the specification of abstract concepts, a formal mathematical proof of the correctness of the abstract program may be feasible. If the abstract program should fail to exhibit good structure, it is generally a result of its including too much, a clear signal to the programmer to go back and review the abstractions used to express the program.

Only after the abstract program has been written do we turn to the question of how to implement the abstract concepts it uses. Eventually each of the abstract concepts will have a direct analog in the target programming language; when this is true of all the abstractions, the program is finished. Often, however, this will not easily be the case, and we will need to find some way to implement the abstract concepts. Note, however, that two important things have happened.

- First, we have divided the original problem into a number of smaller, precisely defined subproblems, and each of these subproblems can be worked on in relative isolation.

- Second, each subproblem is similar to the original one in that the same approach may be used to solve it. In particular, the subprogram may still be conceptually too distant from the target programming language to permit an *understandable* direct implementation. There will, however, be abstract concepts that are natural and specific to the subproblem, and that suffice to express the solution. We can then write another abstract program, specify the abstractions it uses, verify it, implement its abstract concepts, and so on.

Organization and Content of this Book

This textbook is divided into four major parts. Part One is principally concerned with what we call *control structures*; Part Two, with *data structures*; and Part Three, with the interaction of the two. Part Four is a collection of chapter-length examples that illustrate the application of the material in realistic programming contexts.

Within the first three parts we have established parallel structures.

- First, an introduction of mathematical tools and models appropriate to a deeper understanding of the major topics.

- Second, a discussion of the central topics as they appear in programming languages.

- Third, a discussion of *representation*, a concept often foreign to new programmers, but one of the most important concepts in program design and analysis.

∎ Finally, ways to reason about the correctness and efficiency of programs: about what programs do (specification and validation), and about how well they do it (performance analysis).

Following this introduction is a matrix that shows the relationships between the various topics covered. Each part of the book is displayed as a row in the matrix; corresponding chapters as columns. The entries of the matrix are illustrative topics from the chapters. Within the first three parts there is a strong relation between the topics of any one row or any one column. We have chosen to order the material by discussing first all the topics under control, then all topics under data, etc. However, except for our desire to introduce practical issues earlier, we might equally well have discussed mathematical models for control, data, and their interaction before presenting anything about programming languages—that is, we could have placed primary structural emphasis on the topics of the columns rather than the rows.

Although we intend this book to be read in the order it is presented, you should keep in mind the relationships exhibited by the matrix. For example, as the representations of control and data are not *really* independent—they interact in important and, sometimes, subtle ways—you will learn the most from this textbook, by reflecting on earlier issues about representing control when reading the chapter on the representation of data.

Topic Organization Matrix

	Basic mathematical models	Basic programming language concepts	Extended programming concepts	Representation	Reasoning about correctness	Reasoning about efficiency
PART ONE: Control	**1** ■ Finite state models: Finite state machines, Regular expressions. Equivalence of the models ■ Limitations of the models	**2** ■ Flow charts ■ Simple control constructs ■ Equivalence with finite state models	**3** ■ Extended control constructs for Selection, Iteration, Routines.	**4** ■ Representation of basic constructs ■ Representation of extended constructs	**5** ■ Specifications ■ Weakest preconditions	**6** ■ Order arithmetic ■ Experimental cost measures ■ Cost preconditions
PART TWO: Data	**7** ■ Models of memory ■ Data types ■ Specification techniques	**8** ■ Scalar types, including references ■ Structured types: Vectors, Records. ■ Other issues	**9** ■ Nonelementary types: Stacks, queues sets, graphs . . .	**10** ■ Representational techniques ■ Alternative representations of types from Chapter 9	**11** ■ Verification of abstract data types ■ Semantics of pointers	**12** ■ Static space analysis ■ Dynamic space analysis
PART THREE: Control-data interactions	**13** ■ Unbounded state models: Push-down automata, Turing machines, Production grammars. ■ Computability and limitations of the models	**14** ■ Dynamic data types ■ Recursion ■ Recursive programming techniques	**15** ■ Scope ■ Extent ■ Binding, parameter mechanisms	**16** ■ Run-time representation of programs: Activation records, Static vs. stack vs. heap allocation, Displays	**17** ■ Inductive proofs of recursive programs	**18** ■ Solving recurrence relations

PART FOUR: Case studies

(19 A Lexical Analyzer) (20 A Fast Implementation of Sets) (21 Iteration over Sets) (22 Symbolic Differentiation) (23 Symbolic Simplifications of Expressions)

Part One

Fundamental Structures of Control

This, the first major part of this textbook, is concerned with the notion of *control* in programs. One of the most fundamental properties of an algorithm is its arrangement in steps to be carried out in some prescribed order; that ordering is commonly called *control* or *control flow*. We begin our study by considering several mathematical models for control, showing that all these models are "equivalent" and that we may use whichever model is most convenient for a particular problem. We then discuss two well-known programming concepts—flow charts and control statements from programming languages. These are also "equivalent" to the mathematical models of control and this equivalence permits us to use them to design and reason about real programs.

Later chapters in Part One address the issues of representation, correctness, and efficiency. Chapter 4, about representation, shows how higher-level programming-language control constructs can be represented in machine language. Chapter 5, about correctness, shows how the semantics, or meaning, of programs can be formally defined, and how the correctness of a program can be demonstrated mathematically. Chapter 6, the final chapter of Part One, shows how program performance can be analyzed.

Chapter 1

Finite State Models

This chapter discusses a simple class of computations, broadly described by the term "finite state." There are two standard ways of describing the results of such a computation: first as an idealized machine that can perform it, and second as the set of acceptable inputs—called a "language." Although these two appear radically different, they are in fact closely related. Furthermore, for our present purpose, the study of control, both kinds of description have the advantage that they do not explicitly involve the notions of "memory," "storage," or "variable."

1.1 Finite State Machines

Consider a program that has no variables, procedure calls, or function calls. The only varying information that a computer must keep around to execute this program is that associated with the *control flow* of the program—the sequencing of its steps. Specifically, a computer only needs to remember the current step in the program. The program itself, which never changes, provides all the other information necessary to determine what to do next. Generalizing this simple program a bit, we get the computer-like devices called *finite-state machines*, *finite automata*, or *sequential machines*.

Consider a device such as that shown schematically in Fig. 1.1. Since we don't know its internal structure, we call it a "black box." The device has one input mechanism and one output mechanism. It consumes input data one item at a time, producing for each a corresponding output item.

We refer to the sequence of data presented to the input mechanism as the *input stream* or *input tape.* Similarly, we refer to the sequence of results as the *output stream* or *output tape.* The individual input and output items we call *symbols*; the set of all possible input items is the *input alphabet*, and the set of all possible output items is the *output alphabet.* For the moment we'll restrict both the input and output alphabets to the set {0, 1}, but that's not essential.

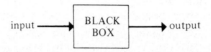

Fig. 1.1 A "black box" device, M_1

Suppose we watch this device (we'll call it M_1) in operation for a period of time and observe the following behavior.

input: 1 1 0 1 1 1 0 1 0 1 0 0 0 · · ·

output: 0 1 0 0 1 1 0 0 0 0 0 0 0 · · ·

Although this sequence of observations is too short to draw any definite conclusions, a pattern of behavior is emerging. Note that the output is a 1 in all cases where both the current input and the immediately preceding input are 1s. In all other cases the output is a 0.

If this is indeed a correct characterization of what the device does, then it has a very interesting property: its output is a function of both the current input *and* the past history of inputs. That is, the device "remembers" something about the sequence of input values. In this case it remembers very little past history, namely the immediately preceding input value, but we can imagine a similar device that might have to remember much more.

Now, since we still can't describe the device's internal mechanism, let's see if we can find some way to describe its external behavior more precisely. One way is to say that the machine, however it is constructed, can be in either of two conditions, or states. It is in one state when the previous input was a 0, and in the other state when the previous input was a 1. Let's call the first state LIZ (last input zero) and the second state LIO (last input one). Using these terms, we can characterize the device's behavior as follows:

- When the device is in state LIZ and the current input is 0, the output is 0 and the state remains LIZ.

- When the device is in state LIZ and the current input is 1, the output is 0 and the state changes to LIO.
- When the device is in state LIO and the current input is 0, the output is 0 and the state changes to LIZ.
- When the device is in state LIO and the current input is 1, the output is 1 and the state remains LIO.

Note that we haven't proved this is the way the machine is constructed; it's merely a description that accounts for all the information we have (such a description is often called a *model*). Also note that each state in our model can be thought of as either a step in a computation or as a situation (condition). The notion of state in this particular model is related to an intuitive notion of memory; each individual state corresponds to a single condition, and the set of all possible states corresponds to the set of all situations, or conditions, that can be distinguished.

The number of possible states of M_1, and hence the amount of information it implicitly stores, is finite. Therefore, M_1 is an example of what we usually call a *finite state machine* (FSM) or *finite state automaton* (FSA). Such machines are very common. A telephone switching circuit, for example, accepts a sequence of inputs, "remembers" them until they are all collected, and finally produces an output (establishes a connection) based on the sequence of inputs.

A vending machine is another practical device that can be modeled as an FSM. Ours accepts only nickels and dimes, and thus its input alphabet consists of various coins, buttons, and error signals (indicating foreign coins, pennies, faucet washers, and the like.) Its output alphabet consists of various lights, clicking noises, change, items for sale, and so forth. We can describe its behavior by assuming the existence of various states and state transitions. For example, there might be one state for each quantity of money that can be deposited:

- ZERO: the initial state; no money has been deposited.
- NICKEL: a total of five cents has been deposited.
- DIME: a total of ten cents has been deposited.

.
.
.

It should not be too hard to see the kinds of state transitions that will happen for various inputs. For example, suppose we start in state ZERO; dropping a nickel into the coin slot will cause a transition to state NICKEL. A dime will cause a transition from state ZERO to state DIME. If the device is in state NICKEL, inserting a nickel will cause a transition to state

DIME. The outputs in each of these cases will probably just be some reassuring clicking noise. In some of these states, however, pushing one of the selection buttons will be a legal input—and the result will be the output of the selected item, as well as a return to the initial state.

The concept *state* is one of the essential ideas for much of what follows throughout this textbook and in all of computer science. It is worth dwelling on for a moment to be sure that we have a good intuitive grasp of its meaning. Informally, we mean a summary of a condition or situation. So, for example, the U.S. President annually delivers a State of the Union address to Congress in which he summarizes the nation's condition as he sees it. A state in a program is also its "condition," consisting of the values of all the program's variables plus the point at which it is executing. In the following paragraphs, we give more formal definition to the concept, but this formal definition is based on our more common, intuitive one.

1.1.1 STATE TRANSITION FUNCTIONS

We can make descriptions of machines like M_1 considerably more concise by introducing some notation. Let us denote the set of possible states of a machine M by Q, where

$$Q = \{q_0, q_1, q_2, \cdots, q_p\}$$

if M has $p+1$ states. The q_i's are merely the names of these states. Thus, for example, $Q = \{LIZ, LIO\}$ in the first example. One state (typically called q_0 in this textbook) is designated the *initial state*.

The machine M starts in the initial state, q_0, and then repeatedly performs an *execution cycle*. On each cycle, M "looks at" an input x and its current state q; on the basis of these, it outputs something, say y, and switches to a new state, q' (it is acceptable to have $q = q'$). Both q' and y depend on x and q; that is, we can characterize the ouput and next state as functions of the current state and input:

$$y = OUT(q, x)$$

$$q' = NEXT(q, x)$$

where "input" and "output" denote the finite input and output alphabets. We call *NEXT* the *state transition function* and *OUT* the *output function*. Together with Q and q_0, they completely characterize the behavior of an FSM. It is common in computer mathematics to describe functions such as *OUT* and *NEXT* in the following form:

    OUT: state × input → output

    NEXT: state × input → state

This notation emphasizes the nature of the domain (inputs) and the range (output) of the functions. In this case, for example, the domain of *OUT* is the cartesian product* of states and inputs, and the range of *OUT* is the set of outputs.

    Note, by the way, that it is permissible for *OUT* or *NEXT* to be undefined for some pairs of inputs and states. Should the machine find itself in a situation where its next state or output is undefined, it *jams*; that is, it stops and cannot proceed. (This is rather analogous to the situation in which an unexpected input, like chewing gum, would, in all senses, gum up the works of a vending machine.)

    The functions *OUT* and *NEXT* for "black box", M_1, are defined in Fig. 1.2.

	state				state	
NEXT	LIZ	LIO		*OUT*	LIZ	LIO
input $\begin{cases} 0 \\ 1 \end{cases}$	LIZ LIO	LIZ LIO	input	$\begin{cases} 0 \\ 1 \end{cases}$	0 0	0 1

Fig. 1.2 *OUT* and *NEXT* for M_1

Let's consider a second device; we'll call it M_2. Suppose we wish M_2 to be an FSM that will look at a stream of 1s and 0s and output the *parity* of the stream it has scanned so far. The parity is to be 1 if the stream contains an odd number of 1s and 0 otherwise. We give our FSM two states, q_0 and q_1, corresponding respectively to "the state of having scanned a stream with even parity" and "the state of having scanned a stream with odd parity." We can then illustrate M_2 as in Fig. 1.3. For example, we can tell that if we are currently in state q_0 (even parity) and a 1 signal is input, then the new state, $NEXT(q_0, 1)$ is q_1 and the output, $OUT(q_0, 1)$ is 1.

    To summarize, then, an FSM has a finite set of states, a finite set of possible input and output items, a pair of functions that for each cycle define the next state and next output, and a specified initial state. We can formalize description as follows:

---

* The cartesian product of two sets, $A \times B$, is the set of all *pairs* of elements from A and B. Thus, if $A = \{a_1, \cdots, a_n\}$ and $B = \{b_1, \cdots, b_m\}$, then $A \times B = \{<a_1, b_1>, <a_1, b_2>, \cdots, <a_n, b_m>\}$.

	state	
NEXT	q_0	q_1
input $\begin{cases} 0 \\ 1 \end{cases}$	q_0	q_1
	q_1	q_0

	state	
OUT	q_0	q_1
input $\begin{cases} 0 \\ 1 \end{cases}$	0	1
	1	0

Fig. 1.3 *OUT* and *NEXT* for M_2

Definition 1.1. An FSM is a *4-tuple* $(Q, q_0, NEXT, OUT)$, where

- Q is a finite set of possible *states*;
- q_0 in Q is the *initial state*;
- *NEXT* is a function that maps <state, input symbol> pairs to states; and
- *OUT* is a function that maps <state, input symbol> pairs to output symbols. □

Remember, this definition is merely a formalization of our intuitions. It introduces nothing new. However, this definition, and ones like it later in the text, allow us to be precise—and hence to have more confidence in our reasoning.

1.1.2 STATE TRANSITION DIAGRAMS

Often it is more convenient to describe an FSM by a *state transition diagram* than by the function definitions of *OUT* and *NEXT*. A state transition diagram is a directed graph with labeled arcs. Figs. 1.4 and 1.5 are state transition diagrams for M_1 and M_2.

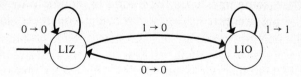

Fig. 1.4 State transition diagram for M_1

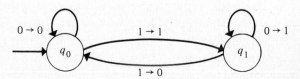

Fig. 1.5 State transition diagram for M_2

Each node, or circle, represents a state and has an arc, or arrow, leading out of it for each legal input in that state. Each arc points to the node representing the state to which the machine will transfer when it reads the appropriate input. Each arc is labeled with a phrase of the form $x \rightarrow y$, where x is the input value that causes the state transformation and y is its corresponding output value. Generally, in such diagrams the initial state is indicated by an arrow that "comes from nowhere."

The state transition diagram for an FSM is often both more compact and more understandable than its formal description as a 4-tuple. For example, the tuple describing the vending machine we discussed earlier would obscure some interesting properties—such as the set of ways to deposit twenty cents—that are quite clear in the diagram in Fig. 1.6.

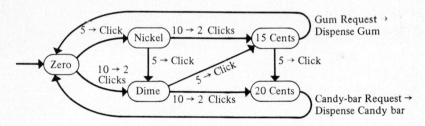

Fig. 1.6 Vending machine state diagram

1.1.3 SPECIAL FSM'S: RECOGNIZERS AND GENERATORS; NULL TRANSITIONS

FSM's, as we have defined them so far, always produce sequences of outputs. We have sometimes used the output to indicate only whether the portion of the string processed thus far has some property—odd parity, for example. One common and useful variety of FSM does not have an output stream. Instead, a certain subset of Q is designated as the set of final or accepting states. We call such a machine a *finite state recognizer* or FSR.

Definition 1.2. A *finite state recognizer* (FSR) is a 4-tuple $(Q, q_0,$ *NEXT, F*), where Q, q_0, and *NEXT* are as for FSM's, and $F \subset Q$ is a set of final states. The operation of an FSR is identical to that of an FSM, except that it produces no output. If, after running through the input tape, the machine is in one of the states in F, it is said to have *accepted* or *recognized* its input. Otherwise, it is said to have *rejected* its input. The machine also rejects the input if it jams (hits an undefined state transition). □

Note that an FSR accepts its input if it is in a final state *after* running through the input tape. The FSR may be in one of the final states any number of times before reaching the end of the input—this says nothing about acceptance or rejection. For example, suppose we throw away the output function in M_2 and designate $\{q_1\}$ as the set of final states. We now have a finite state recognizer for strings of odd parity. If we designate $\{q_0\}$ as the set of final states, we have a recognizer for strings of even parity. These two versions of M_2 are shown in Fig. 1.7. Note that we label the arcs with tags such as $x \rightarrow$ to emphasize that there is no output. In such cases, we also may say that the machine produces the *null output*. We represent the final states in the diagrams with double circles to distinguish them from other states.

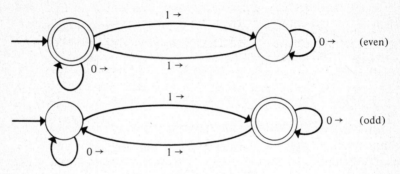

Fig. 1.7 FSR's for M_2

Now let's consider another special kind of FSM. Just as we can dispense with the output of an FSM, we can dispense with the input. In particular, consider a class of machines that take no input at all, but write a fixed sequence of outputs. These machines are called *finite-state generators*, or FSG's, because they produce, or generate, an output sequence. In addition, we call their state transition that consumes no input a *null transition.*

As an example, consider the machine in Fig. 1.8. It will produce a repeating pattern of two zeros followed by a one.

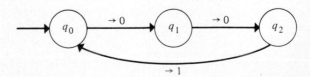

Fig. 1.8 A simple FSG

With the intent to emphasize that no input is expected or required, we've labeled the arcs in Fig. 1.8 with the notation $\rightarrow y$. These transitions, since they consume no input, are null transitions. A fully automatic traffic light is a practical example of an FSG; it consumes no input, but it cycles through a series of states that, in turn, display red, yellow, and green lights, turn signals, walk lights, warning bells, and so on.

By allowing ϵ (the null string) as a possible result of the *OUT* function and as a possible argument to both *OUT* and *NEXT*, we can build FSM's which are "mixtures" of FSR's and FSG's. That is, these FSM's sometimes consume input without producing output or sometimes produce output without consuming input. In effect, an FSR is an FSM (together with a set of final states) that always outputs ϵ, and an FSG is an FSM whose *OUT* and *NEXT* functions are defined only on ϵ.

Of course, nothing prevents certain transitions from producing no output *and* consuming no input. The arc for such a transition would simply be labeled with an arrow, $\rightarrow$. We can always eliminate these transitions from an FSM, FSR, or FSG, as demonstrated in Exercise 1.13. Nonetheless, sometimes they may be useful in constructing machines.

1.1.4 BIGGER INPUTS AND OUTPUTS

So far, our FSM's, FSR's, and other devices have all had inputs and outputs from the set $\{0, 1\}$. In general, however, the input and output *alphabets* (that is, the sets of possible inputs and outputs) can be arbitrary finite sets. For example, Fig. 1.9 shows a machine for reading strings of a's, b's, and c's; the machine prints 0, 1, or 2, depending on the number of b's scanned. Specifically, the number of b's are divided by three: a 0 is output when there is no remainder; a 1 when the remainder is one; and a 2 when it's two. We call this device a "Mod 3 Machine" because its output is equal to the number of b's *modulo 3* (that is, the output minus the number of b's is evenly divisible by 3).

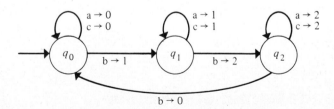

Fig. 1.9 Mod 3 machine

The relaxation of restrictions on input–output alphabets—seems to have increased what our machine can do. In fact, we have added nothing to the essential power of our model, because we can represent any finite alphabet by strings of 1s and 0s. For example, representing the inputs a as 00, b as 01, and c as 10, and the outputs 0 as 00, 1 as 01, and 2 as 10, we can create an alternative to the "Mod 3 Machine." Fig. 1.10 shows one such possible alternative.

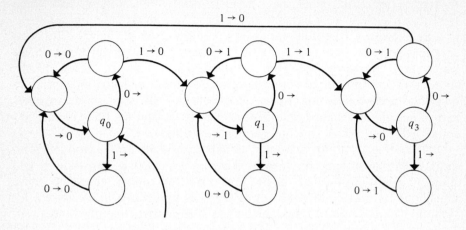

Fig. 1.10 Mod 3 machine with binary alphabet encoding

Note the use of null transitions to output two characters for each input character and the use of null outputs to "wait" for complete input pairs. True, we now have a machine that reads 0s instead of a's and writes 01s instead of 1s; but by appropriately interpreting the inputs and outputs, we see that the machine is essentially the same as that of Fig. 1.9.

Although expanding the alphabet does not essentially expand the types of computations we can do, it does improve the convenience with which we can express them. A comparison of Figs. 1.9 and 1.10 shows that the simpler alphabet forces us to use many more states, and hence leads to a more complex machine. Furthermore, logic circuits are perfectly capable of processing multiple bits in parallel, so that Fig. 1.9 is actually a more accurate representation of most *sequential circuits*, the name given to electronic realizations of FSM's.

1.2 Nondeterministic FSM's

Each of the preceding generalizations of an FSR—adding null transitions and larger input–output alphabets—made it more convenient to describe some class of computation; but none of them added any power to the

original model. In this section we'll define yet another generalization, and
it too will make things more convenient but won't add power. One of the
principal advantages of these new machines is that they provide a
convenient way to combine other FSM's. Consider the two FSR's in Fig.
1.11 that take inputs from the singleton set {1}:

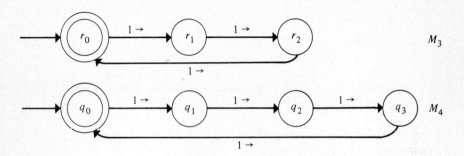

Fig. 1.11 M_3 and M_4

    Convince yourself that machine M_3, in addition to the null input,
accepts any input that is some multiple of three characters long (for
example, it accepts the null input, 111 and 111111, $\cdots$). Similarly,
convince yourself that M_4 accepts any input that's a multiple of four
characters long.

    Suppose we wish to build a machine, M, that accepts inputs that are
multiples of either three characters or of four characters long. It would be
nice if we could construct M directly from M_3 and M_4. Unfortunately, our
current model doesn't make this easy. We can, however, solve the
problem by extending the model as we do in Fig. 1.12 with machine M', a
nondeterministic finite state recognizer (NDFSR). Note the difference
between this and an ordinary FSR: Both arcs leaving s_0 carry the same
label, namely $1\rightarrow$. We seem to have indicated that M' may go to *either* q_1
or r_1 from s_0. In fact, that is precisely what we mean. M' accepts an input
string iff* *any* of the possible paths through the states indicated by the arcs
and their labels leaves the machine in a final state.

    For example, confronted with the input 111, there are two possible
paths that M' can follow through its states.

       $s_0 \rightarrow q_1 \rightarrow q_2 \rightarrow q_3 \rightarrow q_0$

---

* *iff* is an abbreviation for "if and only if".

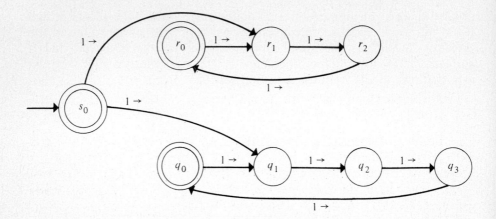

Fig. 1.12 M', an NDFSR

and

$$s_0 \rightarrow r_1 \rightarrow r_2 \rightarrow r_0 \rightarrow r_1$$

The second path does not leave M' in a final state. The first does, however; so M' accepts the input.

In effect, we have turned the state–transition function *NEXT* into a *relation* between pairs of inputs and states (q, x) and possible next states. That is, the relation *NEXT* exists between the state–input pair (q, x) and the state s iff it is possible to go from state q to state s given the input x. Alternatively, you may view *NEXT* as a function that produces a *set* of states (the possible next states) instead of a single next state.

As with FSM's, we define NDFSR's more formally, as follows:

Definition 1.3. An NDFSR, N, is a 4-tuple $(Q, q_0, NEXT, F)$, where

- Q is a finite set of states;
- q_0 in Q is the initial state;
- $F \subset Q$ is a set of final (accepting) states; and
- *NEXT* is a function defined on certain pairs (q, a) of states and inputs (a may be ϵ), and yields sets of possible next states (subsets of Q). If *NEXT* is defined for the pair (q, ϵ), then it is undefined for (q, b), where b is any input symbol.

A computation of N is a sequence of states $s_0, s_1, \cdots, s_n$, where

- $s_0 = q_0$;
- s_{i+1} is in $NEXT(s_i, \epsilon)$ when that is defined;

- s_{i+1} is in $NEXT(s_i, b)$ where b is the next (as yet unconsumed) input;

- The input is consumed either at or before s_n. (We say "or before" because there may be null transitions. Note that this does not imply that all null transitions must necessarily be taken.)

N accepts input x iff *some* computation of N ends with an accepting state. $\square$

From a mechanical point of view, there are several ways to explain M'. When M' comes to a state and input where it has a choice of next states, imagine M' as creating a copy of itself for each possible alternative, with some referee looking on to see if at least one of them completes successfully. You might also think of the state–transition diagram for the machine as containing several movable markers, one on each state that M *might* be in. At each state transition, each token or marker advances (or is destroyed if the machine jams). Whenever more than one arrow leaving a state matches the immediate input, a token or marker on that state "follows" *all* arrows by sending a copy of itself down each of them.

It should be clear that an NDFSR can do anything an FSR can do (an FSR is just an NDFSR that doesn't have any choices). Remarkably enough, an FSR can do anything an NDFSR can. That is,

Theorem 1.1. For any NDFSR, there is an FSR that accepts exactly the same set of inputs. $\square$

The idea behind the proof is quite simple; given an arbitrary NDFSM, we construct an FSM equivalent. To do this we define the states and state–transition function of the FSM. This is the point of the proof that requires some clever insight: at any time, an NDFSR can be in one of a *set* of possible states; Although the next state is *not* unique, the set of the next possible states *is*. Therefore, the FSR we construct to correspond to a given NDFSR uses as its own states the *sets of states* of the NDFSR.

Proof of Theorem 1.1. Let $N = (Q, q_0, F, NEXT)$ be an NDFSR. Without loss of generality, we assume (see Exercise 1.13) that N has no null transitions. Construct a corresponding FSR $N' = (Q', q_0', F', NEXT')$ as follows.

- Let
  $$Q' = 2^Q,$$
  the set of all subsets of Q. (When N' is in state $q' = \{q_1, \cdots, q_r\}$ at some point, it will mean that N is in state $q_1, \cdots,$ or q_r after reading the same input).

- Let

$$q_0' = \{q_0\}.$$

(By definition, the only state that N can be in at the outset is q_0).

- Let

$$NEXT'(\{q_1, \cdots, q_r\}, a)$$
$$= NEXT(q_1, a) \cup NEXT(q_2, a) \cdots \cup NEXT(q_r, a).$$

If any of the $NEXT(q_i, a)$ is undefined, substitute $\{\}$ for it. (That is, $NEXT'$ gives the set of all states that N could reach from some state $q_1, \cdots, q_r$ when the input is a).

- Let

$$F' = \{q' \text{ in } Q' \mid q' \cap F \neq \{\}\}.$$

(That is, F' consists of all subsets of Q containing at least one final state. We define F' this way because an NDFSR is defined to accept a string if it *could be* in a final state after reading the string).

With this construction, the rest of the proof is a straightforward exercise, showing that after reading any string accepted by N, N' will be in a state that contains one of the final states of N, and that any string N rejects is also rejected by N'. $\square$

For example, consider machine M'' of Fig. 1.13.

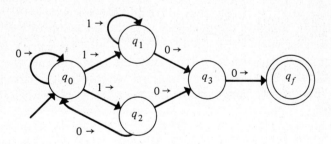

Fig. 1.13 M''

Since $Q = \{q_0, q_1, q_2, q_3, q_f\}$,

$$Q' = \{ \{\}, \{q_0\}, \{q_1\}, \{q_0, q_1\}, \cdots, \{q_0, q_1, q_2, q_3, q_f\} \}.$$

We define

$$NEXT'(\{q_0\}, 0) = \{q_0\},$$
$$NEXT'(\{q_0\}, 1) = \{q_1, q_2\}$$
$$NEXT'(\{q_1, q_2\}, 0) = \{q_0, q_3\}$$
$$NEXT'(\{q_1, q_2\}, 1) = \{q_1\}$$

(Note that $NEXT(q_2, 1) = \{\}$)

$$NEXT'(\{q_0, q_3\}, 0) = \{q_0, q_f\}$$

.

.

$$F' = \{ \{q_f\}, \{q_0, q_f\}, \{q_1, q_f\}, \{q_0, q_1, q_f\}, \cdots, \{q_0, q_1, q_2, q_3, q_f\} \}.$$

Thus, we are free to build NDFSR's to solve problems, confident that we can always convert them into "real world" FSR's. This is nice because NDFSR's are often smaller than FSR's (M' had only eight states as opposed to the twelve required by its deterministic cousin) and was easier to design. Furthermore, mathematicians find many theorems easier to prove with NDFSR's, as we will see in Section 1.4.

1.3 Languages and Regular Expressions

We now turn from FSM's to languages and grammars. The change is really less drastic than it might appear, because the kinds of languages we consider in the following paragraphs are closely related to FSM's.

1.3.1 LANGUAGES

We use the word "language" in a very restricted technical sense, specifically

> **Definition 1.4.** A *language* is a set of strings of symbols. Each string from the language is called a *sentence* (or sometimes a *word*) of the language. The set of symbols of the language is called its *alphabet* or *vocabulary*. We refer to languages (sentences, words) *over* a particular alphabet when we must be specific. In this text, we always take the alphabet to be finite. □

For example, according to Definition 1.4, each of the following sets is a language over the alphabet $\{a, b\}$.

$$L_1 = \{a, b, ab\}$$
$$L_2 = \{aa, aba, abba, abbba, abbbba, \cdots\}$$
$$L_3 = \{ab, abab, ababab, abababab, \cdots\}$$

Each of the strings in L_2, "*aa*," "*aba*," etc., is a sentence of that language.

Note that lacking from the definition is any hint of such concepts as "meaning" or "interpretation" of sentences. Of course, in all practical languages—whether they are natural (like, English) or artificial (like, FORTRAN)—meaning is of paramount importance. For our present purposes, however, we are concerned only with the *form*, or *syntax* of sentences. For example, if our object of study were English syntax or grammar, we would not concern ourselves with the fact that the sentence "Colorless yellow ideas sleep furiously" does not *mean* anything in the ordinary sense of the word. Instead, we simply note that it is grammatically sound, or syntactically correct.

Almost all languages of interest to us have an infinite number of sentences. These languages cannot be defined by explicit enumeration. The language L_1 above happens to be finite, so we were able to list (enumerate) all its sentences. Languages L_2 and L_3, however, were not finite, and we were forced to use "$\cdots$" in their definitions. The usual way to define infinite languages is to supply a grammar and to say that the language is the set of all strings that are grammatically correct according to the given grammar. Informally, for example, the sentences of L_2 can be defined by, "an *a* followed by some number (possibly zero) of *b*'s followed by another *a*." This is a particularly simple example of a grammar, but it illustrates how a grammar can define a language.

1.3.2 REGULAR EXPRESSIONS

One way to define a language is by a very simple type of grammar called a *regular expression* (RE). An RE is, in effect, a formula that specifies a set of strings. The following is a formal definition, with examples, of several RE's.

Definition 1.5. Let $A = \{a_1, a_2, a_3, \cdots, a_n\}$ be an alphabet. Then a *regular expression* (RE) over A denotes a language over A and is defined by the following set of rules:

1. *Atom:* Any single symbol of A is a valid regular expression. The language it denotes consists of the single string consisting of that symbol. The symbol ϵ is a valid RE, denoting the language consisting of the empty string (as opposed to the language containing no strings).

Examples: a_1 a_2

Denoting: $\{a_1\}$ $\{a_2\}$

2. *Alternation:*   If R_1 and R_2 are both regular expressions then $(R_1 + R_2)$ is also a valid regular expression. The language it denotes is the union of the languages denoted by R_1 and R_2.

Examples: $(a_1 + a_2)$ $((a_1 + a_2) + a_3)$

Denoting: $\{a_1, a_2\}$ $\{a_1, a_2, a_3\}$

3. *Composition:*   If R_1 and R_2 are both regular expressions then $(R_1 R_2)$ is also a valid regular expression. The language it denotes is the set of all strings formed by concatenating a string from the set denoted by R_2 to the end of one in the set denoted by R_1.

Examples: $(a_1\ a_2)$ $((a_1\ a_2)\ a_3)$ $((a_1 + a_2)\ a_3)$

Denoting: $\{a_1 a_2\}$ $\{a_1 a_2 a_3\}$ $\{a_1 a_3,\ a_2 a_3\}$

4. *Closure:*   If R_1 is a valid regular expression, then $(R_1)^*$ is also a valid regular expression. The language it denotes consists of all strings formed by concatenating zero or more strings in the language denoted by R_1. This operator $(^*)$ is often called the *Kleene star.* The notation $(a)^+$ is shorthand for $(a(a)^*)$.

Examples: $(a_1)^*$ $(((a_1)^*\ a_2))^*$

            $((a_1 + a_2))^*$

Denoting: $\{\epsilon,\ a_1,\ a_1 a_1,\ a_1 a_1 a_1,\ \cdots\}$ $\{\epsilon,\ a_2,\ a_1 a_2,\ a_1 a_1 a_2, \cdots\}$

           $\{\epsilon,\ a_1,\ a_2,\ a_1 a_1,\ a_1 a_2,\ a_2 a_1,\ a_2 a_2, \cdots\}$

5. Nothing else is a regular expression.

If R is an RE and L is the language it denotes, we will (by slight abuse of notation) write $R = L$ and $L = R$. $\square$

Note that the plus, asterisk, epsilon, and parentheses symbols used in the regular expressions are part of the RE-notation, not part of the languages being defined.

Also, note that Definition 1.5 is *recursive.* That is, Rules 2, 3, and 4 define RE's in terms of RE's. Such a definition would be circular except for Rule 1, which allows us to generate "primitive" RE's. Once we have derived two RE's from Rule 1, we can use Rules 2, 3, and 4 to construct bigger and better ones. Rule 5 makes the definition complete.

Although the rules in Definition 1.5 define the set of valid regular expressions completely, to avoid redundant parentheses, we also use the common rules of precedence and omit parentheses if the result is unambiguous. The standard precedence rules are that in the absence of parentheses, closure takes precedence, followed by composition, then alternation. The expression $ab + c^*$, then, is shorthand for $((ab) + (c)^*)$, and ab^* is shorthand for $(a (b^*))$. The following are some examples of regular expressions and the languages they define. (We are often able to give two equivalent RE's, or RE's that define the same sets of strings. Because some of these languages are infinite we indicate only a few representative strings from the language.)

$$(b + c) \quad = \{b, c\}$$

$$ab \qquad = \{ab\}$$

$$ab^* \qquad = \{a, ab, abb, abbb, \cdots \}$$

$$a(b + c) \; = (ab + ac) = \{ab, ac\}$$

$$(a + b)^* \; = (a^* + b)^* \; = \; (a + b^*)^* \; = \; (a^* + b^*)^*$$

$$\qquad \qquad = \{\epsilon, a, b, aa, ab, ba, bb, aaa, aab, \cdots\}$$

$$a^* b^* \qquad = \{\epsilon, a, aa, aaa, \cdots, b, ab, aab, \cdots, bb, abb, \cdots\}$$

$$\text{(but not } aba, ba, bab, \cdots)$$

Note that we have eliminated unnecessary parentheses; for instance in the last example, the strict definition requires us to write $((a)^*(b)^*)$. We leave out the parentheses because there is no possible ambiguity.

You will, perhaps, already know that at least some portions of common programming languages can be defined by RE's. For example, one common convention for writing *identifiers* is given by

$$(A + B + \cdots + Z) (A + \cdots + Z + 0 + 1 + \cdots + 9 + \_)^*$$

That is, an identifier is simply a letter followed by some sequence of letters, digits, or underscores.

A slightly harder example is the definition of *unscaled floating point numbers* (as in PASCAL). These consist of a sequence of digits (at least one) containing exactly one decimal point. Your first choice for an RE might be

$$d^*.(d)^*$$

where

$$d = 0 + 1 + 2 + 3 + 4 + 5 + 6 + 7 + 8 + 9.$$

However, this would permit there to be no digits both before *and* after the decimal point; A lone decimal point would be, therefore, legal by this definition. PASCAL requires at least one digit both before and after the decimal point; thus we must use

    $dd^*.dd^*$

Some programming languages permit either the digit preceding the decimal point, or the one following it to be omitted—but not both. For such a language we must use one of these more complex expressions:

    $(d^*.d(d)^*) + (d(d)^*.(d)^*)$
    $d^*((d.) + (.d)) (d)^*$
    $(dd^*.d^*) + (.dd^*)$

1.4 Equivalence of Regular Expressions and FSM's

In this section we prove the following, perhaps surprising, theorem:

> **Theorem 1.2.** Any language that can be denoted by a regular expression can be recognized by an FSR. □

As it happens, we could prove even more—that the class of languages denoted by RE's is *identical* to that recognized by FSR's. To do so requires that we also prove the converse of Theorem 1.2. That is, we must show that for any FSR, there is an RE describing *exactly* the language that FSR accepts, which—for our purposes, however—is not very interesting and is rather involved.

Theorem 1.2 has two practical implications. First, it's very easy to build a machine or a program to recognize sentences generated by regular expressions (see Chapter 19). Second, some languages cannot be specified with regular expressions; in particular, regular expressions are too "weak" to describe entire programming languages. (Later we'll concern ourselves with a more powerful tool for specifying languages; and at that time, we look more deeply into the limitations of RE's.)

Before moving on to the proof of Theorem 1.2, let's reflect a moment on the proof technique that we used for Theorem 1.1 and that we'll use again for Theorem 1.2. Both theorems, in a loose sense, state the existence of something: one asserts that for every NDFSR there is a deterministic FSR that will do the same job; the other asserts that the languages generated by an RE can be recognized by an FSR—that is, that there exists an FSR that will recognize the sentences generated by the RE. As is common for these *existence theorems*, both proofs demonstrate a *construction* that will produce an object with the claimed properties. In effect, the theorem and proof say:

> I assert that for every A there is B that will do the same job. I'll prove it to you by, after defining an A, constructing a B that does A's job. I won't claim this B is the *best* B for the job, only that it does the same thing as A.

Such *constructive proofs* are an important part of your programmer's tool kit; it's a good idea to learn how to use them as we do here:

Proof of Theorem 1.2. Suppose that R is an RE over the alphabet A. We will show how to build an NDFSR model that recognizes exactly the language denoted by R. This theorem, then, follows from Theorem 1.1. Our argument is essentially an induction on the size of R and follows the definition of RE. That is, we construct NDFSR's for the smallest RE's and then show that we can construct an NDFSR for an arbitrary R, *assuming* that we can construct NDFSR's for each of its smaller constituent parts. Each NDFSR constructed in this process has exactly one final state.

For the purposes of this discussion, assume that we will be constructing NDFSR's out of other NDFSR's that have been constructed previously. For consistency, we'll always call these previously constructed NDFSR's F_1 and F_2, and will assume that they have initial states q_0 and r_0 and the final states q_f and r_f, as illustrated by Fig. 1.14. Before any construction is performed, we will assume that the states q_f and r_f are first changed to nonfinal (nonaccepting) states. Then each construction provides a new final state.

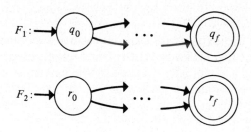

Fig. 1.14 F_1 and F_2

1. Suppose R is the single symbol a in A or is ϵ. Then clearly, the NDFSR's of Fig. 1.15 recognize precisely the language denoted by R. This step establishes that we can build a corresponding NDFSR for any RE of length 1.

2. Suppose R is the expression $(R_1 + R_2)$, where R_1 and R_2 are RE's. Assume further that the NDFSR's F_1 and F_2, respectively, accept the

Fig. 1.15 NDFSR's for the RE's a and ϵ

languages denoted by R_1 and R_2 (That is, F_1 and F_2 have also been constructed previously. This is the inductive hypothesis). Then F_3 of Fig. 1.16 accepts the language denoted by $(R_1 + R_2)$.

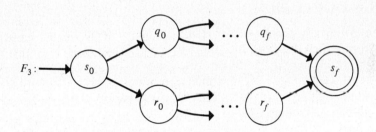

Fig. 1.16 F_3: RE for $(R_1 + R_2)$

Note that F_3 has exactly one final state (s_f). Note also our use of null transitions (from s_0, q_f, and r_f) conveniently "glues" F_1 and F_2 together. These null transitions, of course, make F_3 nondeterministic.

3. Suppose R is the expression $(R_1 \ R_2)$. Letting F_1 and F_2 be as above, the NDFSR F_4 of Fig. 1.17 accepts the language denoted by R.

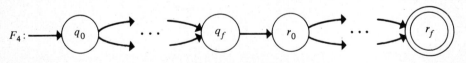

Fig. 1.17 F_4: RE for $(R_1 \ R_2)$

Again, note the null transition connecting F_1 and F_2.

4. Suppose R is $(R_1)^*$. Letting F_1 be as previously described, the NDFSR F_5 of Fig. 1.18 accepts the language denoted by R. There are no further possibilities for R; the construction is complete. $\square$

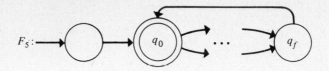

Fig. 1.18 F_5: RE for $(R_1)^*$

1.4.1 EXAMPLE: CONSTRUCTION OF NDFSR FROM RE

To illustrate the construction used as proof of Theorem 1.2, we derive the FSR for the RE

$$R = (1(0)^* + 101)$$

Following the proof, we see that R is of the form $(R_1 + R_2)$, so that its corresponding NDFSR has the form given in Fig. 1.19. We get the NDFSR shown in Fig. 1.20 for 101. For $1(0)^*$, we have the skeletal NDFSR shown in Fig. 1.21, and then the construction for closure gives the NDFSR in Fig. 1.22 for $(0)^*$. Finally, putting Figs. 1.19, 1.20, 1.21, and 1.22 together gives Fig. 1.23.

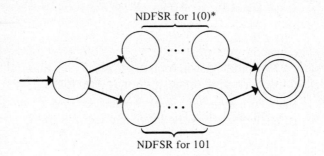

Fig. 1.19 Skeleton NDFSR for R

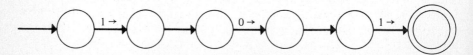

Fig. 1.20 FSR for recognizing 101

The NDFSR of Fig. 1.23, having been constructed in a formal and mechanical manner, has many more states than necessary. The null transitions, in fact, are all unnecessary (see Exercise 1.29). Nonetheless,

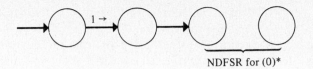

Fig. 1.21 Skeleton NDFSR for $1(0)^*$

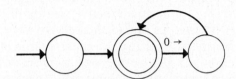

Fig. 1.22 NDFSR for $(0)^*$

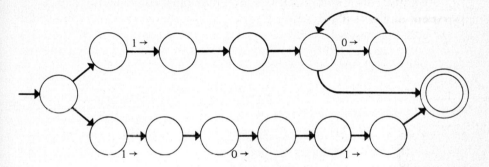

Fig. 1.23 NDFSR for $(1(0)^* + 101)$

by allowing their introduction, we have simplified the description of a conversion process. This is just one of the ways in which the constructions in an existence proof can be unrealistic in practice. The point of such proofs is *not* to provide a *practical* way of building something—but to prove that it's *feasible* to build it.

1.5 Characterizing the "Power" of Computational Models

Both here and throughout the remainder of the text we will occasionally be concerned with the "power" of a computational model such as the FSM. By the power of a model we will simply mean the class of computations that can be performed by a machine conforming to the constraints of the model.

We argued in Section 1.2 that an FSR can do anything that an NDFSR can. In the next section, 1.5.1, we will show you a language no FSM can recognize. Both of these subjects are, in effect, concerned with the power of the FSM model. The first shows that NDFSR's are no more powerful than FSM's, and the second shows that the FSM model is not powerful enough to compute certain mathematical functions.

It is often convenient to have alternative ways of discussing the power of a model. Section 1.4 gave us one such alternative; by showing that the class of languages generated by a regular expression is identical to the set of strings recognized by a FSR, we are free to characterize the power of finite state models in terms of the languages they can recognize. In Chapter 13 we introduce computational models that are more powerful than FSM's. We also introduce corresponding classes of languages that can be recognized by these models.

Since the power of a model is sometimes most conveniently expressed as the set of functions that can be computed by the model, and sometimes as the class of languages it can recognize, we use phrases such as "the class of languages recognized by $\cdots$" and "the set of functions computable by $\cdots$" pretty much interchangeably.

1.5.1 A LIMITATION OF FSM'S

One of the more surprising conclusions in the body of theory that has developed around computers is that it is not possible for computers (and less possible for people) to compute all mathematical functions. In this section we illustrate a simple example of this limitation as it applies to FSM's. A more complete treatment is in Section 13.2 with another model (the Turing Machine, which is much more like real computers).

Consider the set, NMN, of all sequences of 0s and 1s consisting of a sequence of 1s followed by a sequence of 0s, and finally followed by another sequence of 1s of the same length as the first. Some strings in this set are

101

1001

11011

11100111

By extending the notation we used for RE's, we can describe these strings more succinctly as a limited form of repetition; in particular, we use a superscript on an RE to denote a fixed number of repetitions:

$$R^n = R R \cdots R \ (n \text{ times})$$

where n is some nonnegative integer.   It should be clear that this extension does not add any power to RE's; anything written in the extended notation could have been written in the original one by simply repeating some formula a fixed number of times.   Using this extended notation, the set NMN is

$$\text{NMN} = \{1^n 0^m 1^n \mid n, m > 0\}$$

Theorem 1.3. No FSR can recognize exactly the elements of NMN. □

Intuitively, this is not too surprising.   Since the number of initial and trailing 1s must be the same, the FSR must remember the number of initial 1s while scanning the 0s. The only way that an FSR can remember any one thing is by having a state to represent exactly that one thing. Therefore, the FSR requires a state to keep track of each of the possible numbers of initial 1s.   There is an infinite number of possibilities; hence, the FSR would require an infinite number of states. If we were able to construct it, it might look like the machine in Fig. 1.24.

Proof of Theorem 1.3. Suppose that M is an FSR recognizing exactly NMN. Consider the set

$$S = \{1^n \mid 1 \leq n \leq K + 1\}$$

where K is the number of states of M. There must be at least two distinct members of S, say 1^p and 1^q $(p \neq q)$, such that M is in the same state after reading 1^p as after reading 1^q (because S has $K + 1$ members, and M has only K states).   But the set of strings that M accepts following any string depends *only* on M's state after reading that string. Since this is the same state for 1^p and 1^q, and since by assumption M accepts $1^p 0^m 1^p$, M must also accept $1^p 0^m 1^q$.   But this is not in NMN, contrary to the hypothesis. □

For Theorems 1.1 and 1.2, we used an important classification of proof—*constructive proof*—to show that two concepts were equivalent. The proof for Theorem 1.3 illustrates another important classification—*proof by contradiction.* In this proof, we assume something is true, then show that this assumption leads to an impossible conclusion. We will see further examples of such proofs later in the text.

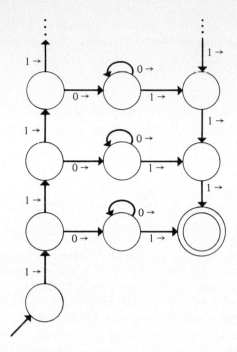

Fig. 1.24 An infinite device to recognize NMN sequences

1.6 Further Reading

There is a well-developed body of knowledge about finite automata and regular expressions. The principal theories and conclusions, as well as a bibliography to most other work, can be found in Minsky (1967), Hopcroft, and Ullman (1969) and Salomaa (1973). In Chapter 19 of this textbook you will find a description of one of the classical applications of finite automata theory to programming—lexical analysis. Further discussion of finite automata and lexical analysis can be found in Aho and Ullman (1977).

Exercises

1.1 Construct an FSM with input alphabet {0,1} and output alphabet {0,1} that behaves as follows: The output is 1 if the current input character and the last two input characters are identical; the output is 0 otherwise. The first two outputs must be 0, since there will not have been two previous inputs. For example, input sequence: 1100001110111

output sequence: 0000110010001

1.2 Consider the FSM of Fig. 1.25.

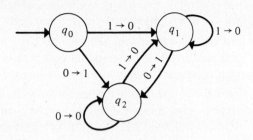

Figure 1.25

a) Write the *NEXT* and *OUT* functions for this machine.
b) What is its output for each of the following inputs?
    i) 1100
    ii) 1010
    iii) 11111100000

c) Describe the function of this machine in English.
d) Describe each state (that is, give the intuitive meanings of the states).

1.3 Construct a deterministic FSM to accept R^*, where R is defined by:

$$R = \{x \text{ in } \{0, 1\}^* \mid x \text{ contains exactly two "00" sequences}$$
$$\text{and no "000" sequence}\}$$

1.4 Given the languages L_1 and L_2 defined by:

$$S = \{p, q\}$$
$$L_1 = \{x \text{ in } S^* \mid x \text{ contains an odd number of p's}\}$$
$$L_2 = \{x \text{ in } S^* \mid x \text{ contains an even number of q's}\}$$

a) Construct an FSM to accept $L_1 \cap L_2$.
b) Construct an FSM to accept $L_1 \cup L_2$.

1.5 Write the tables for the transition functions that describe the FSM of Fig. 1.6.

1.6 Expand the vending-machine example to provide a change-making capability.

1.7 Construct an FSR that acts like a telephone circuit with the following rules.

> a) If the first digit is 0, the call is operator assisted. A single 0 is a legal input, as is a 0 followed by any other legal input.
>
> b) If the first digit is not 0 and the second digit is not 0 or 1, the call is local and any five other digits complete a legal input, for a total length of seven digits.
>
> c) If the first digit is not 0 and the second digit is 0 or 1, the call is long distance. To complete a legal input you need one more digit (any one) followed by a local or seven-digit number.

1.8 Construct an FSR that recognizes (accepts) any string over the alphabet {a, b} containing fewer than four a's.

1.9 Construct an FSR that accepts all and only input strings over the alphabet a, e, b, c, d, comma (,), period (.), and arrow($\leftarrow$), in which

> a) The string ends with a period.
>
> b) Every comma is preceded by a letter and followed by at least one arrow ($\leftarrow$).
>
> c) Every contiguous group of letters contains at least one a or e.

1.10 Assume that you have been hired by a traffic-light manufacturer to design an "intelligent" traffic light. Inputs to your system will include signals from various timers, from sensors that detect the presence of cars in the left-turn and through lanes, and from pedestrian "walk" buttons. The outputs set the lights (including left-turn and pedestrian signals), ring a bell for blind pedestrians, and reset the timers.

> a) Write an English specification of what your intelligent traffic-light system is to do.
>
> b) Construct an FSM to perform to these specifications.
>
> c) Encode this FSM in a form suitable to be simulated in Exercise 1.11.

1.11 The following exercises are related to the construction of an "FSM simulator"—a program that reads in the definition of an FSM and then, when told to start, will behave like the FSM whose definition it just read. In each part you may make reasonable assumptions about the input alphabet, number of states, etc. We suggest that you proceed in several steps:

a) Design an internal data structure to represent the *NEXT* and *OUT* functions.

b) Construct a simulator program that will read the definitions of *NEXT* and *OUT* from an external source and store them in the internal data structures developed in part (a). Be sure the simulator checks the definitions for validity as it reads them.

c) Construct a procedure to print the *NEXT* and *OUT* functions in some easily-readable form.

d) Construct a procedure to print *NEXT* and *OUT* in the form of a state–transition diagram. (If you wish, you may assume that the number of states is small.)

e) Construct a procedure to "simulate" the FSM represented by these internal structures—that is, the procedure should read input symbols, choose the appropriate next states, and output the appropriate symbols. The program should be able to draw a "trace" of what is happening.

f) Test your procedures on all the examples and exercises in this Chapter.

1.12 Give several "interesting" examples of FSR's, FSG's, and FSM's from everyday life. For each, describe the set of inputs, at least some of its states and state transitions. Argue convincingly that it has *finite* memory.

1.13 Show that for any FSM, FSR, or NDFSR, with null transitions that produce no output, there is another FSM, FSR, or NDFSR, with no null transitions that produces the same output (or recognizes the same language). To do so, give a general method for transforming any such FSM into one without those transitions.

1.14 Construct an FSG to generate each of the following strings:

a) 100100100···

b) 1100110011001100···

c) 101101011010110···

1.15 What are the *NEXT* and *OUT* functions for the machines of Figs. 1.9 and 1.10?

1.16 What does the machine of Fig. 1.9 do with each of the following inputs?

a) ab

b) abcba

c) accacccac

d) bbbbacacacbbbb

e) abcabcabcabcabc

1.17 What does the machine of Fig. 1.10 do with each of the following inputs?

a) 0001100100

b) 00101100010010

c) 011001110001

1.18 Design an FSM to convert letters to pairs of bits, as indicated in the following table. Design another FSM to convert pairs of bits to letters. Use null input and output transitions to deal with the differences in the lengths of the input and output streams. The conversions are

$$A \leftrightarrow 00, \quad B \leftrightarrow 01, \quad C \leftrightarrow 10, \quad D \leftrightarrow 11$$

1.19 Formally define a type of FSM that includes the possibility of N simultaneous inputs—that is, a type of FSM that "reads" N input symbols simultaneously and uses all N values to determine the next state and next output. Suggest a scheme for labeling the arcs in the state–transition diagram for such a machine. Show that such a machine is equivalent to an ordinary FSM.

1.20 The language of *propositional logic* is just the set of logical expressions formed out of simple logical variables (such as p or q), parentheses, and the operators "$\wedge$" (and), "$\vee$" (or), and "$\sim$" (not). Informally, an *inference rule* in logic is a rule for taking a logical expression and producing a second expression that must be true whenever the first is true. Consider FSM's with input alphabet = output alphabet = $\{p, q, r, (,), ., \sim, \wedge, \vee\}$. It is possible to construct FSM's over this alphabet that accept sentences in propositional logic and produce other sentences in propositional logic as output. For example, here is a machine that performs the inference

Given: $A \wedge B$

Conclude: A

where A and B stand for any of the propositional variables p, q, or r.

Figure 1.26

The input must be followed by a period. The machine writes a period at the end of its output if its input is in the form required by the inference rule.

    a) What does this machine do to the input "$p \wedge r.$"? To "$q \wedge r \wedge p.$"?

    b) Modify the machine to perform the same deduction, but with A and B being any logical sentences enclosed in parentheses in such a way that the input parentheses are never nested more than two levels deep.

1.21 A sentence in propositional logic (see Exercise 1.20) is said to be in *conjunctive normal form* (CNF) if it has the form

$$A_1 \wedge A_2 \wedge \cdots$$

where each of the A_i are of the form

$$(a_1 \vee a_2 \cdots)$$

and each of the a_i is either a simple variable or the negation ($\sim$) of a simple variable. For example,

$$(p \vee \sim q) \wedge (q \vee r) \wedge (\sim p)$$

is in conjunctive normal form. A sentence is said to be *satisfiable* iff there is some set of values for the propositional variables in it (each is either true or false) that makes the sentence true. Thus, the preceding example is satisfiable (let p and q be false and let r be true), and so is $(p \vee \sim p)$, but

$$(p \vee q) \wedge (\sim p) \wedge (\sim q)$$

is not satisfiable.

    a) Design an FSR with input alphabet $\{p, q, r, \wedge, \vee, \sim\}$ that accepts exactly the set of logical sentences over that alphabet that are in CNF.

    b) Write an RE that describes the same set.

    c) Design an NDFSR that accepts exactly the satisfiable logical sentences in CNF over the alphabet $\{p, q, r, \wedge, \vee, \sim\}$. This is only possible if we assume that all variable names are a single character. Why? [Hint: The problem is to find a proper set of values for p, q, and r. NDFSR's can "guess."]

1.22 Construct a deterministic machine that is equivalent to M'—the nondeterministic machine in Fig. 1.12.

1.23 Argue convincingly that machines M_3 and M_4 accept exactly strings of 1s whose lengths are multiples of 3 and 4 respectively.

1.24 Design a deterministic FSR that will accept strings of lengths that are multiples of either 3 or 4.

1.25 Prove that the FSR constructed in the proof of Theorem 1.1 actually does recognize exactly the same strings as the given NDFSR.

1.26 Give an upper bound on the number of states required for an FSR that accepts the same language as an NDFSR with n states.

1.27 Consider the NDFSM shown in Fig. 1.13.

a) Write the *NEXT* and *OUT* functions for this machine.

b) What sequence of states does the machine go through for each of the following inputs?
   i) 00011100
   ii) 01010100
   iii) 0110110110
   iv) 10111000

c) Describe the behavior of this machine in English.

d) Exhibit an FSM that is equivalent to it.

1.28 What set of strings is accepted by the NDFSR below:

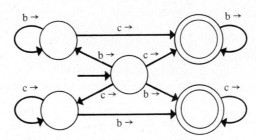

Figure 1.27

Design an FSR that accepts the same set. Note: the construction given in Proof of Theorem 1.1 will certainly work, but you should first try to build the FSR "from scratch," knowing just what strings of inputs it is supposed to accept.

1.29 Simplify the NDFSR of Fig. 1.23 as much as possible, eliminating all null transitions and minimizing the number of states.

1.30 The construction of Section 1.4 depends on the fact that at every step, each of the component machines has exactly one starting state and one final state. Prove that it is always possible to modify the component machines to have this property without affecting the strings they accept.

1.31 Bovik Railways, Inc. owns a section of track that looks like this:

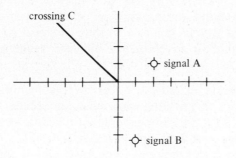

Figure 1.28

Trains travel north and west through C (but not at the same time!). There are four possible events:

W_A—Westbound train arrives at signal A.

W_C—Westbound train crosses at C.

N_A—Northbound train arrives at signal B.

N_C—Northbound train crosses at C.

These events occur repeatedly in a sequence constrained as follows.

- Only one train may be at A or at B at a time. Thus, after W_A occurs, it may not occur again until W_C occurs; likewise, N_A may not occur again until N_C occurs.
- A train must arrive at A or B before it can cross at C.
- If a train arrives at a signal, it must eventually cross, but it may wait before doing so. Thus, after W_A, an N_A or several N_A's and N_C's may occur before W_C.
- If both a westbound and a northbound train are at their respective signals, either may cross first, regardless of arrival

order.

For example, the sequences

$$W_A W_C N_A N_C W_A W_C W_A W_C$$
$$W_A N_A N_C W_C N_A W_A N_C N_A N_C W_C$$
$$W_A W_C$$
$$W_A N_A W_C N_C$$

are legal, and the sequences $W_A N_A N_C$ (westbound train never crosses), $W_A N_C W_C$ (northbound train crosses without passing signal), and $W_A W_A W_C W_C$ (two arrivals before a crossing at signal A) are illegal. Construct an FSR to recognize all valid sequences of the events W_A, W_C, N_A, N_C.

1.32 In the interests of safety, a certain car manufacturer devised a system of detectors for making sure that all occupants' seat belts are fastened. One detector determines whether a seat is occupied. A second detects whether the seat belt on that seat is fastened. A third detects whether the key is in the ignition. The basic idea is that if the key is in the ignition, the seat is occupied, and the seat belt is not fastened, the system turns on a buzzer. Furthermore, to make sure that the occupants do not try to trick the system, it also turns on the buzzer if it detects any suspicious *sequence* of events. For example, if the belt is fastened without the seat first being occupied, the system assumes that somebody is attempting subterfuge for the sake of peaceful comfort, and it buzzes.

Design an FSM whose input alphabet is the set {KI, KR, SO, SU, BF, BU}, standing, respectively, for "key just inserted," "key just removed," "seat just became occupied," "seat just became unoccupied," "belt has just been fastened," and "belt has just been unfastened." The FSM's output alphabet is {ON, OFF}, for "turn (or keep) the buzzer ON" and "turn (or keep) the buzzer OFF." The FSM must turn on the buzzer for all illegal or suspicious sequences and turn it off only when law and order is restored. Identify all suspicious sequences, and state your reasons for thinking them so.

1.33 Construct a program that accepts program text as input and writes a "documentation file" as output. A documentation file contains *only* those lines of the input file that contain:

- procedure or function headers
- **var**, **const** or **type** declarations
- comments.

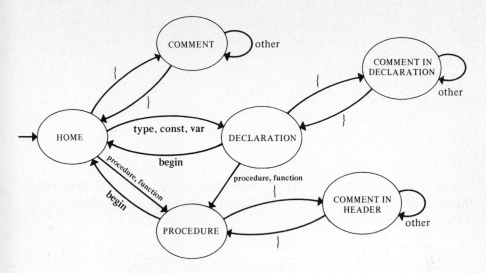

Fig. 1.29 Scheme for Exercise 1.33

Organize the program along lines suggested by Fig. 1.29.

1.34 For each of the following RE's, show at least four "typical" strings generated by the expression. Also give a concise (and accurate) description in English of the set of strings accepted. The description should *not* merely be a restatement of the RE.

    a) b* (b* + a + ab*a + ab*ab*a) b*

    b) b*cb* + c*bc*

    c) (a(c+d))* + (b(e + f))*

1.35 For each of the following pairs of RE's, determine whether both define the same set of strings. Prove your answers (that is, either argue convincingly but concisely that they are the same, or give a counter-example—a string generated by one, but not the other).

    a) (a* + b*)*
       (a+b)*

    b) (a*b*)*
       (a+b)*

    c) (ab)*
       a*b*

    d) a*
       a**

1.36 Use the precedence rules to restore parentheses to the following RE's. Restore *all* the parentheses so that the RE's correspond exactly to the definition.

    a) ab + c

    b) (a + b)*

    c) b*

    d) ab + cd + efg

    e) a*b*c*

1.37 Determine which of the following are valid laws for manipulating RE's. Either prove the law by analyzing the sets of strings described by each RE, or give a counterexample. Let X, Y, Z be RE's.

    a) *Associativity of alternation:* $X + (Y + Z) = (X + Y) + Z$

    b) *Associativity of composition:* $X(YZ) = (XY)Z$

    c) *Commutativity of alternation:* $X + Y = Y + X$

    d) *Commutativity of composition:* $XY = YX$

    e) *Distributivity of composition over alternation:*
       $X(Y + Z) = XY + XZ$

    f) *Distributivity of alternation over composition:*
       $X + (YZ) = (X + Y)(X + Z)$

    g) *Distributivity of closure over alternation:* $(X + Y)^* = X^* + Y^*$

    h) *Distributivity of closure over composition:* $(XY)^* = X^* Y^*$

    i) *Composition with empty string:* $X\epsilon = X\epsilon = X$

    j) *Alternation with empty string:* $X + \epsilon = X$

1.38 Use the notation R^+ to get shorter definitions of unscaled floating-point numbers, with and without a leading (trailing) digit required (see Section 1.3).

1.39 Design an RE to describe scaled floating-point numbers (i.e., those in PASCAL, FORTRAN, etc.—reals optionally followed by a signed integer exponent).

1.40 For each of the following languages, given an RE, an NDFSR, and an FSR, all of which define the same language:

    a) {ab, aab, abb, aaab, aabb, abbb, $\cdots$ }

    b) {ϵ, ab, c, abc, abab, ababc, cc, ababcc, $\cdots$}

    c) {a, ab, ac, abc, abb, acc, abbc, abcc, abbb, $\cdots$}

1.41 Design an FSR (possibly an NDFSR) that will accept all strings from the language defined by each of the following RE's:

    a) $(0+1)^*(000 + 111)$

    b) $(a + bc)^*$

    c) $ab^*c + (b + a^*)c$

    d) $x(((xy)^* + z) + ((x + y)z)^*)y$

1.42 Find a regular expression that describes the language accepted by each of the following FSM's. Try to simplify the RE as much as possible.

a)

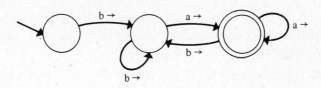

Figure 1.30

b)

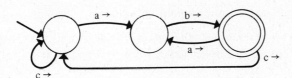

Figure 1.31

c)

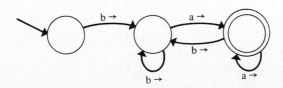

Figure 1.32

1.43 Here is a transition table for an FSR with input alphabet $\{0, 1\}$, initial state q_0, and final state q_2:

	input	
state	0	1
q_0	q_0	q_2
q_1	q_2	q_1
q_2	q_1	q_0

a) Draw the graph representing this FSR.

b) Give an equivalent RE.

c) Consider the set of binary strings, or language, this FSR accepts. Each of these strings can represent a positive integer. Exactly what integers does this FSR accept? (If you can't describe them mathematically, try listing them in order.)

1.44 Prove the converse of Theorem 1.1—that is, show that for any FSR, F, there is an RE that defines the same language as is accepted by F.

1.45 Is it possible to construct an FSG that can output the following sequence?

1010010001000010000010000001 $\cdots$

If it is possible, do so. If not, explain why not.

1.46 Consider a subset of the class of arithmetic expressions that involves just addition, multiplication, and parentheses. For example, "$a + b$," "$a* (b + c)$," and "$(a + b)* (c + d)$" belong to this class. Is it possible to build an FSR to recognize this class of expressions? If so, do it. If not, why not?

Chapter 2

More Models of Control: Flowcharts and Programs

In this chapter we consider two more models of control—models that more closely resemble the kinds of control constructs usually encountered in programming. One model is flowcharts; the other is a subset of the control structures of many programming languages. Suitably restricted, these models are equivalent to the FSM and RE models of Chapter 1. Because of this equivalence, we are able to use the model most convenient for analyzing a particular problem.

2.1 Flowcharts

Although the flowchart is historically rooted in programming it is, like the FSM, simply another representation of a computation as a graph. You are probably already familiar with flowcharts such as that shown in Fig. 2.1, at least informally. They have been used for many years as design and documentation aids in practical programming situations. This use is declining, however, and is not our reason for discussing them; rather, we are interested in investigating their computational power.

As we said previously, a flowchart representation of a program is similar to its representation in conventional programming languages; the major difference is the use of a graph to denote the control relations. The nodes of the graph represent steps in the computation and the arcs represent the sequence of these steps. It's common to define flowcharts and the computations they perform only informally. Here, however, we

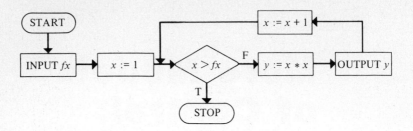

Fig. 2.1 An example of a flowchart

provide a formal definition in order to show the computational power of the flowchart model.

> **Definition 2.1.** A flowchart, F, consists of a labeled directed graph; a set of variables, $x_1, \cdots, x_n$, each of which may take on values from a corresponding domain, $D_1, \cdots, D_n$; a set of initial values, $i_1, \cdots, i_n$ for the variables; and input and output tapes, identical to those for FSMs. The directed graph (see Fig. 2.2) contains:
>
> a) One entrance node labeled START, having no predecessors and exactly one successor;
>
> b) One or more exit nodes, each labeled STOP, and having no successor(s);
>
> c) Zero or more computation nodes, each having exactly one successor and labeled with a sequence of assignment statements whose only variables are the x_i;
>
> d) Zero or more decision nodes, each labeled with a logical expression whose only variables are the x_i, each having exactly two exiting arcs, one labeled T (for true), the other F (for false);
>
> e) Zero or more input nodes, having exactly one successor labeled INPUT x_i, where $1 \leq i \leq n$ and D_i (the domain of values of x_i) contains the input alphabet;
>
> f) Zero or more output nodes, having exactly one successor labeled OUTPUT x_i, where $1 \leq i \leq n$ and the output alphabet contains D_i;
>
> A *finite flowchart* is a flowchart with a finite number of nodes whose domains $(D_1, \cdots, D_n)$ are finite sets. □

The flowchart shown in Fig. 2.1, for example, is legal according to this definition.

In addition, a flowchart defines a computation to be performed on an input as follows.

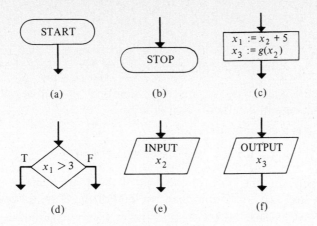

Fig. 2.2 Flowchart nodes

Definition 2.2. For a given input tape, the computation defined by a flowchart, F, is a sequence of *snapshots*, s_0, s_1, $\cdots$, and a corresponding sequence of output symbols (including ϵ), o_0, o_1, $\cdots$. Each snapshot is an $n + 1$-tuple (where n is the number of variables) of the form

$$(\text{node}, v_1, v_2, \cdots, v_n),$$

where "node" is a flowchart node and the v_i are variable values. A snapshot indicates the current node in the flowchart and the current values of the variables. The sequences $<s_i>$ and $<o_i>$ are defined as follows.

- If e is the entrance node, then the initial snapshot and output symbol are

$$s_0 = (e, i_1, \cdots, i_n)$$
$$o_0 = \epsilon$$

(where the i_j, again, are the initial variable values supplied with the flowchart's definition).

- Assume that

$$s_i = (p, v_1, \cdots, v_n) \, .$$

Then,

- If p is the entrance node, then

$$s_{i+1} = (\text{successor}(p), v_1, \cdots, v_n)$$

- If p is a computation node, then

$$s_{i+1} = (\text{successor}(p), v_1', \cdots, v_n')$$

where values of v_i' are the values of x_i that result from the assignment statements that label p.

- If p is a decision node, then

$$s_{i+1} = (q, v_1, \cdots, v_n) \, ,$$

where q is the node at the end of the T arc if p's label evaluates to *true*, and q is the node at the end of the F arc otherwise.

- If p is labeled INPUT x_i, and the next input symbol is y, then

$$s_{i+1} = (\text{successor}(p), v_1, \cdots, v_{i-1}, y, v_{i+1}, \cdots, v_n)$$

- If p is labeled OUTPUT x_i, then

$$s_{i+1} = (\text{successor}(p), v_1, \cdots, v_n), \quad o_i = v_i$$

In all other cases,

$$o_i = \epsilon$$

- If p is an exit node, or p is an input node and the input is exhausted, then s_i is the last snapshot in the computation. □

You can see that the definitions above correspond to the usual notion of what a flowchart is and how it specifies a computation. You will also find definitions in which the only input and output occurs through the initial and final values of the variables. This leads to a slightly simpler model than ours, but one inconvenient for our discussions of finite flowcharts.

Looking back at the flow chart in Fig. 2.1 we see that the variables are fx, x, and y. Thus, assuming that the initial value of all these variables is zero (which doesn't really matter in this case) and that the input will be 10, we get the following sequence of snapshots:

s_0: (START, 0, 0, 0)

s_1: (INPUT fx, 0, 0, 0)

s_2: ($x := 1$, 10, 0, 0)

s_3: ($x > fx$, 10, 1, 0)

s_4: ($y := x * x$, 10, 1, 0)

s_5: (OUTPUT y, 10, 1, 1), $o_5 = 1$

s_6: ($x := x + 1$, 10, 1, 1)

s_7: ($x > fx$, 10, 2, 1)

s_8: ($y := x * x$, 10, 2, 1)

s_9: (OUTPUT y, 10, 2, 4), $o_9 = 4$

.

.

.

2.1.1 THE RELATION BETWEEN FSM'S, RE'S, AND FLOWCHART PROGRAMS

Because FSM's and flowcharts denote computations that consume inputs and produce outputs, we may, therefore, compare their input–output behaviors.

> **Definition 2.3.** Two programs are considered input–output equivalent iff for any input tape, they produce the same sequence of output symbols after each input symbol read. □

We say "· · · after each symbol read" rather than "produce the same output tape," because we wish to emphasize that the two programs produce the same outputs at the same relative points in their reading of the input stream.

Flow charts might seem to describe a richer set of computations than FSMs. It turns out, however, that despite the flowchart's appearance of greater computational power, the two models are actually equivalent.

> **Theorem 2.1.** For any FSM, there is an input–output equivalent finite flowchart. For any finite flowchart, there is an input–output equivalent FSM. □

The restriction to *finite* flow charts (a finite set of nodes, variables, and domains) is crucial. We will see part of the reason for this in the way the proof of the theorem is constructed, but a fuller explanation must wait until the discussions about more general computational models in Chapter 13.

Proof of theorem 2.1. This proof consists of two parts, each of which is a construction corresponding to one of the statements in the theorem. The first part is easy, and is left as an exercise (see Exercise 2.9). Suppose, then, that we need only prove the theorem's second assertion—that, given a flowchart we can construct an input–output equivalent FSM.

Assume that F is a finite flow chart with variables $x_1, \cdots, x_n$, initial values $i_1, \cdots, i_n$, domains $D_1, \cdots, D_n$, and nodes $p_1, \cdots, p_r$. Our goal is to construct an FSM, M, that is input–output equivalent to F. Since the set of nodes and the domains are all finite, the set, Q, of all possible snapshots is finite. This is the central observation of the proof (indeed, you might first try completing the proof yourself before reading further). Essentially, we now use Q as the set of states of the FSM we are constructing; that is, there will be one state in this FSM corresponding to each possible set of variable values and positions in the flowchart. Therefore, we can build an FSM, $M = (Q, q_0, NEXT, OUT)$, where

- If e is the entrance node, then

$$q_0 = (e, i_1, \cdots, i_n);$$

- If p is a computation, decision, entrance, or output node and $(p', v_1' \cdots, v_n')$ is the snapshot that follows $(p, v_1, \cdots, v_n)$ according to Definition 2.2, then

$$NEXT(\ (p, v_1, \cdots, v_n), \epsilon) = (p', v_1', \cdots, v_n');$$

- If p is labeled INPUT x_i,

$$NEXT(\ (p, v_1, \cdots, v_n), y)$$
$$= (successor(p), v_1, \cdots, v_{i-1}, y, v_{i+1}, \cdots, v_n);$$

- $NEXT$ is undefined otherwise;
- If p is labeled OUTPUT x_i,

$$OUT(\ (p, v_1, \cdots, v_n), \epsilon) = v_i;$$

- OUT is undefined otherwise.

With a little thought, it should be clear that M and F are input–output equivalent. There is a state in M corresponding to each possible configuration in a snapshot of F; moreover, the $NEXT$ function has been defined so that these states occur in the proper order. □

Note that what we treat as data (variable values) in a flow chart we treat as control (position in a program) in an FSM. This is but one example of duality between control and data; we will see other examples later. Further, in other contexts, when we refer to the "state of a computation" we will mean, as in this proof, both the current variable values and the current control values—our current location in the program. At the same time, we also have a concise way of describing related states of an FSM and of defining the state transition function as a function of the variables that "contain" the state.

2.2 Simple Programming Language Control

Historically, flowcharts were used as graphic representations of programs. More specifically, they have been used to display the control flow of programs—the order of statement execution. Therefore, it is hardly surprising to find that the class of computations that can be present in the flowcharts of Section 2.1 can also be presented in a large class of programs written in typical programming languages. We demonstrate this in the sections to follow, first by defining a family of hypothetical programming languages called FCL/1, FCL/2, etc. (each of which is slightly different from the others), then by showing their essential equivalence to each other and to flowcharts. In doing so, we also explore alternative ways to design the control constructs of a language—those statements that describe control flow.

2.2.1 FCL/1

The statements of FCL/1 (for Flowchart Language 1) include all of the assignment statements, INPUT statements, and OUTPUT statements allowed in a flowchart. All statements in FCL/1 may have labels. In addition, FCL/1 has two control statements:

- **goto** *L*, for *L* a statement label, and
- **if** *C* **then goto** *L*.

Any FCL/1 program has a set of variables and respective domains, as does a flowchart. Statements in the program (separated by semicolons, as in PASCAL) are executed left to right, except when directed otherwise by one of the control statements.

FCL/1 programs describe the same set of computations as flowcharts. That is, any FCL/1 program is input–output equivalent to a flowchart program and vice–versa. Indeed, any FCL/1 program is simply a linear (or one-dimensional) representation of a flowchart. This result is sufficiently clear that we will not bother with a formal proof here (see Exercise 2.10).

The **goto** construct of FCL/1 and similar constructs in other languages are known generically as unconditional transfers. The **if-then-goto** construct is an instance of a conditional transfer construct. One final control construct—sequencing—is implicit in the semantics of FCL/1; that is, two adjacent statements are by default executed separately in the order written.

2.2.2 FCL/2

The simple statements of FCL/2 will be the same as those of FCL/1 and flowcharts (assignment, INPUT, and OUTPUT).   Its three control constructs, however, are the following:

- Simple sequential execution is represented by the construct

  begin $S_1; \ldots ; S_n$ **end**

  and is treated as a single statement whose execution consists of executing the statements S_i separately and in order.
- Conditional execution is represented by the construct

  if C **then** S_1 **else** S_2,

  where C is a logical expression and S_1 and S_2 are statements (including **skip**, meaning "do nothing").   The meaning of this construct is, as in PASCAL: If C is true, S_1 executes; otherwise S_2 executes. The construct is an instance of a *selection construct.*
- Iterative execution is represented by the construct

  while C **do** S,

  where C is a logical expression and S is a statement. The meaning of this statement is also as in PASCAL: If C is false, the statement has no effect; and if C is true, S is executed and the entire process is repeated. The statement is an instance of an iteration or looping construct.

2.3 Equivalence of Flowcharts and Programs

While it may have been obvious that FCL/1 programs and flowcharts are input–output equivalent, it's not so obvious that the same is true of FCL/2 programs. What should be clear, however, is that for any FCL/2 program there is an input–output equivalent flowchart.   Any simple statement corresponds, by assumption, to a flowchart node.   Any sequence, for example,

$$S_1; \ S_2; \ \cdots \ ; \ S_n$$

corresponds to a flowchart as the sequence above does to the flowchart in Fig. 2.3. The statement

  if C **then** S_1 **else** S_2

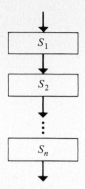

Fig. 2.3 Flowchart for a statement sequence

corresponds to the flowchart in Fig. 2.4.

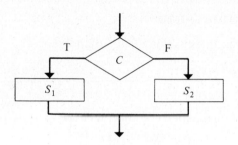

Fig. 2.4 Flowchart for an **if** statement

Finally, the statement

    while C **do** S

corresponds to the flowchart in Fig. 2.5.   Moreover, any flowchart is input–output equivalent to some FCL/2 program.   That is, the control constructs of FCL/2 are adequate for expressing any computation expressible by a flowchart program.   Consider any flowchart and number its nodes with integers from 1 to, say, n.   Introduce a new variable, $NEXT$, distinct from all others and having as its domain the set $\{0, \cdots, n\}$.   We construct the following FCL/2 program:

    $NEXT := 1;$
    while $NEXT \neq 0$ **do**
        if $NEXT = 1$ **then** S_1

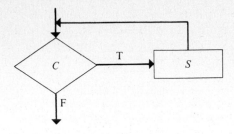

Fig. 2.5 Flowchart for a **while** statement

else if $NEXT = 2$ **then** S_2

$$\cdot \qquad \cdot \qquad \cdot$$
$$\cdot \qquad \cdot \qquad \cdot$$
$$\cdot \qquad \cdot \qquad \cdot$$

else if $NEXT = n$ **then** S_n **else** $NEXT := 0$

where S_i is determined as follows.

- If node i of the flowchart is STOP, then S_i is

  $NEXT := 0$

- If node i is a decision node with label C whose T successor is node j and whose F successor is node k, then S_i is

  if C **then** $NEXT := j$ **else** $NEXT := k$

- Otherwise, if node i contains some other simple statement(s), S and has successor j, then S_i is

  begin S; $NEXT := j$ **end**

Note that if we start with a finite flow chart, we end up with an FCL/2 program with all variables having finite domains.

We can summarize these results about flowcharts and linearized programs in the following theorem.

Theorem 2.2. For any flow chart program, there are input–output equivalent FCL/1 and FCL/2 programs. For any FCL/1 or FCL/2 program, there is an input–output equivalent flowchart program. Hence, for any FCL/1 program, there is an input–output equivalent FCL/2 program and vice versa. □

From Theorem 2.1 we therefore conclude

Theorem 2.3. For any FSM there are input–output equivalent FCL/1 and FCL/2 programs all with finite variable domains. Conversely, for any FCL/1 or FCL/2 program that has only finite variable domains, there is an input–output equivalent FSM. □

Although FSMs, FCL/1, FCL/2, and finite flow charts are equivalent in power, it would be a mistake to conclude that they should be used interchangeably. Real programming languages include some or all of the constructs of FCL/1 and FCL/2; they include other features as well. The selection of which of these constructs to use in a particular program is a design decision, and the mark of a good programmer is intelligent, tasteful use of the features provided by his language.

2.4 Further Reading

The equivalence of finite automata, flowcharts, FCL/1 and FCL/2 is an issue of some mathematical and philosophical interest. As a whole the subject has not been of much practical interest, owing primarily to the inappropriateness of the level at which finite automata resemble real digital computers. Later, in Chapter 13, we discuss another computation model that is more realistic.

The subject of flowcharts is of waning interest in programming and is not of much value for further study in this textbook. The fact that FCL/1 is equivalent to flowcharts is not really surprising because of the obvious power of the **goto**. Boehm and Jacopini (1966) first showed that all flowchart programs could be written using only sequential, conditional, and iterative control, if extra variables were allowed (which implies the equivalence of FCL/1 and FCL/2).

Exercises

2.1 Consider the following program that reads a value for x ($x \geq 2$) and computes the smallest power of 2 that is larger than x.

```
begin
var x, y: integer;
read(x);
y := 2;
while y ≤ x do y := y + y;
write(y);
end.
```

a) Draw a flowchart, *F*, for this program.

b) Show the computation performed by this program on the input of $x = 13$ as a series of snapshots and outputs.

c) Construct an FSM, *M*, that is input–output equivalent to your flowchart *F*. You may assume $2 \leqq x \leqq 6$.

2.2 Suppose the definition of "flowchart" (Definition 2.1) is changed to

- restrict computation, decision, input, output, and exit nodes to exactly one predecessor, and

- introduce a new type of node, *MERGE*, that does no computation but has two predecessors and one successor.

Represent a *MERGE* node as a small circle with two arcs entering and one leaving. All nodes with more than one predecessor will, thus, be designated *MERGE* nodes. Show that the new kind of flowchart is equivalent to the one in Section 2.1.

2.3 Examine the informal definitions of flowcharts in several introductory texts. Give a more formal definition of them. In what ways do these definitions differ from the one used in this text? Are they input–output equivalent to the one given here? (Prove your answer.)

2.4 Define a "nondeterministic" flowchart and a corresponding concept of a computation. Prove that finite nondeterministic flowcharts have no more power than the deterministic ones.

2.5 Suppose that the statement **if** *C* **then hop** replaces **if** *C* **then goto** *L* in FCL/1, but the unconditional **goto** remains. (The statement **hop** means "ignore the next statement and continue on from there.") Would the new FCL/1 still have the same power as the old one? Prove your answer.

2.6 Suppose the statement **if** C **then** *L*1, *L*2 replaces both the **goto** and the **if** *C* **then goto** *L* in FCL/1. The new statement has the effect of transfering control to *L*1 when *C* is false and to *L*2 when *C* is true. Would the new FCL/1 still have the same power as the old one? Prove your answer.

2.7 Suppose the statement **if** C **then** S_1 **else** S_2 (from FCL/2) replaces both the **goto** and the **if** *C* **then goto** *C* in FCL/1. Would the new FCL/1 still have the same power as the old one? Prove your answer.

2.8 Why does Theorem 2.1 not work for nonfinite flowcharts?

2.9 Prove the first part of Theorem 2.1—that for any FSM there is an input–output equivalent flowchart.

2.10 Write an algorithm for converting a flowchart into an FCL/1 program, and vice versa. Suppose we had defined flowcharts to be connected graphs. Would Theorem 2.2 be affected? Why or why not? How would the changed definition affect your algorithm?

2.11 Define an FCL/1 program "snapshot". In doing this, you will need to introduce some systematic naming of the statements. Show traces of the execution of several FCL/1 programs, including the programs of Exercises 2.16 through 2.18.

2.12 The definitions of FCL/2 control constructs permit statements to contain statements. Consider the following FCL/2 program:

```
begin
x := 10;
while x > 5 do
    if odd(x) then x := x−1 else x := x/2;
x := x+3
end
```

a) Prove that this program was constructed according to the rules for FCL/2 (that is, parse the program).

b) Draw a flowchart for this program using the correspondences in Figs. 2.3 through 2.5.

c) Write an equivalent FCL/1 program.

d) Draw a flowchart for the FCL/1 program.

2.13 The FCL/2 **if–then–else** statement is, in fact, redundant. Obtain its result using only sequencing and the **while** statement.

2.14 In this exercise you will analyze, write, and rewrite a program that reads in sets of three coefficients A, B, and C and solves the quadratic equation

$$Ax^2 + Bx + C = 0$$

Your program must explicitly recognize and label the special cases that arise for complex roots and when certain coefficients are zero. The program should have a loop that prompts the user to see if more tests are to be made.

a) Your analysis of special cases should reveal a small number of distinct states that the program might be in after preliminary inspection of the input data. Illustrate the analysis that determines the state in the FSM style, but label tranition arcs with data tests instead of possible input values. (The result should resemble a flowchart.)

b) On the basis of the analysis in (a), generate a set of test cases for checking the program. Include the values you expect the program to compute. Explain why you think these tests will convince you that the program works. Do this *before* you attempt to execute the program.

c) Write the program using *only* FCL/2 control constructs. Test it with the data you generated in (b).

d) Rewrite the program using *only* FCL/1 control constructs. Test it, too. Include comments to show the correspondence of the FCL/2 program to the FCL/1 program of (c).

2.15 Redo Exercise 2.14 for the problem of reading x, y, and z, the lengths of the sides of a triangle, and the triangle classifications *equilateral*, *isosceles*, and *scalene*, or the identification *not a triangle*.

2.16 You are given the task

read x; compute and print $x!$, where $x! = x \cdot (x - 1) \cdots 3 \cdot 2 \cdot 1$.

Do the following:

a) Draw a valid flowchart for a program that performs the task.

b) Show the computation performed by the flowchart on a typical input series of snapshots and outputs.

c) Write an FCL/2 program to perform the same task.

d) Write an FCL/1 program to perform the same task.

2.17 Redo Exercise 2.16 for the task

read x and n; compute and print x^n.

2.18 Redo Exercise 2.16 for the task

read x; print x, x^2, x^3

2.19 A "planar" graph is one that can be drawn on a sheet of paper without ever having two arcs cross.

a) Prove that every FCL/2 program has a planar flowchart.

b) Give an example of a nonplanar flowchart and an FCL/1 program that corresponds to it.

c) Translate your nonplanar example of (b) into FCL/2.

2.20 Write an FCL/2 program that is an input–output equivalent to the flowchart in Fig. 2.6.

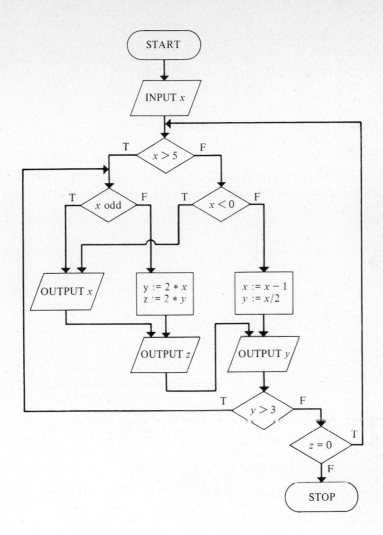

Fig. 2.6 Flowchart for Exercise 2.20

2.21 Let *test*(*c*) be a boolean function that returns *true* if *c* is the current input character and returns *false* otherwise. Let *more*() be a Boolean function that returns true only if the input has not been exhausted. Let *scan*() advance the input to the next character. Then the regular expression (*a* + *b*)* would be recognized by the program

```
while more() do
    begin
    scan();
    if not test('a') and not test('b') then error;
    end;
```

Write a program that accepts the regular expression

$$(a + b)(a + b + 1 + 2)^*,$$

which is one way to express rules for constructing legal identifiers in a programming language (of course more letters than just a and b and more digits than just 1 and 2 are typically allowed). Write programs for the two RE's $(a^*(bc + d)^*)^*$ and $ba^*(c + b^*)$.

Chapter 3

Additional Control Structures

In the last chapter we introduced two simple programming languages, FCL/1 and FCL/2. The languages differed only in the control constructs provided: FCL/1 provided sequencing, and **goto** and conditional **goto** statements; FCL/2 provided sequencing, and statements, **if–then–else**, and **while do**.

The three constructs of FCL/2 are generally thought to be "more expressive" than those of FCL/1. The notion of one construct being more expressive than another is not very precise; generally, however, it means "closer to the kinds of things we usually want to say." For example, the effect of the FCL/2 construct

   if $x > y$ **then** $min := y$ **else** $min := x$

seems clearer than that of the equivalent in FCL/1

```
     if x > y then goto xbig;
     min := x;
     goto next;
xbig:
     min := y;
next:
```

The control structures of FCL/2 are, as a group, less primitive than those of FCL/1; they more closely match the kinds of things we want to write in typical programs. However, there are many instances in which the constructs of FCL/2 are less than convenient. For example, suppose we want to sum a sequence of positive numbers as they are read in; in FCL/2 we must write something like the following:

> *INPUT x*;
> **while** $x > 0$ **do**
>   **begin** $y := y + x$; *INPUT x* **end**;

Note, we must repeat the INPUT statement because of the placement of the test in the **while** statement. This is a place, in fact, where FCL/1's **goto** would have been rather handy. Alternatively, we could search for another control construct that more closely reflects this particular style of processing; if this kind of loop is sufficiently common, it would be nice to express it directly rather than to construct it from either **goto**'s or **while**'s.

A good deal of research and experimentation has been done to find control constructs that, when added to a language, increase the convenience and clarity of programs. In this chapter we define some of these constructs.

3.1 Selection Constructs

The **if** statement allows the selection of precisely one of two control paths. In many programming situations, however, it's necessary to select one of several paths. Two proposals have been made (and implemented in some languages) that permit such a selection. Unfortunately, although the two constructs are somewhat different from each other, they have both been given the name "**case** statement." To avoid confusion we will call them the **case** and **numberedcase** statements, even though these terms differ from common usage.

3.1.1 THE "CASE" STATEMENT

The syntax of the **case** statement is generally of the form:

> **case** E_0 **of**
> $E_1: S_1$;
> $E_2: S_2$;
>
> .   .
>
> .   .
> $E_n: S_n$;
>   **otherwise** S_*
>   **end**

where $E_0, \cdots, E_n$ are all expressions of the same type and the S's are statements. Semantically, this is equivalent to:

```
begin
var t: Etype;
t := E₀;
if        t = E₁ then S₁
else if   t = E₂ then S₂
          .

          .

          .
else if   t = Eₙ then Sₙ
else      S*
end
```

where *Etype* denotes the type of expressions E_0 through E_n. As you can see, exactly one of S_1–S_* will be executed. In particular, if expressions E_1 through E_n have different values, and if E_0 equals E_i, then S_i will be executed. If none of E_i are equal to E_0, then S_* will be executed. Hence, the **case** statement expresses a case analysis of the computation based on possible values of E_0.

3.1.2 THE "NUMBEREDCASE" STATEMENT

The **numberedcase** statement is a special form of the **case** statement in which E_0 is an integer and the other E_i are implicit. Typically, it has a syntax such as

    numberedcase E **of** S_1; S_2; S_3; $\cdots$; S_n **end**

where E is an integer-valued expression and S_1 through S_n are arbitrary statements. The **numberedcase** statement is equivalent to the following **case** statement.

```
case E of
    1:  S₁;
    2:  S₂;

        .

        .

        .

    n:  Sₙ
end;
```

Thus the effect of executing the **numberedcase** statement is to execute S_E, and only S_E, so long as the value of E lies in the range 1 to n. If the value of E is outside this range, no statement is executed. This is considered an error in some languages.

A minor variation of the **numberedcase** statement provides an **otherwise** clause and a statement, S_*, to be executed when the value of E lies outside the range 1 to n. Typical syntax is

> **numberedcase** E **of** S_1; S_2; S_3; $\cdots$; S_n; **otherwise** S_* **end**

and means

> **case** E **of**
>   1: S_1;
>   2: S_2;
>   .
>   .
>   .
> n: S_n;
>   **otherwise** S_*
> **end**;

The **numberedcase** statement may seem a trivial modification of the **case** statement. It is. Furthermore, it can lack readability; if there are fifty cases, for example, one cannot easily tell which case corresponds to which number. Historically, though, the **numberedcase** statement was invented before the **case** statement. This invention was largely due to a particularly elegant implementation that was carried over from machine language programming.

3.2 Iterative Constructs

There are many variations and elaborations of the **while** statement, but we'll discuss only a few to illustrate them.

3.2.1 THE "REPEAT" AND "LOOP" STATEMENTS

The **while** statement places the termination test before the loop body. Thus, if the Boolean expression evaluates to false the first time, the body will never be executed. Sometimes it is convenient to have the test placed after the body, thus guaranteeing that the body will be executed at least once. The PASCAL syntax for a construct with this property, for example, is

> **repeat** S **until** B,

which is semantically equivalent to

S;   **while** $\sim B$ **do** S.

We use $\sim B$" here instead of B", since repeating the body *until* B is true is the same as repeating it as long as B is false.

The **while** statement places the termination test before the loop body; the **repeat** statement places it after the body. There is, however, no overwhelming reason for the termination test to occur at either of these places. Some languages, in recognition of this fact, provide two related constructs—the **loop** and **exitif** statements. The statement

    loop S_1; $\cdots$; S_m **end**

repeats the execution of S_1 through S_m indefinitely. The statement

    exitif B

causes the innermost loop in which it occurs to be terminated if the boolean expression, B, is true. Thus, for example, we may write

    loop
       S_1; $\cdots$; S_k;
    exitif B_1;
       S_{k+1}; $\cdots$; S_l;
    exitif B_2;
       .
       .
       .
    end;

This loop may terminate from an interior point for any number of reasons.

3.2.2 THE "FOR" STATEMENT

The **for** statement expresses a common kind of loop in which a program segment, called a body, is executed repeatedly for various values of a quantity called a loop control variable. Most often, the values of the loop control variable are an ascending sequence of integers. If there were no special construct for this form of iteration we might write it as

    $i := E_1$;
    while $i \leq E_2$ **do**
       begin S; $i := i + 1$ **end**

where i is an integer variable, E_1 and E_2 are integer-valued expressions, and S is the statement to be repeated.

Since iteration is so common, most programming languages provide a special construct for it. It generally looks something like

> **for** $i := E_1$ **to** E_2 **do** S;

where the subscripts on the E's were chosen to match their use in the preceding **while** statement. The "meaning" of the **for** statement, however, is not exactly the same as that of the **while** statement. Instead, its execution is more like that of the sequence (possibly empty) of statements

$$S(\alpha); \ S(\alpha + 1); \ \cdots ; \ S(\beta)$$

where α is the initial value of E_1, β is the initial value of E_2, and $S(k)$ is the statement S with k substituted for all instances of the control variable, i.

Unlike the **while** version, in **for** statements the expression E_2 is computed only once, prior to any execution of the loop body. Thus, any changes by the loop body to the variables in E_2 will not affect the number of times the loop is executed. In PASCAL and FORTRAN, in fact, such changes are illegal. Nothing that the loop body does can affect the number of times the body will be executed. To understand the effect of a loop, we can separate the control of the loop from the computation that the body performs on each iteration.

Although the intended use of **for**-like statements is roughly the same in all languages, this is one place where language designers' imaginations have run wild. Here are some common variations.

- The termination value, E_2, may be re-evaluated on each iteration.
- Many languages allow the specification of a "step size," so that the control variable may be incremented by values o her than 1.
- Some languages allow one to count either up or down; that is, the step size can be either positive or negative, and the termination test is determined accordingly.
- In some languages the control variable is an ordinary variable and has a defined value upon loop completion. In others, it is an ordinary variable, but its value on termination of the loop is not defined.
- In some languages the control variable must be of type *integer*; in others it may be type *real*; in yet others it may also be any of a large number of other types.
- In some languages there is a Boolean **while** termination test, as well as a counting test.

All of these variations are useful in certain situations; however, the simple form provided by PASCAL is sufficient for the vast majority of loops.

3.3 Routines

The **case**, **numberedcase**, **repeat**, and **for** statements are not strictly necessary because we can express each in terms of other control constructs. However, they are all extremely useful because they serve to focus the attention of the programmer on a segment of code that is of comprehensible, but not trivial, size. Furthermore, the syntax of these constructs are direct indications of the programmer's intent and someone reading a program with these constructs does not have to decipher (or guess) what's going on.

By way of analogy, consider a half-tone photograph of the kind used in newspapers. When you look at it closely with a magnifying glass, you see collections of irregularly placed dots. The dots seem to run into one another in random patterns and it's hard to recognize the objects photographed. Putting down the magnifying glass and moving back, you begin to see the shapes formed by patterns of dots. At some particular distance, which may vary slightly for different people, the dots "disappear" and you see only a picture. Stepping farther back from this point will obscure details that are important visual clues; and, of course, at a very great distance, you can't see the picture at all.

The same phenomenon occurs in a program that is read by others. Programs composed of very many adjacent primitive operations are as difficult to understand as the magnified isolated dots of a half-toned photograph. In general, if primitive operations are grouped together into a smaller number of less primitive ones (which might therefore be called "macro-operations"), you perceive more easily their significance and their relationships to one another. This observation is true of built-in macro-operations such as the control constructs we have been discussing. However, it is also true of programs that users write.

Programming languages provide various facilities for constructing macro-operations, but by far the most important is the *routine*. (Particular kinds of routines are often called *procedures, functions,* or *subroutines.*) In its simplest form, a routine is merely a named group of statements. This in itself is very useful for two reasons. First, if we use an operation in several places, it saves us a certain amount of effort to be able to reuse the text that implements the operation. Likewise, we may save program space if the compiler can contrive to keep just one copy of the machine code implementing a certain piece of repeated text. Second, and perhaps more important, we increase the clarity of our programs if we write concise and descriptive statements rather than unnecessarily detailed ones. At some point, of course, we must write the code that implements these operations. In the places where we use the operations, however, these implementation details can be irrelevant and confusing.

3.3.1 ROUTINE INVOCATION

The control action associated with routines is called *routine invocation* (we also use the synonymous terms *invocation, routine call,* and *call*). A declaration of routine provides a *body*—a segment of program text. When a routine is called, the effect is as if that body of text—suitably modified—replaces the call statement. As a routine is supposed to encode a macro-operation, this should be more or less the definition you would intuitively expect.

The complexities surrounding a definition of routine invocation in most languages (and those complexities are considerable) all hinge on the phrase "suitably modified." One necessary modification is obvious: the identifiers defined within a routine's body are supposed to be distinct from those defined in the text surrounding its call. For example, in

> **var** *i: integer*;
>
> **procedure** *skip*(*n: integer*);
> **var** *i: integer*;
>   **begin**
>   **for** *i* := 1 **to** *n* **do** *Write*(' ')
>   **end**;
>   .
>
>   .
>
>   .
> *i* := 5; *skip*(10)

the call *skip*(10) is equivalent to the statement

> **for** *i'* := 1 **to** 10 **do** *Write*(' ')

where *i'* represent some new identifier used nowhere else in the program. Thus, part of suitably modifying the routine's body involves renaming. Another, and major part, of a suitable modification involves the treatment of parameters, which we discuss in Section 3.3.2.

Technically we have been a little inaccurate in our definition of invocation. Languages such as PASCAL or FORTRAN do not, in general, allow bodies of code in all the contexts in which a routine call may appear. In particular, certain routines (generally called functions) can be invoked inside an expression—even though the body is a statement, and statements are not allowed inside expressions. For example, PASCAL does not allow us to replace *sin*(*x*) in

> *y* := *sin*(*x*) + 1

with the body of the sin routine, no matter how it is modified. However, we can replace the entire assignment statement with

$S; \quad y := t + 1.$

where S is the body of *sin*, modified to work on the value of x and to put its result in the new variable t. Therefore, we continue to define an invocation as a "suitably modified replacement," with the understanding that the details of this replacement can get rather complicated.

3.3.2 PARAMETERS

If routines provided only the ability to abbreviate a single fixed body of text, their utility would be rather limited. For example, the definition

procedure *QSolve*;
    begin
      $x := (-b + sqrt(b * b - 4. * a * c))/(2. * a)$
    end;

does allow us to write

    QSolve;

wherever we might want to write

    $x := (-b + sqrt(b * b - 4. * a * c))/(2. * a)$

but this is not as useful as it might be. In particular, the fact that all the variables involved are those defined at the definition site of the procedure is an awkward limitation.

Hence, most languages allow the introduction of some systematic modification of the body of a routine as part of its invocation. Specifically, they allow programmers to provide routines with formal parameters— identifiers that are systematically substituted as part of routine invocation. For example, n in the definition of *skip* in section 3.3.1 is a formal parameter (or, for short, a *formal*) of *skip*. The constant 10 in the invocation *skip*(10) is called an actual parameter, (or for short, an *actual*).

There are various rules by which one might define the systematic substitution of actuals by formals. We call the substitution process "parameter passing" and the rules for the substitutions, "parameter passing rules" or "parameter passing mechanisms." The rule used for n in the skip procedure is known as call-by-value and is defined by the steps

1. Create a new variable, named, say, n', of the type indicated in the list of formal parameters (*integer*, for instance, in the case of routine *skip*).

2. Replace all occurences of n in the routine body with n'.

3. Insert the statement

$n' := a;$

where a is the actual corresponding to n, at the beginning of the routine body.

Thus, $skip(10)$ becomes

$n' := 10;$
for $i := 1$ **to** n' **do** $Write(' ')$

Call-by-value is by no means the only conceivable mechanism for passing parameters to routines. Another important one is call-by-reference, used for **var** parameters in PASCAL. Call-by-reference is fully explained in Section 15.3; but intuitively, we can understand that a formal parameter name is replaced with an actual parameter name.[*] Call-by-reference parameters can, therefore, be altered by the routine body.

Not only are parameters "identifiers that get replaced at each invocation," but they are also "quantities on whose values computations depend" and thus, by extension, "variables modified by computation (or output parameters)." In any case, the term parameter can refer to variables not explicitly included in a formal parameter list. We call these global variables. For example, in

procedure $ResetLine$;
   begin
   $LinePos := 0;$
   end;

$LinePos$ is a global variable.

3.3.3 VALUE-RETURNING AND VALUE-LESS ROUTINES

Value-returning routines are those routines whose invocations may appear in expressions, where something denoting a value is expected. They are often used to implement mathematical functions, hence the frequent use of the keyword **function** to introduce a definition.

Value-less routines (which we often call *procedures*) are actions performed on a program's variables (including their input and output streams). The invocations of these routines are individual statements rather than expressions.

---

[*] There are some subtle issues here involving conflicting names, which is one of the reasons we have deferred a fuller discussion to a later section. For now, assume all the variable names are unique.

In most languages that provide both types of routine, it is possible, in fact, to specify a value-returning routine that also requires an additional action. For example, routines that return pseudorandom numbers not only return values, but also alter variables in a way that guarantees that the next call will produce a different value.*

Actually, the difference between value-less and value-returning routines is smaller than it might appear. Mathematically, both are functions from their inputs (including any global variables) to their output parameters (again, including globals). A value-less routine *does*, as we've shown, return values in the form of the output parameters it modifies. Further, a procedure invocation such as

> *increment*(x)

could be rewritten

> $x :=$ *increment*(x),

if *increment* were suitably redefined. If we had a language allowing multiple, simultaneous assignment statements, such as

> $(x, y, z) := (3, z, x + y)$,

we could even write routines that modify several parameters. By the same token, we may rewrite any value-returning routine as a value-less routine (with, however, considerable loss of clarity).

3.3.4 RECURSIVE ROUTINES

During an invocation of a recursive routine f, f itself may be invoked. A routine, f, is called *recursive* if f may be invoked during another invocation of itself. Recursive routines are the natural implementations of many algorithms.

Consider, for example, the *Comb* routine in Fig. 3.1, which computes the number of combinations of n things taken k at a time in almost the same way as standard mathematical textbooks. *Comb* is recursive because its body, specifically its **else** clause, contains not only one invocation of itself, but in fact, two such invocations.

It's worthwhile, if you haven't done it before, to see how a routine such as *Comb* is evaluated for specific argument values. Consider, for example, the call "*Comb*(3,2)". Since k is neither 0 nor n, the **else** clause

---

* There is a school of thought that holds that writing such routines is poor programming practice—precisely because routines are not functions in the mathematical sense.

will be executed, giving rise to two more invocations of *Comb*—*Comb*(2, 2) and *Comb*(2, 1). The initial invocation of *Comb* cannot be executed until the latter two *Comb's* are evaluated: Its execution is suspended for the moment.

The invocation *Comb*(2, 2) has the property that $k = n$, so the **then** clause of its body will be executed and will immediately return the value 1. The invocation *Comb*(2, 1), however, will execute the **else** clause and give rise to two more invocations, *Comb*(1, 1) and *Comb*(1, 0). Again, the current invocation—*Comb*(2, 1)—is suspended until the two new *Comb's* have been executed and have returned values. Before *Comb*(1, 1) and *Comb*(1, 0) are executed, we have two suspended invocations, *Comb*(3, 2) and *Comb*(2, 1).

As you can see, both of the newest invocations of *Comb* will immediately return the value 1. Their completion allows us to resume execution of *Comb*(2, 1), which returns the value 2. Completion of *Comb*(2, 1), in turn, allows us to resume execution of *Comb*(3, 2), which finally returns the value 3.

```
function Comb(n, k: integer): integer;
   begin
   if (k = 0) or (k = n)
      then  Comb := 1
      else  Comb := Comb(n − 1, k) + Comb(n − 1, k − 1)
   end
```

Fig. 3.1 *Comb*: A recursive routine to compute the number of combinations of *n* objects taken *k* at a time

Tracing the execution of a recursive routine is a good way to develop a feel for what's going on in a program, but its *not* a good way to reason about recursive routines in general. Actually, the organization of *Comb* is an example of a more general divide-and-conquer programming strategy that is very useful in reasoning about most recursive routines. The divide-and-conquer strategy simply involves splitting a large problem into two or more simple problems of the same kind as the original—and then providing an explicit solution to the simplest form of the problem.

Of course, that's exactly what the *Comb* routine does. We know an explicit solution for the simple cases: if $k = n$ or $k = 0$, then *Comb*=1. In more complicated cases we simply use the known property that

$$Comb(n, k) = Comb(n - 1, k) + Comb(n - 1, k - 1)$$

to divide the problem into two simpler ones of the same kind. We know that if the resulting problem is simple enough it will be solved directly. If it is still too large it will be split into two more smaller subproblems until each one can be solved directly.

Note, by the way, that a routine need not contain a direct call on itself to be recursive. It may also call a second routine which, in turn, calls the first. For example,

> **procedure** *Term*;
>   **begin** $\cdots$; *Factor*; $\cdots$ **end**;
> **procedure** *Factor*;
>   **begin** $\cdots$; *Term*; $\cdots$ **end**;

are both recursive (we say mutually recursive, because they invoke each other). In fact, we call a routine recursive if there is any chain of calls beginning with that routine that will eventually reinvoke itself.

We were careful, when defining routine invocation, to specify insertion of a suitably modified copy of the routine body *at the time the invocation is executed.* This is important in recursive routines. Had we specified that the substitution take place before the program started to execute, we would have had to deal with infinite programs. It is possible to do so, in fact; but we prefer to remain on more familiar ground.

3.3.5 THE POWER OF ROUTINES

As we have defined them, routines may seem simply a convenience so far as computational power is concerned. This is the case, in fact, if we restrict ourselves to nonrecursive routines. We will do so for now but return to recursive routines at greater length in Section 14.2.

Meanwhile, let's try another language, FCL/3, which extends FCL/2 with nonrecursive routine definitions and invocations. We'll assume a strictly call-by-value parameter mechanism, although this restriction is not really necessary. We now have

> **Theorem 3.1.** For any FCL/3 program, there is an input–output equivalent flowchart program (and, hence, input–output equivalent FCL/1 and FCL/2 programs). If all variable domains in the FCL/3 program are finite, the flowchart is finite. □

For proof, note that the necessary modifications of the routine body for any FCL/3 routine invocation result in a finite program. The lack of recursion allows us to replace all routine invocations with equivalent FCL/2 text.

3.4 Summary: Equivalence, Power, and Convenience

In this chapter and in the previous two chapters we proved that a large number of things are equivalent to each other. Consider the following partial list:

- FSM's with an alphabet of $\{0, 1\}$ are equivalent to FSM's with larger (but finite) alphabets.
- FSR's are equivalent to NDFSR's.
- Languages generated by RE's are identical to those that can be recognized by FSR's.
- FSM's are input–output equivalent to finite flowcharts.
- FCL/1, FCL/2, and FCL/3 are input–output equivalent to flowcharts and to each other.

You might well be wondering whether everything is going to be equivalent to everything else. Be assured, it won't.

The important points to be made at this juncture are

- The equivalence of all these formalisms allows us, without sacrificing generality, to reason about a problem, using whatever formalism best suits the situation.
- Equivalence is the equivalence of computational power—not of expressiveness. In any specific case, one of the equivalent formalisms will generally be more convenient than another.

3.5 Further Reading

The proliferation of syntax for control structures in programming languages is immense. Fortunately, most are similar in behavior to those discussed in this chapter; hence we need not go into a detailed analysis of control structures on a language-by-language basis. Certain languages stand out as either pioneers or unifiers. Among the most prominent are ALGOL 68 by van Wijngaarden, et al. (1975) and on PASCAL by Wirth (1971). One of the earliest languages to abolish the **goto** statement in favor of higher level control structures was Bliss as discussed in Wulf, et al. (1971).

An excellent article by Knuth (1974) addresses the issue of using appropriate control structures without being unduly biased by impassioned cries for a set of constructs that would, by themselves, eliminate the "software crisis."

Exercises

3.1 Suppose each **else** in the translation of the **case** statement in Section 3.1.1 were changed to a semicolon. Would this significantly change the meaning of the **case** statement? If so, in what way(s)?

3.2 Suppose the definition of the **case** statement were changed to require $E_1, \cdots, E_n$ (but not E_0) to be constants. Would this significantly change the meaning of the **case** statement? If so, in what way(s)?

3.3 Why wasn't the behavior of the **numberedcase** statement described as

> **if** $E = 1$ **then** S_1 **else**
> **if** $E = 2$ **then** S_2 **else**
>
>   .
>
>   .
>
>   .
>
> **if** $E = n$ **then** S_n

instead of the description given?

3.4 Write a definition of the **loop** construct using FCL/2. Suppose you are limited to one **exitif** statement. Why does this limitation simplify the definition?

3.5 Section 3.2.1 shows a **repeat** loop in terms of a **while** statement. In a similar manner (a) write a **while** loop in terms of a **repeat** statement and (b) a **loop** statement in terms of a **while** statement.

3.6 Redo Exercise 3.4, but use FCL/1.

3.7 Show that a language with the **loop** construct does not strictly need **while** and **repeat** statements.

3.8 Consider the statement

> **for** $i := E1$ **step** $E2$ **until** $E3$ **do** S

where $E1$, $E2$, and $E3$ are integer-valued expressions. Construct FCL/1 programs that are equivalent to this statement under the following assumptions about its meaning:

a) The loop variable, i, is local to the loop; the expressions $E1$ through $E3$ are evaluated only once. The expression $E2$ is assumed to be positive and the loop "counts up" from E1 to E2.

b) The loop variable, i, is local to the loop; the expressions $E1$ through $E3$ are evaluated only once. Expression $E2$ may be either positive or negative. If it is negative the loop "counts down" from $E1$ to $E3$; if it is positive, the loops "counts up" in the usual fashion. (Be careful about the end conditions.)

c) The loop variable is again local, and the expression $E2$ may be either positive or negative as in (b). In addition, however, the expressions $E1$ through $E3$ are re-evaluated on each iteration.

d) The loop variable is *not* local; otherwise the loop is as in (c). What is the value of the loop variable on exit?

3.9 Do each of the parts of Exercise 3.8, above, but use FCL/2.

3.10 Describe the characteristics of the **for** statement (or its equivalent) in the programming language you are using for this course.

3.11 Section 3.3.1 describes the effect of a routine call in terms of replacing the call with a modification of the procedure body. Show the effect of this substitution for the *skip* procedure in the following code segment:

```
skip(5);
a = 7;
skip(a);
a := a + 2;
skip(a);
```

3.12 Change procedure *ResetLine* so that its expected result is obtained through a formal parameter list instead of through a global variable.

3.13 Describe or define a general scheme for converting a value-returning routine to a value-less routine, and for converting all calls in the value-returning version of the routine to calls in the value-less version.

3.14 Carefully trace the execution path of the call *Comb*(4, 2) to determine the value computed. Be prepared to show your work.

3.15 Not all functions that are described recursively are best computed recursively. Consider the Fibonacci numbers, defined

$$F_0 = 1$$

$$F_1 = 1$$
$$F_n = F_{n-1} + F_{n-2}.$$

These can be computed recursively, or they can be computed iteratively using temporary variables.

 a) Write a recursive procedure to compute $F(n)$.

 b) Write an iterative procedure to compute $F(n)$.

 c) Compare the number of additions the two procedures use to compute F(6).

3.16 Write a program that reads English text and outputs that text along with a count of the number of occurrences of each word. The text consists of words, blanks, and punctuation. Identify the macro-operations; write a main program that invokes them as procedures; and then write the procedures themselves, if necessary iterating this kind of decomposition until detailed code is called for.

3.17 Fill in the details, and prove Theorem 3.1.

3.18 Strictly speaking, a language with recursive procedures doesn't need a loop construct. For example, consider

   while B **do** S;

this can be coded as

   procedure WBS; **begin if** B **then begin** S; WBS **end end**

Show how the same thing can be done with **repeat**, **loop**, and **for** statements.

3.19 Suppose you want to allow procedure parameters in a procedure declaration—that is, you want to permit a formal parameter to be specified to be a **procedure**. Consider, for example, the following program fragment:

   procedure $P1$; $\cdots$
   procedure $P2$; $\cdots$

   procedure $XMP(P\!: \textbf{procedure})$;
       begin $\cdots$ P; $\cdots$ **end**;

   $XMP(P1)$;
   $XMP(P2)$;

In this program, the first call on XMP will cause a call on $P1$; the second call on *XMP* will cause a call on *P2*. Now, consider the **for** statement variation of Exercise 3.8.

Write one recursive procedure for each of these variations. Each procedure must be equivalent to one of the variants and accept a procedure parameter for the loop body.

3.20 Write a procedure that reads a character from the input and, if the character is A, B, C, D, or E, invokes procedures *Acommand, Bcommand, Ccommand, Dcommand,* or *Ecommand* respectively, but if the character is anything else, invokes procedure *error.*

3.21 Consider the following argument.

> There is a contradiction in this book. Look, it says
>
> a) that FSM's and regular expressions and so on are all equivalent. In particular, it says that all programs you can write using the primitive control constructs are equivalent to FSM's.
>
> b) that adding other control constructs (like **for**, **case**, and so on) doesn't add any power.
>
> c) that there are some problems you can't solve with FSM's. Section 1.5.1 gives the example of recognizing the NMN set—the set of strings with *N* ones followed by *M* zeros followed by *N* ones again.
>
> Okay, putting those three things together, I ought to be able to conclude that I can't write a program that recognizes *NMN*. But you can plainly see that's false—in fact it's a trivial Pascal program—just count the initial ones, skip any zeros, count the trailing ones, and then compare the two counts.

Is this argument right or wrong, and why?

3.22 For several years following 1968, there raged a debate about whether **goto** statements should be included in a programming language (or used if they are included). This exercise is an excerpt from that saga. Consider the following program (*g* and *f* are procedures, and *A* and *B* are Boolean expressions, all of which may act on the global variable *s*):

```
var s: integer;
begin
    Read(s);
L1: if A then
        begin  g; goto L1  end
    else if not B then g
    else
        begin
        g;
L2:     if B then
```

                begin f; **goto** $L2$ **end**
            else if not A **then** f
            else begin f; **goto** $L1$ **end**
            end
     end

It has been shown that for some f, g, A, and B, this program cannot be rewritten without **goto** statements if we make the following restrictions:

- The only control statements allowed (other than **goto**) are **while** and **if–then–else**.
- Additional variables and routines may not be introduced.
- The variable s may only appear in the initial *Read* and in A, B, and the bodies of f and g.

Answer the following questions about this program:

a) If you were allowed to use other control statements from this chapter, with the exception of routines, could you recode this program without a **goto** statement? Why or why not?

b) If you were allowed to introduce one new variable, say x, could you recode this program without a **goto** statement? Why or why not?

c) On the basis of your answers so far, do you think that the existence of this program says anything significant about the necessity of **goto** statements? Be prepared to defend your answer.

Chapter 4

The Representation of Control

In this chapter we examine some ways to represent the control constructs we used previously. The language we use for these representations, SMAL, resembles FCL/1, but is even more primitive; its statements resemble those of a typical machine language. (See Appendix D for a full explanation of machine and higher-level languages.) The only features of SMAL that concern us here are as follows.

- All statements must be very simple; the expressions on the right hand side of an assignment, for example, must have at most one operator. Parenthesized expressions are not permitted.

- The only control statements (other than inplicit sequencing) are an unconditional and a conditional transfer, as in FCL/1. These are written

  goto < *place* >

  and

  if < *simple Boolean* > **then goto** < *place* >,

  respectively. A < simple Boolean > is either a single Boolean variable or a single relationship between two arithmetic variables. A < *place* > is either < *label* > or @< *variable* >; the meanings of these two alternatives follow.

- Statements may be labeled, and the label of a statement may be thought of as the name of the place in the program whose statement is labeled L. (In a real computer, $\#L$ is the memory location where that statement is stored.) The statement

$x := \#L;$

where L is a label, stores the memory location name of the statement labeled L in the variable x. The statement

goto $@x$

transfers control to the statement whose memory location name is stored in the variable x.

SMAL is exceedingly primitive, so primitive that ordinarily we wouldn't use it for programming interesting problems. However, in this chapter its simplicity is its virtue. By showing how other control constructs can be represented in this very simple language, we implicitly show one way to represent them in more sophisticated languages.

4.1 Representation of FCL/2 Constructs

The following paragraphs describe how to represent each of the three FCL/2 control constructs—sequencing, **if–then–else**, and **while do**. In presenting these representations we will use the notation "$rep(S)$" to denote the representation of S.

Definition 4.1. The notation

$rep(S)$

denotes the representation of the statement, phrase, or value, S. The language of the representation depends on the discussion context. □

In defining the representation of a construct we will often use the *rep* of its components. For example, since complicated expressions are not permitted in SMAL, we might define

$rep(x := E_1 / E_2) =$
    $rep(t1 := E_1);$
    $rep(t2 := E_2);$
    $x := t1 / t2;$

where E_1 and E_2 are arbitrary expressions that, in turn, may be defined using the *rep* function.

Note that the definition of $rep(S)$ may involve the introduction of new temporary variables (such as $t1$ and $t2$ above) and labels. By convention, we assume that the names of these variables and labels are distinct from the program's names and from each other.

4.1.1 THE REPRESENTATION OF SEQUENTIAL CONTROL

Because a computer is basically a sequential processor, there is nothing special involved in translating successive high-level language statements into SMAL. The rule is simply

$$rep(S_1; S_2; \cdots ; S_n) = rep(S_1); rep(S_2); \cdots ; rep(S_n)$$

That is, the representation of a sequence of statements is the sequence of representations of the individual statements.

4.1.2 THE REPRESENTATION OF "IF-THEN-ELSE"

In order to represent the general **if** statement we translate it into statements involving, at most, assignment and transfers.

$rep(\textbf{if } B \textbf{ then } S_1 \textbf{ else } S_2) =$
  $rep(t := B)$;
  if t **then goto** L_1;
  $rep(S_2)$;
  goto L_2;
L_1: $rep(S_1)$;
L_2:

This translation should be familiar if you've used FORTRAN or BASIC, neither of which has the full FCL/2 conditional construct. Note the use of $rep(B)$, $rep(S_1)$, and $rep(S_2)$ to indicate where the appropriate representation must be substituted.

4.1.3 THE REPRESENTATION OF "WHILE"

The representation of the **while** statement is quite simple:

$rep(\textbf{while } B \textbf{ do } S) =$
L_1: $rep(t := \textbf{not } B)$;
      if t **then goto** L_2;
      $rep(S)$;
      goto L_1;
L_2:

So, for example, if we expand

  $rep(i := 0; s := 0; \textbf{while } i < 100 \textbf{ do } s := s + i;)$

we obtain

$$i := 0;$$
$$s := 0;$$
$$L_1:\ t1 := i < 100;$$
$$\quad t2 := \mathbf{not}\ t1;$$
$$\quad \mathbf{if}\ t2\ \mathbf{then\ goto}\ L_2;$$
$$\quad s := s + i;$$
$$\quad \mathbf{goto}\ L_1;$$
$$L_2:$$

In more complicated cases, of course, even greater expansions occur.

4.2 Representation of Other Control Constructs

Earlier we introduced a number of extensions to the three basic control constructs, namely **numberedcase**, **case**, **repeat**, and **for**. We defined each of these extensions by giving a representation of it in terms of other constructs. Therefore, we could obtain a representation in SMAL by "plugging in" the representations of Section 4.1. In many cases, however, doing so would result in a rather inefficient representation. In the following sections, therefore, we'll consider only the more efficient representations of these constructs.

4.2.1 REPRESENTATION OF "REPEAT"

The **repeat** statement,

repeat S **until** B

was defined as

begin S; **while** $\sim B$ **do** S **end**

which seems to imply duplicating the code for S. What's intended, of course, is simply that the evaluation and test of B be placed after the body of the loop. So, a better way to represent the construct in SMAL is

$rep(\mathbf{repeat}\ S\ \mathbf{until}\ B) =$
$\quad L:\ rep(S);$
$\qquad rep(t := \mathbf{not}\ B);$
$\qquad \mathbf{if}\ t\ \mathbf{then\ goto}\ L;$

4.2.2 REPRESENTATION OF "NUMBEREDCASE"

Each of the other extended constructs may be similarly optimized; most are quite obvious and are left as an exercise to the student. One, however, is not so obvious—the use of a transfer vector in the implementation of the **numberedcase** statement.

As we defined it, the statement

numberedcase E **of** S_1; S_2; $\cdots$; S_n; **otherwise** S_* **end**

seemed to require a long sequence of comparisons of the value of E against (in the worst case) each of the numbers 1 through n. We can avoid the sequence of explicit comparisons by constructing a vector of addresses of the code for statements $S_1 \cdots S_n$, using E as an index into the vector, and transferring control (or branching) *indirectly* through this vector. Branching through a variable is not common in high-level languages, yet it pervades the machine level.

As noted in an earlier section, SMAL allows constructs of the form

> $loc := \#L$;
> **goto** @loc;

The notation $\#L$ means "that place in the program whose statement is labeled L." Therefore, these two statements assign loc a value that represents the statement labeled L. The statement **goto** @loc means "branch to the label indicated by the contents of loc"; hence this program is equivalent to

> **goto** L;

Now, suppose that we initialize the vector $TRV[1 \colon n]$ so that

> $TRV[1] := \#L_1$;
> $TRV[2] := \#L_2$;
>
> $\qquad \cdot \qquad \cdot$
> $\qquad \cdot \qquad \cdot$
> $\qquad \cdot \qquad \cdot$
>
> $TRV[n] := \#L_n$;

Note that $\#L_1$, etc. are constant values, so that we can do this initialization before the program begins running. The representation of **numberedcase** then becomes

> $rep(\textbf{numberedcase } E \textbf{ of } S_1; \cdots ; S_n \textbf{ otherwise } S_* \textbf{ end}) =$
> $\qquad rep(t := E)$;
> $\qquad$ **if** $t < 1$ **then goto** L_*;
> $\qquad$ **if** $t > n$ **then goto** L_*;
> $\qquad$ **goto** @$TRV[t]$;
>
> $\qquad L_1: rep(S_1)$;

> $\quad$ **goto** *Lexit*;
> L_2: *rep*(S_2);
> $\quad$ **goto** *Lexit*;
>
> .   .
>
> .   .
>
> .   .
>
> L_n: *rep*(S_n);
> $\quad$ **goto** *Lexit*;
> L_*: *rep*(S_*);
>
> *Lexit*:

4.3 Representation of Routines

Our discussion in Section 3.3.1 suggests an obvious implementation for routine invocation. For example, if procedure *P* is declared

$\quad$ **procedure** $P(<$*formal parameters*$>$); $\quad$ *BODY*;

then we define the representation of an invocation of *P* to be

$\quad$ *rep*($P(<$*actual parameters*$>$)) $=$ *rep*($BODY'$)

where *BODY'* is the "suitably modified" version of the body *BODY*. This representation is called an *in-line expansion*. We also call a procedure such as *P*, whose invocations are "expanded in-line," an *open procedure*.

In-line expansion has the virtue that it involves execution of no more instructions than those required to represent *BODY'*. It is a time-efficient representation that has two disadvantages: It leads to multiple copies of *rep*(*BODY'*), and it cannot be used for recursive routines.

The *closed routine* representation, however, avoids these problems, but at the expense of time efficiency. That is, the closed routine representation allows us to have just one copy of *rep*(BODY') stored, and allows recursive calls, but requires us to add a sequence of SMAL instructions, known as a *calling sequence*, at each invocation. Essentially, this calling sequence keeps track of what is to happen after the body of the invoked routine finishes—it stores the point in the calling program to which control returns.

For now, we'll consider only parameterless procedures, starting with an incorrect representation, but one that illustrates some basic concepts. Then we'll point out the flaw in the mechanism and correct it.

Consider a program containing a declaration of a procedure, *P*, and a call on *P*

> **procedure** P;
> $BODY$;
>
> $\cdots$; P; $\cdots$

We might try the following representations for the declaration and call:
The declaration is:

> $rep($**procedure** P; $BODY) =$
>   **var** $Preturn$: $word$;
> $Plabel$:
> $rep(BODY)$;
>   **goto** @$Preturn$,

The call is:

> $rep(P) =$
> $Preturn := \#Lnext$;
>   **goto** $Plabel$;
> $Lnext$:

The label $Lnext$ must be different for each invocation statement, or *call site*. Each call site stores the address of the *return point* (the label of the next statement to be executed) in the variable $Preturn$. After the procedure completes, it transfers control to the location whose address is stored in $Preturn$; that is, it branches to the statement that the call site designates the next to be executed.

As we said earlier, this representation is incorrect–but it has the essential property of all correct ones: the call site must save the return point in a variable, and the procedure body must return to that point by branching indirectly through this variable.

The error in the representation arises when we have recursive calls. Suppose that someone calls a recursive routine, F, and that F then calls itself before returning. The first call on F executes the calling sequence

> $Freturn := \#L1$;
>   **goto** $Flabel$;
> $L1$: $\cdots$

The representation of procedure F is

> $Flabel$:
> $BODY\ OF\ F$
> $Freturn := \#L2$;
>   **goto** $Flabel$;
> $L2$: $\cdots$

But when the body of F executes $Freturn := \#L2$, it loses track of the fact that it must eventually return to $L1$. A correct solution will require different locations for storing the return points for the two calls–indeed, for all potentially nested calls.

There is an elegant representation of procedure calls that uses a *stack*, a data structure we discuss in Section 9.1.3. Here we use a special case of a stack that meets our needs for a restricted routine-calling mechanism. We first assume that the representation of every program contains two declarations,

> **var** *Stk*: **array** [1..*N*] **of** *word*;
> **var** *PStk*: *integer*;

that N is "big enough," and that *PStk* is initialized to zero.

The following code shows how to use *Stk* and *PStk* to keep track of the return points. The declaration is:

> *rep*(**procedure** *P*; *BODY*) =
> *Plabel*:
>   *rep*(*BODY*);
>   **goto** @*Stk*[*PStk*];

The call is

> *rep*(*P*) =
>   *PStk* := *Pstk* + 1;
>   *Stk*[*PStk*] := #*Lnext*;
>   **goto** *Plabel*;
> *Lnext*:
>   *PStk* := *PStk* − 1

Each time that P, or any other closed routine, is called, the variable *PStk* increases by one and the return point is stored in *Stk*[*PStk*]. This, in effect, allocates a new variable to the job of holding the return point each time a routine is called. Each time a routine returns, the value of *PStk* decreases by one, thus simultaneously freeing a location for reuse and ensuring that *Stk*[*PStk*] holds the return point for the calling procedure. With this scheme, a collection of routines may invoke each other in any order, as long as the number of invocations in progress at the same time is less than or equal to the size of *Stk*.

4.4 Further Reading

The representation of one control structure by another is a topic usually associated with compiling. Discussions of the machine-level (SMAL) representations of programming languages can be found in Gries (1971) and Aho and Ullman (1977).

Exercises

4.1 Assume that we have added a *print* command to SMAL. What does the following SMAL program print?

```
    var L, M: word;
    L := #R;
P:
    M := #P
    print('A');
    goto @L;
Q:
    L := #S;
    print('B');
    goto @M;
R:
    L := #Q;
    print('C');
    goto @M;
S,
    print('D');
```

4.2 Assume again that we have added a *print* command to SMAL. What does the following SMAL program print?

```
    var i: word;
    var V: array [1..3] of word;
    V[1] := #R;
    V[2] := #S;
    V[3] := #T;
    i := 0;
R:
    print('A');
    i := i + 1;
    goto @V[i];
S:
    print('B');
    goto @V[i - 1];
```

T:

　　print(' *C*');
　　i := *i* − 1;
　　goto @ *V*[*i*];
Q:

4.3 Assume that input and output commands have been added to SMAL. Rewrite the programs of the following exercises in SMAL:

　a) Exercise 2.1

　b) Exercise 2.14 (assume also that a square root routine is primitive)

　c) Exercise 2.15

　d) Exercise 2.16

　e) Exercise 2.17

　f) Exercise 2.18

　g) Exercise 3.15 (the iterative version)

4.4 Perform a careful and complete expansion of each of the following into SMAL. Show all steps of the expansions.

　a) *rep*(*x* := *y* ∗ *z* + *w*)

　b) *rep*(**if** *a* < 0 **then** *b* := *c* ∗ *d* + *e* ∗ *f* **else if** *g* **then** *f* := *g*)

　c) *rep*(**while** *a* < 0 **and** *i* > 5 **do begin** *i* := *i* − 1; *a* := *a* + 1 **end**)

4.5 Represent the following program segments in SMAL. That is, write program fragments in SMAL that have the same effect as the following program fragments.

　a) *y* := (*w* ∗ *z* − *p* ∗ *q*)^3;

　b) *i* := 0; *j* := 1;
　　　for *k* := 1 *to* 10 **do**
　　　　begin
　　　　i := *i* + *k*; *j* := *j* ∗ *k*;
　　　　end;

　c) **if** *x* > 0
　　　　　then begin if *y* = 5 **then** *z* := 1 **end**
　　　　　else begin if *y* = 4 **then** *z* := 2 **else** *z* := 3 **end**

　d) **numberedcase** *x* ∗ 21 − 5 **of**

$$x := x + 1;$$
$$y := 32;$$
$$\textbf{begin } x := x + y;\ y := y + 1 \textbf{ end};$$
   end;

e) **case** i **of**
   $5: j := j + k;$
   $6: j := j * k;$
   $8: j := j / k;$
   $4: j := 0;$
   end;

4.6 The statement

numberedcase E **of begin** S_1; S_2; $\cdots$; S_n; **otherwise** S. **end**

may be represented in SMAL either as a sequence of **if–then–goto**'s or with a transfer vector.

a) Implement **numberedcase** in SMAL without using indirect **goto**'s.

b) A clever programmer from our Computer Center says that if $n > 3$, one should always use a transfer vector to implement a **numberedcase** statement because "otherwise you have to do all those tests." Is he right? Can you think of a time when you would use the implementation you found in (a) rather than a transfer vector? What are your criteria for choosing one implementation rather than the other? Why? State any assumptions you make. Be as quantitative as you can. If you can find formulas for determining the costs of the two methods, so much the better.

4.7 Define a representation for the **loop** and **exitif** constructs defined in Section 3.2.

4.8 Recall the **for** statement introduced in Exercise 3.8,

   for $i := E1$ **step** $E2$ **until** $E3$ **do** S;

Give a representation in SMAL for each of the variant interpretations discussed in that problem.

4.9 Consider the simple PASCAL **for** statement

   for $i := E1$ **to** $E2$ **do** S;

The most obvious representation for this loop will place the test for termination at the "top" of the loop body. An alternative implementation is slightly faster; it places the test at the bottom of the loop so that an extra **goto** can be avoided.

    a) Give the suggested "faster" representation for the PASCAL **for**.

    b) This representation requires an extra **goto** in the test "the first time." Discuss under what circumstances this **goto** can be avoided.

4.10 The simplest form of ALGOL 68 iteration statement is

$$\textbf{do } S_1; \cdots; S_n \textbf{ od}$$

This is just like a PASCAL **loop**—it causes statements S_1 through S_n to repeat indefinitely. However this can be prefixed by a **for** clause or a **while** clause or *both*. For example,

$$\textbf{for } i \textbf{ from } 1 \textbf{ to } 10 \textbf{ while } x > 0 \textbf{ do } \cdots \textit{od}$$

will perform the loop body with increasing values of i—but only as long as $x > 0$.
Construct a SMAL representation for the ALGOL 68 **for** statement as illustrated here. (If you're ambitious, you may want to explore other properties of the ALGOL 68 **for**—notably the consequences of defaulting the values of the initial, final, and step sizes.)

4.11 Boolean expressions involving relational tests can often be represented in either of two ways—directly in terms of operations on Boolean values, or by conditional jumps. For example, the statement

    if $x > 5$ **or** $y = 3$ **then goto** L;

could be represented as

    var $t1, t2$: *word*;
    $t1 := x > 5$;
    $t2 := y = 3$;
    $t1 := t1$ **or** $t2$;
    if $t1$ **then goto** L;

or, more simply, as

    if $x > 5$ **then goto** L;
    if $y = 3$ **then goto** L;

where the **or** operator never explicitly occurs and is, instead, implicit in the control flow. Use both representations for the following program fragments

a) **if** $x > 5$ **and** $y = 3$ **then goto** L;

b) **if** $(x > 5$ **and** $y = 3$ **and** $z < 2)$ **or** $w < 12$ **then** $S1$ **else** $S2$;

c) **while** $w = 0$ **and** $x > 46$ **do** S;

4.12 The Boolean operators **cand** and **cor** behave just like **and** and **or**, respectively, except that they do not evaluate the second (right) subexpression if the truth or falsehood of the whole expression can be determined after evaluating the first (left) subexpression. Consider the statements

$x := y$ **cand** z;   **if** y **cand** z **then goto** L;

$a := b$ **cor** c;   **if** b **cor** c **then goto** L;

a) Rewrite these statements in FCL/2.

b) Rewrite these statements in SMAL.

c) It is clear that **cand** and **cor** can save time if the right subexpression is complex. Under what other circumstances are they particularly useful?

4.13 Suppose the statement

when C **do** S

means that the next occurrence of the statement

signal C

will result in the execution of S, followed by a return after the **signal** statement. For example,

when *thud* **do** *print*(x);
$x := 2$; **signal** *thud*;
$x := 3$; **signal** *thud*;
when *thud* **do** *print*$(x + 1)$;
signal *thud*;

will print 2, then 3, then 4.

What are *rep*(**when** C **do** S) and rep(**signal** C)?

4.14 Give an open representation in SMAL of

begin
var x, y: *word*;
procedure $P(a$: *word*$)$;
    begin $a := a * 3$ **end**;
$x := 3$;
$P(x)$;

```
y := 4;
P(y);
end
```

4.15 Give both open and closed representations in SMAL of

```
begin
var x, y: word;
procedure P;
    begin x := x + y * 3; y := 4 * x + 5 * y end
x := 3;
y := 5;
P;
x := 7;
y := 9;
P;
end
```

4.16 A *coroutine* is a special kind of closed routine that, when invoked, begins execution from where it last left off. For example, the following will print 1, 4, 2, 5, 3, 4, 1, 5, · · · in response to **resume** *A*:

```
coroutine A;
    begin
    while true do
        begin
        print(1);
        resume B;
        print(2);
        resume B;
        print(3);
        resume B
        end
    end;

coroutine B;
    begin
    while true do
        begin
        print(4);
        resume A;
        print(5);
        resume A;
        end
    end;
```

The statement **resume** C, where C is the name of a coroutine, causes that C to take up where it left off the last time it performed a **resume**. If C has not started executing, then **resume** C causes C to begin executing at its beginning. What could be *rep*(**coroutine** C; S) and *rep*(**resume** C)? Assume recursion is not allowed (that is, forget the stack). Also, assume that if a coroutine ever terminates (reaches its **end**), it terminates the entire program. Although the program above has only two coroutines, you should make your solution general enough to handle any number of coroutines.

4.17 Consider again the program in Exercise 3.22. Assuming you are allowed to introduce nonrecursive coroutines (see Exercise 4.16), rewrite the program without using a **goto** statement.

4.18 Devise a representation in SMAL for closed value-returning routines.

4.19 Consider the following recursive program for computing factorials. It computes $n!$ if x is initialized to n and F is initialized to 1.

> **procedure** *Fact*;
>   **begin if** $x \neq 1$ **then** $F := F * x$; $x := x - 1$; *Fact* **end**

a) Write a program that uses this procedure to compute 4! and 6!.

b) Comment on the use of global variables to provide parameters of a recursive procedure.

c) Give a closed representation of *Fact* in SMAL.

d) Discuss what is involved in using open representations of *Fact*.

e) Rewrite *Fact* as a function with one parameter.

f) Use the result of Exercise 4.18 to represent this function in SMAL.

4.20 Consider the routine *Comb* of Fig. 3.1, Section 3.3.4. Translate *Comb* into SMAL using the stack implementation of Section 4.3. Trace the execution of this SMAL program for the invocation

$$x := Comb(4, 2).$$

In general, how much space must be left on the stack for the call $Comb(x, y)$ to work?

4.21 The "incorrect" solution for procedure invocation (the one that didn't use a stack) is incorrect *only* because it doesn't support recursion. It is perfectly acceptable for languages such as FORTRAN that don't permit recursion. Advocates of languages without recursion sometimes charge that recursion is very expensive.   Assuming that each SMAL statement requires one unit of time, compare the costs of the recursive and nonrecursive representations. Relate this cost to the size of the body of the subroutine.   Then examine several programs you have written recently, estimate the size of their SMAL representations, and thus estimate the overhead that recursion would have imposed on your own programs.

Chapter 5

Formal Specification and Proof of Programs

In this chapter we discuss some methods of reasoning about *what* a program does that fall into several related categories.

1. *Specification.* We must be able to state precisely what a program is intended to do; such a statement is called a *specification* of the program. array *M*. specifications are usually written in a natural language (such as English). There is a growing trend, however, to use mathematical notation for specifications which allows us to be both more concise and more precise.

2. *Program proofs.* Given a specification and a program, we must prove that the program does what the specifications say it should do. Since complete testing of a program is usually impossible, and because a program that has merely been "debugged" is likely to contain undetected errors, we need to find other ways to increase our confidence in a program's correctness. By proving that a program conforms to its specifications, we can have greater confidence in it. Further, program-proof techniques are helpful in reading and writing programs (even if we do not carry out formal proof procedures), and in reasoning informally about programs.

3. *Specification of programming languages.* In order to know whether a program performs as specified, we must know what each statement in a program will do. That is, we must have a precise specification of the semantics, or meaning of each programming language construct.

4. *Testing.* In many cases, a complete and formal proof may be

unnecessary. In some cases, such proof is infeasible, usually when complete testing is also infeasible. We can often, however, substantially increase our confidence in a program by careful testing, using the style of reasoning developed for program proofs to guide the testing process. In some instances, we can mix program proofs of some properties with tests of others.

5.1 Specification of Programs

For the moment let's consider only programs, or portions of a program, that have no input or output, (see Exercise 5.25 for proving program statements about input and output). We can express what a program does with a predicate describing the relationship between the values of program variables when a program stops, and their values before it started. Such a predicate is called a *postcondition* of the program. Consider, for example, the simple program

$$y := x / k$$

One possible postcondition of this program is

$$x = y \cdot k.$$

We must stress that this is only one of the program's *possible* postconditions. This particular postcondition is a predicate that captures the fact that y is the quotient of x divided by k. One might very well have chosen a different postcondition if a different property of the program were desired, and we'll see an example of this in a moment.

The postcondition alone, however, is not enough to specify the behavior of this program. In particular, division is not defined for a divisor of zero. We must be able to say that the value $k = 0$ is illegal. We can also express such constraints as predicates called *preconditions*. The combination of preconditions and postconditions permits us to say what a program does.

Definition 5.1. The notation,

$$P \{S\} \ Q,$$

where S is a program* or portion of a program and P and Q are predicates, is a logical (true or false) statement meaning that if P is

---

* Within braces enclosing a program or a portion of a program, the notation will be in program language. For instance, the multiplication operator outside the braces would be a multiplication dot; the operator inside the braces, an asterisk.

true when S begins executing, then S will terminate and Q will then be true. The notation is read "S is strongly correct with respect to precondition P and postcondition Q." The phrase "strongly correct" is often abbreviated to simply "correct." □

The example we've been discussing could also be written

$$k \neq 0 \ \{y := x / k\} \ x = y \cdot k.$$

That is, if k is nonzero and the statement $y := x / k$ executes, then x will be the product of y and k.

Suppose we had chosen a different postcondition for this simple program. In particular, suppose that we had wanted $x = y$ to hold after the assignment is executed. Clearly, for this to be true, k would have to be 1 beforehand. We can express this as

$$k = 1 \ \{y := x/k\} \ x = y.$$

Note the wording of Definition 5.1. We carefully avoided saying anything about what will happen if S executes when P is false. If P is false, S can fail without invalidating $P\{S\}\ Q$. If P is true, and only in that case, S is obligated to produce a state described by Q (that is, a state in which Q is true). In other words, promises about programs (the postconditions) apply only when the programs are used as intended (the preconditions are satisfied).

Also note that the definition states two things: (1) that Q will hold true after the execution of S, and (2) that S will *terminate*—that is, will eventually finish execution.* Not all formulations of the $P\{S\}\ Q$ notations require the second of these properties; these formulations are said to specify the *weak correctness* of the program with respect to P and Q. A further discussion of termination is included in sections about loops (see 5.3.7.3, for instance.).

Let's consider another very simple example that illustrates two important aspects of specifications. Suppose we want to specify the effect of the statement

$$x := x + 1$$

The postcondition must say that the value of x will be one greater than its "previous" value, its value before the statement is executed. We need some way to name the old, previous value of the variable, x, as well as its value after it has been changed by the statement. We do so by appending a prime mark to the variable name (x') when we refer to the old value; without the prime, the variable name refers to its new, or current value.

---

* The program, "**while** *true* **do** $x := y$", for example, will not terminate.

Thus, for our example we write

$$\{x := x + 1\} \; x = x' + 1.$$

The postcondition asserts that the current value of x, (that is, the value after executing the statement) is one greater than its previous value. Formally, the meaning is identical to that of

$$x = x' \; \{x := x + 1\} \; x = x' + 1.$$

The precondition deals only with the values of the variables before the statement is executed. Since it is not necessary to refer to the values of a variable at two different times before execution, primes are unnecessary. Note that an unprimed variable name in a precondition and the corresponding primed name in a matching postcondition will refer to the same value of the variable. In general, we will use names without primes to refer to the values of variables at the point the predicate appears. Another convention about the use of the prime notation is that we use it only for those variables whose values have changed as a result of executing the program—and then only if the prior value of the variable matters. For example, we don't use primes in

$$k \neq 0 \; \{y := x/k\} \; x = y \cdot k$$

because the values of x and k didn't change, and the prior value of y didn't matter.

The example, $x := x+1$, permits us to illustrate another important point. Look at the statement again. Does it need a precondition? If the data manipulated by programs really behaved like mathematical objects, the answer would be *no*. But they don't. In this particular case we must account for the fact that integers in a computer have some fixed maximum size—usually something like $2^w - 1$, where w is the "word size" of the computer. In fact, we can execute the statement $x := x + 1$ only if the resulting value is in the range supported by the machine. Thus, we might write

$$(x + 1) \leqq 2^w - 1 \; \{x := x + 1\} \; x = x' + 1.$$

In practice, we often ignore such details, but sometimes they must be considered. Obviously, if we intend to provide an absolutely unassailable proof that our program works, we cannot ignore them. If there is ever a fully computerized program prover, it may very well be able to check all preconditions. However, until that time we must adopt the more pragmatic policy of being content simply to increase our confidence in a program, without considering every single detail.

Note, by the way, that preconditions are not executable statements, and are not subject to the same restrictions as program statements. The term $x + 1$ makes mathematical sense in a predicate, even if a particular machine might get an overflow in an attempt to compute it. Likewise, notations such as 2^w make sense, even though most programming languages do not allow superscripts. When we use the language of specification, we speak to each other *about* programs. This is not a programming language with which we speak to machines, even though they share some notation and refer to some of the same program variables.

5.2 Program Proofs

In this section we discuss how one proves that a program with preconditions and postconditio s actually satisfies these conditions. Our method uses *predicate transformers.* We start with a postcondition, Q, and a program, S. Depending on S, we then determine how to *transform* the predicate Q into the weakest predicate that, if true before S is executed, would make Q true afterward ("weakest" here meaning "minimal" or "making the fewest restrictions").

> **Definition 5.2.** The *weakest precondition* of S with respect to Q, denoted here as
>
> $wp(S, Q)$,
>
> is the weakest statement (or the minimum condition) that must be true in order that S terminate successfully and that Q be true after S has been executed. □

Consider $P \{S\} Q$, and for the moment, suppose we can find $wp(S, Q)$. By definition we know that $wp(S, Q) \{S\} Q$, so if we could prove

$$P \supset wp(S, Q)$$

we would know that $P \{S\} Q$. That is, if P is strong enough to imply $wp(S, Q)$, then since $wp(S, Q)$ is just strong enough to ensure that Q is true after S has been executed, P will also ensure that Q is true. Here is a two-step approach to proving $P \{S\} Q$.

1. Find $wp(S, Q)$.
2. Prove $P \supset wp(S, Q)$.

Step 2 is a matter of logic and mathematics; we present many examples of such proofs, but the techniques for carrying out these proofs are standard in mathematics and will not be covered in this textbook. Our main task will be figuring out how to do step 1.

Note that it is always the *weakest* precondition that we wish to find, because the weakest precondition is the least restrictive and, therefore, the most accurate. That is, the weakest precondition is true in *all* cases where execution of the statement in question produces the desired result. If we choose a stronger, more restrictive condition, it might exclude some cases in which the statement actually works. For example,

$$k > 0 \ \{y := x \ / \ k\} \ y \cdot k = x$$

is true, but $k > 0$ is unnecessarily restrictive, for the assignment will work for negative, as well as positive, values of k.

Our definition of $P\{S\}\ Q$ notation is nonstandard in one important sense: we have included the requirement that S terminates, that it does not loop endlessly. Other treatments give a weaker definition of the notation, namely that S need not terminate and that Q need be true only if it does. We call this *weak correctness*.

Definition 5.3. We say that S is *weakly correct* with respect to precondition P and postcondition Q iff, whenever P is initially true, S will halt with Q or S will not halt. The notation,

$P\{S\}\ Q$, *weakly*

is used to express this. □

This weaker definition of correctness is appropriate when we find it convenient to prove the termination of a program independently of proving of its weak correctness. The reader should be aware that often in other books and papers, the "$P\{S\}\ Q$" notation refers to weak correctness. Here we will always designate weak correctness explicitly.

5.3 Computing wp: Language Semantics

Up to this point we have discussed only the ways to specify the behavior of programs; before going on, we need to discuss a method of defining programming languages. In defining a programming language, we describe the effect—which is the same as the meaning or *semantics*—of each statement and combination of statements. For example, we say that the effect of the statement

$$x := q \ / \ r$$

is that the current value of q/r (that is, the current value of q divided by the current value of r) is computed and replaces the contents of the variable x. In other words, using our notation,

$q \ / \ r \ is \ defined \ \{x := q \ / \ r\} \ \ x = q \ / \ r.$

Another way we could write this is

$wp(x := q \ / \ r, \ x = q' \ / \ r') \ \equiv \ q \ / \ r \ is \ defined$

That is, as long as the right-hand side of the assignment has a defined value, the effect of the assignment is to set x to the quotient of the values of q and r. Since assigning to x does not affect the values of q and r, we could have eliminated the prime notations in the postcondition and used q and r in place of q' and r'.

Thus, the task of defining wp for all possible cases is the same as that of defining the meanings of all programs. The following subsections of Section 5.3 define wp for several kinds of statements, specifically, for the null (or empty) statement, the assignment statement, the condition **if–then–else**, **while–do** and **for** statements, and for the procedure invocation. In our definitions we use one piece of standard notation, $R(\mathbf{x})$.

> **Definition 5.4.** Suppose that R is either a predicate or a fragment of a program. The notation
>
> $R(\mathbf{x})$
>
> where $\mathbf{x}$ is a list of variables, indicates that R may contain (and, in the case of a program fragment, modify) the variables in $\mathbf{x}$. It also indicates that
>
> $R(\mathbf{z})$
>
> is to be interpreted as the result of substituting the expressions in the list $\mathbf{z}$ for the corresponding variables in the list $\mathbf{x}$ throughout R. □

For example, if

$P \ \equiv \ P(x, y) \ \equiv \ x > y,$

then

$P(z, 3) \ \equiv \ z > 3.$

We also make the important assumption that expressions—such as the right-hand sides of assignments, the actual parameters of a procedure invocation, and the conditions of an **if** or **while** statement–have no *side effects*. That is, we assume that no expression changes the value of any variable, but merely denotes its value. This assumption is not really necessary, but it greatly simplifies the formulas we derive. The following sections develop the weakest preconditions for a representative set of constructs for many of the statements written in typical algebraic languages. Figure 5.5 summarizes the results.

5.3.1 THE NULL STATEMENT

The weakest precondition for the null statement is trivial. The statement does nothing. Thus, in order for any predicate, Q, to be true after the null statement, it must be true before the statement as well, and

$$wp(\textbf{skip}, Q) = Q \tag{5.1}$$

(We denote the null statement **skip** just to make it visible. In PASCAL, for example, the null statement contains no characters.)

5.3.2 SEQUENCING

The weakest precondition for the sequence of statements

$$S_1; S_2$$

is also simple to derive, if we assume that the weakest precondition of each individual statement is available. In order for Q to be true after the execution of S_2, the predicate $wp(S_2, Q)$ must be true immediately after the execution of S_1. But according to the definition of weakest precondition, in order for $wp(S_2, Q)$ to be true after the execution of S_1, the predicate $wp(S_1, wp(S_2, Q))$ must be true before S_1 executes. That is,

$$wp(S_1; S_2, Q) = wp(S_1, wp(S_2, Q)).$$

Note that this is a recursive definition; we have defined the weakest precondition for $S_1; S_2$ with respect to Q in terms of the preconditions for statements S_1 and S_2. There is nothing wrong with this, since we always define a precondition in terms of the wp's of simpler statements. Since S_1 and S_2 are simpler, or shorter than the original statement, the process of expanding the wp's must ultimately terminate.

Applying our reasoning repeatedly, we deduce that

$$wp(S_1; \cdots; S_n, Q) \equiv wp(S_1, wp(S_2, \cdots wp(S_n, Q) \cdots)). \tag{5.2}$$

5.3.3 THE "IF" STATEMENT

Generating the weakest precondition for the statement

> **if** C **then** S_1 **else** S_2

is simply a matter of case analysis; either C is true or it is false. If C is true, then S_1 will execute. Therefore, if Q is to be true after the **if**, then $wp(S_1, Q)$ must be true before. Likewise, if C is false, then S_2 will execute and $wp(S_2, Q)$ must be true before. Therefore,

$$wp(\textbf{if } C \textbf{ then } S_1 \textbf{ else } S_2, Q)$$
$$\equiv (C \supset wp(S_1,Q)) \, \wedge \, (\sim C \supset wp(S_2,Q)).$$

Likewise, the statement

> **if** C **then** S

is equivalent to

> **if** C **then** S **else** *skip.*

Therefore,

$$wp(\textbf{if } C \textbf{ then } S, Q) \equiv (C \supset wp(S, Q)) \, \wedge \, (\sim C \supset wp(skip, Q))$$
$$\equiv (C \supset wp(S, Q)) \, \wedge \, (\sim C \supset Q).$$

Note that none of these equations for wp make sense if C is undefined, so that there is an implicit requirement that C be defined in each of the two formulas.

5.3.4 THE ASSIGNMENT STATEMENT

The assignment statement

> $x := E$

is a bit more subtle. Assume, again, that the evaluation of the expression E has no side effects. If Q does not involve the variable x, then it is not affected by execution of the assignment, and in order for Q to be true after the assignment, it must also have been true before. If, on the other hand, Q does involve x, we don't expect it to be true before execution of the assignment statement, since the assignment changes the value of x.

The only thing changed by the assignment statement is the variable x. Therefore, if the predicate $Q(x)$ is to be true after the assignment, it also must be true before the assignment, but with x replaced with its final value. That is,

$$wp(x := E, Q(x)) \equiv Q(E).$$

Note that we used the notation defined earlier; $Q(x)$ is a predicate, and $Q(E)$ is that same predicate with the value of E substituted for all uses of x. For example, in order for

> $\{x := y + 3\} \, x > 0$

to be true, we must have as a precondition $y + 3 > 0$. This precondition is, indeed, the same as the postcondition, but with x replaced with its value after the assignment. So,

$$wp(x := y + 3, \ x > 0) \equiv y + 3 > 0,$$

just as the rule specifies.

As another example, suppose that we want

$$\{x := x + 1\} \ x \leq n + 1$$

to be true. Computing the wp we obtain

$$wp(x := x + 1, \ x \leq n + 1) \equiv x + 1 \leq n + 1$$

or, if you prefer, $x \leq n$. Thus,

$$x \leq n \ \{x := x + 1\} \ x \leq n + 1.$$

Of course, $x + 1$ must also be defined. In general, then, since the final value of x is E,

$$wp(x := E, \ Q(x)) \ \equiv \ E \text{ is defined } \wedge \ Q(E). \tag{5.3}$$

Here, we have been explicit in requiring the expression E to be defined. This is to avoid erroneous results such as

$$y > 0 \ \{x := 1/0\} \ y > 0.$$

Without that "E is defined" clause, we get

$$wp(x := 1/0, \ y > 0) \ \equiv \ y > 0,$$

which totally ignores the fact that the division by 0 brings proceedings to a grinding halt. Normally, however, we suppress the "is defined" clause when it is clearly true.

As still another example, consider proving

$$x \geq 0 \ \{x := x \textbf{ div } 2\} \ 2 \cdot x \geq x' - 1.$$

Recall that this is the same as

$$x = x' \wedge x \geq 0 \ \{x := x \textbf{ div } 2\} \ 2 \cdot x \geq x' - 1.$$

Using the two-step approach, we must first find

$$wp(x := x \textbf{ div } 2, \ 2 \cdot x \geq x' - 1)$$

and then prove that the precondition implies the resulting predicate. Combining two steps, we must prove that

$$x = x' \wedge x \geq 0 \; \supset \; wp(x := x \, \mathbf{div} \, 2, \; 2 \cdot x \geq x' - 1)$$
$$x = x' \wedge x \geq 0 \; \supset \; 2 \cdot (x \, \mathbf{div} \, 2) \geq x' - 1$$
$$x \geq 0 \supset 2 \cdot (x \, \mathbf{div} \, 2) \geq x - 1,$$

which follows immediately.

5.3.5 TWO EXAMPLES

Before discussing the weakest precondition of the **while** statement, let's consider two examples that make use of several of the definitions introduced so far. Suppose that x is an integer variable and that we wish to prove

$$x \neq 0 \; \{\mathbf{if} \; x < 0 \; \mathbf{then} \; x := -x \, \mathbf{else} \; x := x - 1\} \; x \geq 0$$

Recall that the proof involves two steps: (1) constructing a wp, and (2) proving that the given precondition implies this wp. For this particular program, then,

Step 1. The weakest precondition is

$$
\begin{aligned}
&wp(\; \mathbf{if} \; x < 0 \; \mathbf{then} \; x := -x \, \mathbf{else} \; x := x - 1, \; x \geq 0) \\
&\quad \equiv \; (x < 0 \supset wp \, (x := -x, \; x \geq 0)) \\
&\quad\quad \wedge (x \geq 0 \supset wp \, (x := x - 1, \; x \geq 0)) \\
&\quad \equiv \; (x < 0 \supset -x \geq 0) \wedge (x \geq 0 \supset x - 1 \geq 0) \\
&\quad \equiv \; true \wedge (x \geq 0 \supset x \geq 1) \\
&\quad \equiv \; x \neq 0
\end{aligned}
$$

Step 2. Therefore, we must prove

$$x \neq 0 \supset wp(\mathbf{if} \; x < 0 \; \mathbf{then} \; x := -x \, \mathbf{else} \; x := x - 1, \; x \geq 0)$$

or

$$x \neq 0 \supset x \neq 0,$$

which is true by inspection. Note that we have suppressed the conditions saying that $-x$ and $x - 1$ are defined.

The second example is (again, assuming that x is an integer variable) a proof for

$$x < n \wedge p = 2^x \; \{x := x + 1; \; p := p * 2\} \; x \leq n \wedge p = 2^x. \qquad (5.4)$$

Step 1. We generate the weakest precondition

$$
\begin{aligned}
&wp(x := x + 1; \; p := p * 2, \; x \leq n \wedge p = 2^x) \\
&\quad \equiv \; wp(x := x + 1, \; wp(p := p * 2, \; x \leq n \wedge p = 2^x)) \\
&\quad \equiv \; wp(x := x + 1, \; x \leq n \wedge p \cdot 2 = 2^x) \\
&\quad \equiv \; x + 1 \leq n \wedge p \cdot 2 = 2^{x+1}
\end{aligned}
$$

Step 2. The proof that the given precondition implies the weakest precondition found in Step 1 is trivial:

$$x < n \wedge p = 2^x \supset x + 1 \leqq n \wedge p \cdot 2 = 2^{x+1}$$

5.3.6 EMBEDDED ASSERTIONS

Sometimes it helps both the reader and writer of a program, if the writer inserts comments indicating what should be true at a particular point in the program's execution. For example, these can also help simplify a proof if we wish to show that

$$P \{S_1; \textbf{ assert } R; \ S_2\} \ Q$$

is valid (the **assert** statement would be a comment in most languages), it suffices to show that

$$P \{S_1\} \ R, \text{ and } R \{S_2\} \ Q.$$

In terms of the weakest precondition,

$$wp(\textbf{assert } R, Q) \ \equiv \ R \wedge Q. \tag{5.5}$$

If we can prove that

$$R \supset Q,$$

then Eq. (5.5) reduces to

$$wp(\textbf{assert } R, Q) \ \equiv \ R. \tag{5.6}$$

It is this reduction that makes the **assert** statement useful: It identifies the assertions that are most important to the correctness of a program.

5.3.7 WEAKEST PRECONDITION FOR THE "WHILE" STATEMENT

Iterative statements present a special problem; they may not terminate. Each of the other statements considered so far has the property that if each of its subparts terminates, so must the entire statement. But

while true do $x := 5$

does not terminate even though the **do** part, $x := 5$, does. The possibility of nontermination is linked to another problem; since a **while** statement is equivalent to an arbitrarily long sequence of statements, its weakest precondition may be infinite, an undesirable situation. Let us see what can be done to avoid this situation.

To say that the execution of

while B **do** S

terminates with some predicate Q being true is also to say that

- After executing S zero times, $Q \wedge \sim B$ is true, or
- B is initially true and after executing S once, $Q \wedge \sim B$ is true, or
- B is initially true and after executing S once, B is still true; and after executing S twice, $Q \wedge \sim B$ is true, or $\cdots$

Thus,

$$
\begin{aligned}
wp(\textbf{while } B \textbf{ do } S, \ Q) \equiv \ &wp(skip, \ Q \wedge \sim B) \ \vee \\
&B \wedge wp(S, \ Q \wedge \sim B) \ \vee \\
&B \wedge wp(S, \ B) \wedge wp(S; \ S, \ Q \wedge \sim B) \ \vee \\
&\cdots
\end{aligned}
$$

We note that

$$
wp(S, \ B) \ \wedge \ wp(S; \ S, \ Q \wedge \sim B) \ \equiv \ wp(S, \ B \wedge wp(S, \ Q \wedge \sim B))
$$

to get (see Exercise 5.35),

$$
\begin{aligned}
\equiv \ &[Q \wedge \sim B] \ \vee \ [B \wedge wp(S, \ Q \wedge \sim B)] \ \vee \\
&[B \wedge wp(S, \ B \wedge wp(S, \ Q \wedge \sim B))] \ \vee \ \cdots,
\end{aligned}
$$

an infinite sentence, as suspected. Let us write this as

$$
wp(\textbf{while } B \textbf{ do } S, \ Q) \ \equiv \ H_0 \vee H_1 \vee \cdots \tag{5.7}
$$

where

$$
\begin{aligned}
H_0 &\equiv \ Q \wedge \sim B \\
H_i &\equiv \ B \wedge wp(S, \ H_{i-1}), \textit{ for } i > 0.
\end{aligned}
$$

The trick now is to find some regularity in the H_i so that we can collapse this infinite formula.

5.3.7.1 A Direct Approach

Consider the program fragment in Fig. 5.1, part of a routine to compute x^n.

```
while  i ≦ n do
   begin
      t := t * x;  i := i + 1
   end
```

Fig. 5.1 Loop for computing x^n

Call the fragment F. Let's calculate $wp(F, Q(t, i))$. From Eq. (5.7), we derive

$$
\begin{aligned}
wp(F,\ &Q(t,\ i)) \\
\equiv\ &Q(t,\ i) \wedge i > n\ \vee \\
&i \leq n\ \wedge\ wp(t := t * x;\ i := i + 1,\ \ Q(t,\ i) \wedge i > n)\ \vee \\
&i \leq n\ \wedge\ wp(t := t * x;\ i := i + 1,\ \ i \leq n\ \wedge \\
&\qquad\qquad wp(t := t * x;\ i := i + 1, \\
&\qquad\qquad\qquad\qquad Q(t,\ i) \wedge i > n))\ \vee \\
&\cdots
\end{aligned}
$$

$$
\begin{aligned}
\equiv\ &Q(t,\ i) \wedge i > n\ \vee \\
&i \leq n\ \wedge\ Q(t \cdot x, i + 1) \wedge i + 1 > n\ \vee \\
&i \leq n \wedge i + 1 \leq n \wedge Q(t \cdot x^2,\ i + 2) \wedge i + 2 > n\ \vee \\
&\cdots
\end{aligned}
$$

$$
\begin{aligned}
\equiv\ &[Q(t,\ i) \wedge i > n]\ \vee \\
&[Q(t \cdot x,\ i + 1) \wedge i = n]\ \vee \\
&[Q(t \cdot x^2,\ i + 2) \wedge i + 1 = n]\ \vee \cdots
\end{aligned}
$$

$$
\equiv\ Q(t \cdot x^j,\ i + j), \text{ for } j \text{ the smallest integer} \geq 0 \text{ such that } i + j > n,
$$

and this last line makes the formula finite.

Now if we had originally wanted to prove that

$$
n \geq 0\ \{i := 1;\ t := 1;\ F\}\ t = x^n,
$$

we would compute

$$
wp(F,\ t = x^n)\ \equiv\ t \cdot x^j = x^n,
$$

where j is the smallest integer ≥ 0 such that $i + j > n$,

so that

$$
wp(i := 1;\ t := 1;\ F,\ \ t = x^n)\ \equiv\ 1 \cdot x^j = x^n,
$$

where j is the smallest integer ≥ 0 such that $1 + j > n$.

The correctness of our program is equivalent to

$$
n \geq 0\ \supset\ 1 \cdot x^j = x^n,
$$

where j is the smallest integer ≥ 0 such that $1 + j > n$.

This last is certainly true, since n is the smallest nonnegative number, so that $1 + n > n$.

We can make this a little more general. To find

$wp(\textbf{while } B(\textbf{x}) \textbf{ do } S, Q(\textbf{x}))$,

where **x** stands for the variables modified by S, we find a general formula giving the value, call it $F(\textbf{x}, j)$, of **x** after j iterations, for all values of j. The weakest precondition then becomes

$$wp(\textbf{while } B(\textbf{x}) \textbf{ do } S, Q(\textbf{x})) \equiv Q(F(\textbf{x}, j)), \tag{5.8}$$

where j is the smallest nonnegative value for which $\sim B(F(\textbf{x}, j))$.

If there is no such j, then we take the formula to be false.

In the preceding example,

$$F(\textbf{x}, j) = F(t, i, j) = (t \cdot x^j, i + j).$$

That is, the values of t and i after j iterations will be $t \cdot x^j$ and $i + j$, respectively.

As another example, consider the program fragment of Fig. 5.2, which is supposed to compute the maximum element of an array, A, with elements indexed from 1 to n. The program—we'll refer to it as *MAXFIND*—won't work correctly as it stands. We'll use wp's to find and solve the problem.

```
MAXFIND:
    while i ≤ n do
        begin
        if A[i] > x then x := A[i];
        i := i + 1
        end;
```

Fig. 5.2 Maximum element loop

We see that after j iterations, the variable i will again be j greater than its original value. Furthermore, the variable x will contain the largest of the element values scanned by the loop, which are

$$x', \ A[i'], \ A[i' + 1], \cdots, A[i' + j - 1].$$

(We use the prime notation here to indicate values before the loop starts.) Ignoring problems of array bounds, we find that the entire weakest precondition with respect to any postcondition $Q(x, i)$ is

$$wp(MAXFIND, Q(x, i))$$
$$\equiv Q(max(x, A[i], \cdots, A[i + j - 1]), i + j), \tag{5.9}$$

where j is the smallest nonnegative value for which $i + j > n$, and max is the maximum of its arguments. Since *MAXFIND* is supposed to compute the maximum element of the array A, a reasonable postcondition might be

$$Q(x, i) \equiv x = \max(A[1], \cdots, A[n]).$$

Note that this particular postcondition does not happen to include the variable i. Plugging this postcondition into Eq. (5.8) gives

$$wp(MAXFIND, x = max(A[1], \cdots, A[n])) \equiv \qquad (5.10)$$
$$max(x, A[i], \cdots, A[i + j - 1]) = max(A[1], \cdots, A[n])$$

where j is the smallest nonnegative value for which $i + j > n$.

If we examine Eq. (5.10), we see that in order to satisfy the weakest precondition, we may start the loop with

$$x = A[1] \wedge i = 2 \wedge n > 0.$$

So, assuming that $n > 0$ initially, the initialization statements

$$i := 2; \quad x := A[1]$$

will suffice to make the loop work.

Note that our arguments about the validity of Eq. (5.10) were informal. In addition, we ignored the fact that $A[i]$ is undefined unless the value of i is in the proper range. (In Exercise 5.26, we consider these points in more detail.)

5.3.7.2 Loop Invariants and Weak Correctness

The *MAXFIND* loop of Fig. 5.2 provided an example of a rather useful concept, the *loop invariant*. The word "invariant," of course, means something that doesn't change, or vary. We use it here to refer to a predicate that is always true; in particular, a *loop invariant* is a predicate (about a loop) that is true each time control reaches the point where the predicate appears in the loop body.

Assume that we have done the initialization statements suggested, namely

$$i := 2; \quad x := A[1]$$

have been added to the beginning of the program. Then, by inspecting the loop body, we can see that the range of i will be between 2 and $n + 1$. This is one invariant of this loop; it starts off being true, and the loop stops before it becomes invalid. There may be many invariants for a particular loop. The trivial predicate "true," for example, is an invariant of all loops. The most important invariants, however, are those that describe the way that the loop operates. To see a helpful invariant for the *MAXFIND* loop, observe that at the beginning of each pass through the loop, x contains the largest of the values $A[1]$ through $A[i-1]$. Again, this property starts out being true, and each pass preserves its truth, so it is an invariant. Moreover, it captures the fact that the loop works by checking each element of A in sequence, and by making sure that x always contains the largest value checked thus far.

Note that we can combine the two invariant properties of the *MAXFIND* loop and the resulting predicate,

$$2 \leq i \leq n + 1 \wedge x = max(A[1], \cdots, A[i-1]) \tag{5.11}$$

is also an invariant; it is true before the loop starts executing and remains true each time control reaches the top of the loop body.

When the loop does terminate, the semantics of the **while** statement guarantee that $i > n$. Since we know that the invariant is always true, we have

$$i > n \wedge 2 \leq i \leq n + 1 \wedge x = max(A[1], \cdots, A[i-1]),$$

which implies that $i = n + 1$, giving

$$x = max(A[1], \cdots, A[n]).$$

This program was supposed to find the maximum element of A, and this is precisely what it does.

Note the strong resemblance between Eqs. (5.11) and (5.10). The essential difference between them is that Eq. (5.11) contains some assumptions about the initial values of i and x. Consider the loop

while B **do** S,

for which we want to prove some postcondition, Q. Suppose also that we can find a predicate, *INV*, that summarizes this loop as did Eq. (5.11). That is, suppose that we find *INV* such that

1. *INV* remains invariant whenever B is true and the body executes:

$$(B \wedge INV) \{S\} INV. \tag{5.12}$$

2. When B is false and the loop terminates, *INV* implies Q:

$$(\sim B \wedge INV) \supset Q. \tag{5.13}$$

Then we are justified in concluding that Q will be true when and if the loop terminates, assuming that *INV* is initially true. Another way to look at this is to realize that the loop corresponds to some sequence of executions of S. We know that B is true before each execution of S and that it is false after the last execution of S. Suppose we could show that INV is true before *and* after each execution of S:

> **assert** $INV \wedge B$; S; **assert** $INV \wedge B$; S; $\cdots$; S; **assert** $INV \wedge \sim B$.

Now we only have to prove that

> $(INV \wedge B) \ \{S\} \ INV$

and that $INV \wedge \sim B$ suffices to prove Q (our postcondition).

An invariant such as INV is an *approximation* of the weakest precondition. It does not imply termination, and it may be unnecessarily strong. We say that a loop invariant is too strong when it includes properties that are not true, or are irrelevant to the correct operation of the loop. For example, in the loop of Fig. 5.1,

> $x = 1 \wedge 1 \leqq i \leqq n + 1 \wedge t = 1$

is a loop invariant that satisfies both Eq. (5.12) and Eq. (5.13). However, this invariant may not be true when the loop begins. Whoever wrote the loop, moreover, probably did not intend for it to work only for $x = 1$.

As another example of the use of invariants, consider the loop of Fig. 5.1 once more. Since the purpose of this loop is to compute x^n, we might suspect that the loop invariant will indicate that t contains some partial result. Careful consideration of the loop body, in fact, shows that

> $t = x^{i-1}$

is an invariant of this loop (assuming that t and i start at 1). We also see that

> $1 \leqq i \leqq n+1$

assuming that $n \geqq 0$. These two equations are now our complete loop invariant. According to Eqs. (5.12) and (5.13), if we can show that the invariant is initially true, we need only show, in addition, that

> 1. $(1 \leqq i \leqq n + 1 \wedge t = x^{i-1} \wedge i \leqq n)$
> $\{t := t * x; \ i := i + 1\}$
> $(1 \leqq i \leqq n + 1 \wedge t = x^{i-1})$
>
> 2. $1 \leqq i \leqq n + 1 \wedge t = x^{i-1} \wedge i > n \supset t = x^n,$

which we leave to the reader.

To summarize, any predicate that satisfies the conditions of Eqs. (5.12) and (5.13) is a loop invariant. If INV is such an invariant, then it is also an approximation of the loop's weakest precondition. We can, however, qualify that precondition by adding to it the condition that the loop terminate, a condition we can prove separately as we show in section 5.3.7.3.

5.3.7.3 Termination of Loops: Well-founded Sets

In our discussion of loop invariants we did not address the issue of termination. Proving loop termination, however, is usually an easy task. One method is called the method of *well-founded sets*. To illustrate the method, reconsider the program fragment, Fig. 5.1. Note that the difference, $n - i$, decreases with each iteration. Furthermore, from our invariant, we know that n can never be less than -1. There is only one way both statements can be true: The loop must terminate.

In general, we know that the loop must terminate if we can find an expression, in terms of the program variables, that

- Takes on values from an ordered set, S, with the property that any decreasing sequence of elements from S must reach a smallest element in a finite number of steps,*
- Decreases in value on each iteration of the loop.

Then in the worst case, the program will terminate when the value of the expression reaches the smallest element. In practice, the most common expressions are integer valued and the set, S, is an integer interval.

As another example, consider the standard binary search in Fig. 5.3.

$$
\begin{aligned}
&\textbf{while } L < U \textbf{ do}\\
&\quad \textbf{if } A[(L+U) \textbf{ div } 2] < x\\
&\quad\quad \textbf{then }\ L := (L + U) \textbf{ div } 2 + 1\\
&\quad\quad \textbf{else }\ U := (L + U) \textbf{ div } 2\}
\end{aligned}
$$

Fig. 5.3 Binary search

where the initial values of L and U are non-negative. Here we see that the quantity U minus L constantly decreases and that its value never goes below 0, unless $L \geqq U$ initially, in which case the loop is never executed. Therefore, if the loop is executed a first time, it must also terminate. A formal proof of its termination is

$$0 \leqq L < U \ \{\textbf{if } \cdots \textbf{ then } \cdots \textbf{ else } \cdots .\} \ 0 \leqq (U-L) < (U'-L'),$$

as in Exercise 5.27.

---

* A set with these properties is called a well-founded set.

5.3.7.4 Summary of **while** Statement Semantics

Calculating the weakest precondition for most constructs in a programming language is simple; loops, however, present a problem. Because a loop statement represents an arbitrary number of possible executions of its body, it's possible that the wp for the loop is an infinite expression–each term being a wp for one of its executions. In the most general case it isn't possible to give a "closed form" formula for the wp of a loop. Nevertheless, two general approaches to finding the wp for **while** loops are the "direct approach" and the use of loop invariants.

The direct approach is

$$wp(\textbf{while } B \textbf{ do } S, Q) \equiv H_0 \vee H_1 \vee \cdots$$

where

$$H_0 \equiv Q \wedge \sim B$$
$$H_i \equiv B \wedge wp(S, H_{i-1}), \textit{ for } i > 0.$$

In the direct approach we simply compute H_i until the formula becomes finite.

The use of loop invariants often simplifies a proof of weak correctness, but requires a separate proof of termination. A predicate, INV, is an approximation to the weakest precondition of

while B **do** S

if we can show that

$(B \wedge INV) \{S\} INV$

and

$(\sim B \wedge INV) \supset Q$

If we choose to use loop invariants to prove weak correctness, we can prove termination using well-founded sets. Basically, we find an expression in the program variables that (a) decreases by some amount each time the loop is executed, and (b) cannot become smaller than some fixed quantity.

5.3.8 THE "FOR" STATEMENT

Although the **while** statement is adequate for writing all programs with loops, it is not necessarily convenient. The **for** statement was originally introduced to make it easier to write certain common loops, but we can also use it to make loop verification easier. PASCAL's **for** statement is of the form

for $i := a$ **to** b **do** S

where a and b are side-effect-free expressions yielding integers, and i is a local variable of the loop body.

The **for** statement "executes S for each value of i from a to b." When the program is midway through this execution, just before S executes with i equal to some intermediate value k, we know that "S has been executed for each value of i from a up to, but not including, k." This statement is similar to a loop invariant for the **while** statement, but it has been specialized to the semantics of a **for** statement. It doesn't say anything about the body, S, of course; so, in order to use it to prove anything about the loop, we must still invent an invariant that captures the effect of S. Choosing this latter invariant is just like choosing one for the body of a **while** statement; in both cases we need a predicate that both expresses the effect of the body and is true on each iteration.

In the case of the **for** statement, the invariant generally involves the loop control variable—i in the loop above—so that we can write the "typical" invariant as $INV(i)$. That is, $INV(k)$ is an assertion that is true just before executing the statement S with k as the value of the control variable. Just before the **for** loop begins running, therefore, $INV(a)$ holds. The state immediately *after* the **for** statement finishes is a little tricky. Assuming that $b \geq a$, we would expect $INV(b + 1)$ to be true, as if we were about to execute S with i set just beyond its last value. However, if $b < a$, then the value "just beyond the last value of i" is a little unclear. Since we assume that $INV(a)$ will hold in this case, we will take the value "just beyond" to be a when $b < a$. Putting these two cases together, we find that just after the **for** statement, we have

$INV(max(a, b + 1))$,

where the *max* function returns the larger of its arguments. Therefore, if Q is a postcondition that we want to guarantee when the **for** statement finishes, and if INV is a valid invariant, then it suffices to show that

$$INV(max(a, b + 1)) \supset Q. \tag{5.14}$$

All that remains is to say what it means for INV to be an invariant. We must show that if, first, INV is true when the loop is about to be executed with the current value of i, and, second, that i is between a and b (that is, the loop isn't finished yet), then—after executing S one more time—INV will be true of the next value of i, which is $i + 1$. More formally,

$$INV(i) \wedge a \leq i \leq b \, \{S\} \, INV(i + 1). \tag{5.15}$$

If the invariant INV satisfies the two properties (5.14) and (5.15), it is also a suitable approximation of the weakest precondition. That is,

$$wp(\textbf{for } i := a \textbf{ to } b \textbf{ do } S, Q) \subset INV(a), \tag{5.16}$$

assuming that (5.14) and (5.15) are true. This is because once we establish that *INV* holds just before starting the **for** loop (that is, that *INV(a)* is true), the two conditions (5.14) and (5.15) guarantee that *Q* will hold when the loop finishes. Read the symbol ⊂ in this equation as "is implied by." Intuitively, we want ⊂ to mean "is approximately." Note that we don't have to worry about showing that *i* is incremented and stays in range—the **for** statement does that for us. More important, we don't need to prove termination—the **for** statement guarantees that too (assuming, of course, that *S* always terminates.)

Several assumptions were made in the development of Eq. (5.16); but the PASCAL **for** statement satisfies these assumptions, so we didn't emphasize them. However, the equivalents of the **for** statement in some other languages do not satisfy the assumptions, so it's worthwhile reviewing them explicitly. In particular, we assume that *S* does not alter *a*, *b*, or *i*. PASCAL, in fact, places precisely this restriction on *S* (officially, that is; in practice, this requirement is not always observed.) We again assume that *a* and *b* have no side-effects.

Consider the following simple example:

$$S = 0$$

$$\{\textbf{for } i := 1 \textbf{ to } n \textbf{ do } \ S := S + A[i]\}$$

S is the sum of all elements of *A* indexed 1 to *n*

(For this example, we use an English postcondition where we could use a more succinct mathematical notation to show that the methods in this chapter do not always require formal notation.) The invariant here is simply

$$INV(i) \equiv S \text{ is the sum of all elements of } A \text{ indexed 1 to } i - 1.$$

To prove the program correct, we must first show that

$$INV(max(1, \ n + 1)) \supset$$

$$S \text{ is the sum of all elements of } A \text{ indexed 1 to } n,$$

and that

$$INV(i) \wedge 1 \leq i \leq n \ \{S := S + A[i]\} \ INV(i + 1).$$

In other words—expanding *INV* according to its definition—

- The assertion

    "*S* is the sum of all elements of *A* indexed 1 to $max(1, \ n+1) - 1$"

    implies the assertion

"S is the sum of all elements of A indexed 1 to n," and

- If the precondition

    "S is the sum of all elements of A indexed 1 to $i - 1$ and $1 \leq i \leq n$"

is true and we execute the statement

$S := S + A[i]$

then the postcondition

    "S is the sum of all elements of A indexed 1 to $i + 1 - 1$"

will be true.

Both of these statements are true. In particular, note that if n is less than 1, the sum all elements of A indexed from 1 to n is 0—there are no such elements. If we wanted to be tedious, we could use the assignment rule to convert the second statement to

    The assertion

    "S is the sum of all elements of A indexed 1 to $i - 1$ and $1 \leq i \leq n$"

implies that

    "$S + A[i]$ is the sum of all elements of A indexed 1 to i."

To complete the proof, we show that the given precondition ($S = 0$) implies the (approximate) weakest precondition of the **for** loop:

$S = 0 \supset INV(1)$

or

    "$S = 0$ implies S is the sum of all elements of A indexed 1 to 0."

Since the latter sum is 0 (there are no elements in the range 1 to 0), the statement is true.

5.3.9 THE PROCEDURE CALL

Just as it can be difficult to understand a large, monolithic program, it is even more difficult to prove it. Verification conditions tend to become large very quickly as one tackles larger and larger programs. There is, in short, a practical limit to the size of program we can verify as a single unit. Procedures play an important role in software engineering precisely because they provide a convenient way to break a program down into comprehensible units. By the same token, they can break program proofs into manageable pieces.

If we can devise a succinct and rigorous statement of what a certain procedure, say f, is supposed to do, we can then

1. Prove once and for all that f does, in fact, accomplish its stated objectives, and then,
2. Assuming that f works as stated, use the specifications to prove the correctness of each program that uses it.

Suppose that f is declared

> **procedure** f;
> **pre** $P_f(x)$;
> **post** $Q_f(x)$;
>   **begin**
> S
>   **end**

Here, P_f and Q_f are predicates—the desired specifications of f—and are the necessary precondition (P_f) and resulting postcondition (Q_f) for any call on f. The variable x stands for all global variables—all variables declared outside f—that f may modify. For now, we assume f to have no parameters.

According to our outline, we first verify that f satisfies its specifications:

$$P_f(x) \; \{S\} \; Q_f(x). \tag{5.17}$$

We do so using the methods already described. Having proved this, we forget the body, S, entirely and consider P_f and Q_f to be all that we know about f. Indeed, we could safely change the body of f without worrying about programs that use it, provided that the new body could also be described by Eq. (5.17).

Suppose that we have a procedure call on f in the middle of some program, and we wish to find

$$wp(f, Q(x)).$$

(Note that Q is an arbitrary postcondition, unrelated to Q_f). To find this weakest precondition, we assume that f satisfies its specifications (Step 2 of our outline), so that we are entitled to conclude that $Q_f(x)$ is true after f is executed, as long as $P_f(x)$ was true before. If, at the end, $Q_f(x)$ implies $Q(x)$ and if, initially, $P_f(x)$ is true, then $Q(x)$ will be true at the end. That is, we can define the *procedure call rule*.

$$wp(f,\ Q(x)) \ \equiv\ P_f(x)\ \wedge\ \forall\, z(Q_f(z)\ \supset\ Q(z)) \tag{5.18}$$

where z is any unused variable, and denotes the possible outputs (possible final values of x) from f. Translated into English, this weakest precondition asserts: first, that $P_f(x)$ is true, and second that any possible outcome (any z) that f could produce that would satisfy $Q_f(z)$, that outcome also satisfies Q.

5.3.10 AN EXAMPLE WITH PROCEDURES

Suppose that we declare a procedure, *EXP2*, as in Fig. 5.4.

```
procedure EXP2;
    pre n ≥ 0;
    post p = aⁿ
        var x: integer;
        begin
        x := 0; p := 1;
        while x < n do
            begin
            x := x + 1; p := p * a
            end
        end
```

Fig. 5.4 Program *EXP2*

It's easy to prove that procedure *EXP2* accomplishes its objectives—that is, that the preconditions and postconditions form a correct specification. The only global variable changed by the body of *EXP2* is p (note that x is not a global variable). Thus, we have

$$P_f \equiv P_f(p) \equiv n \geq 0$$

$$Q_f \equiv Q_f(p) \equiv p = a^n$$

Consider now the program, S,

```
a := 1; n := 3; sump := 0;
while a ≤ m do
    begin
    EXP2;
    sump := sump + p; a := a + 1
    end
```

Suppose we wish to prove

$$m > 0 \;\{S\}\; sump = 1^3 + \cdots + m^3$$

It is sufficient to prove

$$a = 1 \wedge n = 3 \wedge sump = 0 \wedge m > 0$$

$$\{\textbf{while} \cdots \textbf{end}\}$$

$$sump = 1^3 + \cdots + m^3,$$

since the first three terms of the precondition are clearly established by the initial assignment statements. After looking at the loop, we see that each iteration adds another term onto the sum. We eventually arrive at the invariant

$$INV \equiv \; sump = 1^3 + \cdots + (a - 1)^3 \; \wedge \; 1 \leq a \leq m + 1$$

$$\wedge \; m > 0 \wedge n = 3.$$

INV is true at the beginning of the loop. If it is true at loop termination, it implies the postcondition. The remaining task is to prove the invariance of *INV*.

$$INV \wedge a \leq m \; \{EXP2; \; sump := sump + p; \; a := a + 1\} \; INV$$

With a little manipulation, this becomes

$$INV \wedge a \leq m \; \{EXP2\} \; sump + p = 1^3 + \cdots + a^3 \; \wedge$$

$$0 \leq a \leq m \wedge m > 0 \wedge n = 3$$

Applying the procedure call rule, we can now write

$$INV \wedge a \leq m \supset$$

$$n \geq 0 \wedge \; z(z = a^n \supset (sump + z = 1^3 + \cdots + a^3 \; \wedge$$

$$0 \leq a \leq m \wedge m > 0 \wedge n = 3)),$$

which is easily shown to be true.

5.3.11 DEALING WITH PARAMETERS

In section 5.3.10, for the sake of simplicity, we considered only procedures without parameters. In effect, *EXP2* above actually has parameters—p, n, and a; but a good program would not pass parameters in this way, as global variables. A procedure that does so would be too rigid; it would work only with parameters having specific names that might well conflict with other program variables.

However, procedure parameters make the proof rule more complex. We omit them from this discussion in order to emphasize the notion of *modularization*—of dividing a program into logically autonomous pieces with simple, well-defined specifications.

The proof rule for a procedure with parameters can be derived from that for procedures without parameters; it is essentially

> Apply the rule for procedures without parameters after replacing occurrences of the formal parameters of the procedure in the specification with the actual parameters supplied.

We will content ourselves for now with seeing how this works with simple value parameters.

Suppose that f is declared

> **procedure** $f(y)$;
>   **pre** $P_f(y, x)$;
>   **post** $Q_f(y, x)$;
>   **begin**
> S
>   **end**

where y represents the pass-by-value parameters of f (so that P_f and Q_f now may depend on y), and where x, as before, consists of any global variables that f may change. Again, we use the methods already described to prove that f meets its specifications. Now suppose that we have a call on f, $f(a)$ in the middle of some program, and wish to find

$$wp(f(a), Q(a, x)).$$

This is given by

$$wp(f(a), Q(a, x)) \equiv P_f(a, x) \land \forall z(Q_f(a, z) \supset Q(a, z)), \qquad (5.19)$$

where x, as before, stands for all the variables that f may change. In other words, the rule is almost the same as for that for procedures without parameters, but now a provision is made for changing the names of the parameters.

There is a slight, but important, technical difficulty with this rule. Specifically, it doesn't work if one of the variables represented by x is also contained in a. For now, we will not consider this problem, but Exercise 5.37 does ask you to do so.

5.3.12 SUMMARY OF WEAKEST PRECONDITIONS

In order to prove that $P\{S\}Q$, we proceed in two steps: First we compute $wp(S, Q)$, then we prove $P \supset wp(S, Q)$. Figure 5.5 contains a summary of the wp's we have treated thus far.

$wp(\textbf{skip}, Q) \qquad \equiv Q$

$wp(S_1; S_2, LQ) \quad \equiv wp(S_1, wp(S_2, Q))$

$wp(\textbf{if } C \textbf{ then } S_1 \textbf{ else } S_2, Q)$
$$\equiv (C \supset wp(S_1, Q)) \wedge (\sim C \supset wp(S_2, Q))$$

$wp(\textbf{if } C \textbf{ then } S, Q)$
$$\equiv (C \supset wp(S_1, Q)) \wedge (\sim C \supset Q)$$

$wp(x := E, Q(x)) \equiv E$ is defined and $Q(E)$

$wp(\textbf{assert } R, Q) \quad \equiv R \wedge Q$

$wp(\textbf{while } B \textbf{ do } S, Q)$
$$\equiv H_0 \vee H_1 \vee \cdots, \text{ where } H_0 \equiv Q \wedge \sim B \text{ and}$$
for $i \geqq 1$, $H_i \equiv B \wedge wp(S, H_{i-1})$

$wp(\textbf{for } i := a \textbf{ to } b \textbf{ do } S, Q)$
$$\subset INV(a), \text{ where } INV(max(a, b+1)) \supset Q \text{ and}$$
$INV(i) \wedge a \leqq i \leqq b \{S\} INV(i+1)$

$wp(f(a), Q(x)) \quad \equiv P_f(a, x) \wedge \forall z (Q_f(a, z) \supset Q(z))$,

where x is the set of variables modifiable by f, P_f and Q_f are the precondition and postconditions on f, and a is the set of actual (by-value) parameters to f.

Fig. 5.5 Summary of Verification Conditions

In addition, we considered two additional topics for **while** loops—the use of loop invariants and separate proofs of termination:

Invariants

If INV is a predicate satisfying

$$B \wedge INV \{S\} INV \text{ and } INV \wedge \sim B \supset Q$$

then

$INV \bigwedge$ loop terminates $\supset$ $wp(\textbf{while } B \textbf{ do } S, Q)$,

providing an approximation to the weakest precondition for the loop.

Loop termination

To show that **while** B **do** S terminates, typically one finds an integer-valued expression, E, such that

- E has some finite lower bound, and
- E decreases each time S is executed.

5.4 Weaknesses in Correctness Arguments

There is a tendency to regard an alleged mathematical proof of anything as the last word on the subject—a guarantee against any possibility of error. Even in pure mathematics, in fact, this is far from the truth. When it comes to program correctness, it is even more risky to trust to the infallibility of proofs: We can make errors in specification and errors in proofs.

5.4.1 ERRORS IN SPECIFICATION

Suppose that we have a program fragment, S, and wish to prove that it sorts a vector, A. That is, we would like to show that the program rearranges the elements of the vector in ascending order. We might write the specification of S as

$n > 0$ $\{S\}$ The elements $A[1]$ to $A[n]$ are in ascending order.

If we assume that this statement is proved, can we conclude that we have proof that S sorts the first n elements of the vector A? You might be tempted to says so; at least the postcondition indicates that the first n elements of A are sorted. But suppose that S is the program fragment

for $i := 1$ **to** n **do** $A[i] := i$

This S certainly satisfies the postcondition, it certainly doesn't correspond to our notion of a sorting program.

What happened is that the formal specification did not correspond to our intent. In particular, we neglected to say that the final values of $A[1]$ through $A[n]$ were to be a permutation or a rearrangement of their initial values. Furthermore we might also point out in the postcondition that the variable n is not changed.

There is no way to guarantee that our formal specifications match our intentions. For that matter, we seldom have a complete picture of what our intentions really are (not many people would have thought to mention that S was not to change n). What we know is that we are confronted with the original verification problem (Does a program do what we want it to?), except that instead of program correctness, we are now concerned about specification correctness.' However, because specifications are written at a "higher level"—expressing desired results rather than details for obtaining them—we considerably reduce the problem of proving that a program does what we want.

5.4.2 ERRORS IN PROOFS

Errors in program proofs consist mostly of overlooking cases.   For example,

$$\{x := x + 1\} \quad x = x' + 1$$

seems obvious, but is technically incorrect; $x + 1$ might be too large for the machine to handle. We usually ignore this possibility, since it adds a considerable burden of detail to our proofs. Sometimes, though, ignoring it can hurt.

In general, there is a host of technical details that we tend to overlook in a program proof (just as in a mathematical proof). In program proofs, we are usually concerned with proving certain "interesting" properties. Unfortunately, the tiniest "uninteresting" detail can have disastrous effects if overlooked. A possible remedy to these oversights: that is currently under study is the development of programs that assist in program proofs, keeping track of and sometimes proving technical details. The programs that now exist, unfortunately, need considerable improvement before they become practical.

Still, even informal proofs (that is, proofs that ignore some details), performed using the methods of this chapter, serve to increase our confidence in a program's correctness. In a sense, program proving is a sort of "wholesale testing" where large numbers of cases are tested analytically.

5.5 Validation: Testing Versus Verification

The entire task of making sure a program works is called *program validation.* The traditional approach to validation is, of course, testing with sample data; we might call this empirical program validation.   To be convincing, such data must test all of the code of the program—including all special cases and error conditions—and, in addition, must test the program on the "extreme" values of correct data. Unfortunately, there is no coherent, comprehensive theory of program testing, and vague statements such as those in this paragraph are just about all we can say on the subject.

In view of the weaknesses of the verification approaches discussed in Section 5.4, it is clear that testing will remain a major source of confidence in program correctness for some time. Still, program verification, which might also be called "symbolic program validation," supplies an alternative, independent method of validation.

Program verification techniques also *fit in with* one empirical validation technique—*instrumentation.* One way to test a program, or to check upon its continuing correct operation when it is in production, is to instrument it. In programming, the term generally means inserting extra statements not concerned with producing results, but rather with monitoring aspects of the computation. Sometimes, these statements are inserted by a programmer, sometimes by a compiler, and sometimes by a separate monitoring program (such as the "debuggers" common in interactive systems). Program verification, with its emphasis on explicit assertions—loop invariants and preconditions and postconditions—provides obvious instrumentation guidance.

5.6 Further Reading

Manna's text, *Mathematical Theory of Computation*, is an excellent introduction to the subject of verification. We attribute the concept of predicate transformers as presented in this Chapter mainly to Dijkstra (see *A Discipline of Programming*, for example).

Exercises

5.1 Below we give informal English specifications for a number of procedures. Formalize these specifications by writing them as appropriate preconditions and postconditions:

a) *sqrt(x)*: Returns the square root of positive real arguments.

b) *max(a, b)*: Returns the larger of *a* and *b*.

c) *P(A, i)*: Sets *i* to the number of nonzero elements of array *A*.

d) $P(A, i)$: Sets i to the value of the largest element in the array A.

e) $P(A, i)$: Sets i to the average of the largest and smallest elements in the array A.

5.2 In Section 5.1 we remark that, in actual practice, we ignore such details as computer word size and the attendant limits on the magnitudes of integers. List some other important details that we typically ignore in programming and may want to ignore in verification. Why do you think we ignore these details when our formalism allows us to handle them?

5.3 The following is the billing procedure description of a certain fictitious but major department store, which we call "The Store."

The customers have the following options each monthly billing period when they receive their bills.

1. To pay the entire New Balance within 30 days of the Billing Date shown on the monthly billing statement; or

2. To pay for each purchase the deferred payment price consisting of a cash price and a Finance Charge.

   a) The Finance Charge will be computed upon the Average Daily Balance of the customer's account in each monthly billing period.

   (1) The Average Daily Balance is determined by dividing the sum of the Balances Outstanding for each day of the monthly billing period by the number of days in the monthly billing period.

   (2) The Balance Outstanding for each day of the monthly billing period is determined by subtracting payments and credits from the previous day's balance, excluding any purchases added to the account during the monthly billing period and also excluding any unpaid Finance Charge.

   b) The Finance Charge will be determined by applying a periodic rate of 1.25% per month to the Average Daily Balance.

   c) When the Average Daily Balance for a monthly billing period is $40.00 or less, the Finance Charge will be $0.50 instead of the amount computed in (b).

   d) No Finance Charge will be assessed:

   (1) In a monthly billing period during which there was no Previous Balance;

   (2) In a monthly billing period during which payments and/or credits equal or exceed the Previous Balance;

   (3) On unpaid Finance Charge; or

   (4) On purchases during the monthly billing period in which they are added to the account.

e) The Store has a security interest under the Uniform Commercial Code in all merchandise charged to the account. If the customer does not make payments as agreed, the security interest allows The Store to accept return of only the merchandise that has been paid in full. Any payments the customer makes will first be used to pay any Finance Charges, and then to pay for the earliest charges on the account. If more than one item is charged on the same date, the customer's payment will apply first to the lowest priced item.

f) The customer has the right to pay the entire balance in full at any time without incurring a subsequent Finance Charge.

a) Can you understand this description? Assuming that you knew all the dates, times, and amounts of purchases and payments, and the billing dates, could you compute the balance in your account each month? Comment on any redundancies, contradictions ambiguities, or confusing presentation in the description Is, for instance, 2(d) redundant?

b) Prepare some test data for a billing program that is to compute bills under The Store's rules. Have several people *separately* read these rules and compute the bills. Compare the results. Your test data should "exercise" any clauses whose meaning seems obscure.

c) Write a program to compute one month's bill to these specifications. The data on purchases and payments are to be in the form of transactions such as

Purchase No. 1 $15.00 4/15/79

Purchase No. 2 $20.00 5/1/79

Payment $10.00 5/10/79

Repossess No. 1 7/15/79

What other data are necessary? How sure can you be that your program is correct?

d) Rewrite the specifications to correct any flaws you found for part (a) of this exercise and to make them as clear and short as possible. Resolve ambiguities as you see fit.

5.4 State the following weakest preconditions. In some cases, the program statements are in English, rather than PASCAL. In these cases, you may have to rely on Definition 5.2 to decide what the weakest precondition must be.

a) $wp(i := i + 1, A[i]$ is the largest element in $A[1], A[2], \cdots,$

$A[i]$)

b) wp(Set i to the index of the first instance of y in the array A, $A[i] = z$)

c) wp(Paint the living room ceiling, The house is all painted)

d) wp(*Count* := *Count* + 1; $i := i + 1$, *Count* is the number of times y occurs in array A between positions 1 and $i - 1$)

5.5 Compute the following weakest preconditions.

a) $wp(x := 2 * x, 0 < x < 20)$

b) $wp(x := x + y, x = -1 \wedge y = 1)$

c) $wp($**if** $x < 0$ **then** $x := x + 2, x > 0)$

d) $wp($**if** $odd(x)$ **then** $x := x + 1, x = 10)$

e) $wp($**if** $y \neq 1$ **then** $y := 1$ **else** $x := -x, y' \neq y \vee x < 0)$

f) $wp(x := 1 - x;$ **if** $x > 0$ **then** $y := 1 - x, 0 \leq x \leq 1 \wedge y \geq 0)$

g) $wp(t := x; x := y; y := t, x = y' \wedge y = x')$

5.6 What is $wp(S, x > y > 0 \wedge w \cdot x < 0)$ for each of the following statements S?

a) $x := y$

b) $x := -w + y$

c) $x := 2 * y; w := -x$

d) $x := -y$

e) **if** $x > y$ **then** $w := 5$ **else** $w := -3$

f) **while** $a > 0$ **do** $a := d * e + f$

5.7 What is $wp(x := 2 * y; y := x - 4, Q)$ for each of the following assertions Q?

a) $x = 0$

b) $y = 10$

c) $y > 6 \wedge x > 0$

d) $a = 37$

e) $x \cdot y < 0$

f) $w \cdot y = z + x$

5.8 Verify the following program. You can simplify the task by discovering and proving a suitable theorem.

true
$\{$**if** $A < B$ **then begin** $T := A$; $A := B$; $B := T$ **end**;
if $B < C$ **then**
  begin
  $T := B$; $B := C$; $C := T$;
  if $A < B$ **then begin** $T := A$; $A := B$; $B := T$ **end**;
  end; $\}$
$A \geq B \geq C$

5.9 We have assumed that expressions have no side effects. Construct some examples that show the importance of this assumption—that is, exhibit some claims of the form

$P \{S\} \ Q$

that are true under the assumption of no side effects, but false when the assumption proves untrue.

5.10 Prove each of the following, first by the direct method, and then by finding a suitable loop invariant and using the method of Section 5.3.7.2. All identifiers are declared to be integer variables.

a) $y > 0$
    $\{z := x$; $n := y$;
    while $n > 1$ **do**
      begin $z := z + x$; $n := n - 1$ **end**$\}$
  $z = x * y$

b) $n \geq 0$
    $\{p := 1$; $k := 0$;
    while $k < n$ **do**
      begin $p := 2 * p$; $k := k + 1$ **end**$\}$
  $p = 2^n$

c) $p > 0 \wedge q > 0$
    $\{m := p$; $k := 0$;
    while $m \geq q$ **do**
      begin $m := m - q$; $k := k + 1$ **end**$\}$
  $k = (p \ \textbf{div} \ q)$

d) *true*
    $\{i := 1$; $x := A[1]$;
    while $i \neq n$ **do**
      begin
        $i := i + 1$;

$$\text{if } A[i] > x \text{ then } x := A[i]$$
$$\text{end}\}$$
$$\forall\, 1 \leq i \leq n\, (x \geq A[i])$$

5.11 Using the method of well-founded sets, prove that each of the program fragments in problem 5.10 terminates.

5.12 Prove that the procedure *EXP2* in Figure 5.4 meets its specifications

5.13 Consider the annotated program

$$n \geq 1$$
$$\{x := 1;$$
$$\quad loop: \text{if } A[x] = y \text{ then goto } exit;$$
$$\quad x := x + 1;$$
$$\quad \text{if } x \leq n \text{ then goto } loop;$$
$$\quad exit;\}$$
$$A[x] = y \lor x = n + 1;$$

Rewrite this program without **goto**'s, find a suitable loop invariant, and prove the rewritten program correct. If you had to prove the original program correct, what assumption would you have to make about the label *loop*?

5.14 Find a statement or sequence of statements, S, that will satisfy each of the following preconditions and postconditions. Show formally (that is, using the methods of this chapter) that your solution works.

a) $true\ \{S\}\ y \geq x \land [(x = x' \land y = y') \lor (y = x' \land x = y')]$

b) $x \neq 0\ \{S\}\ |x| = |x'| - 1$

c) $x = F_n \land y = F_{n-1}$
$\quad \{S\}$
$\quad [x = F_n = F_{n'} + F_{n'-1}] \land [y = F_{n-1} = F_{n'} \land n = n' + 1],$
$\quad$ where F_n is the nth Fibonacci number.

5.15 Prove or find a counterexample:

$$y \geq 0\ \{p := y;\ q := 0;\ r := x;$$
$$\quad \text{while } p > 0 \text{ do}$$
$$\quad\quad \text{if } odd(p) \text{ then}$$
$$\quad\quad\quad \text{begin}$$
$$\quad\quad\quad\quad p := p - 1;\ q := q + r$$
$$\quad\quad\quad \text{end}$$
$$\quad\quad \text{else}$$
$$\quad\quad\quad \text{begin}$$

$$p := p/2; \; r := r * 2$$
end}
$$q = x * y$$

5.16 Prove or disprove with a counterexample:

$x > 0$ {**if** $(x \, \textbf{div} \, 2) * 2 = x$ **then** $x := x/2$ **else** $x := x * x$} x is odd

5.17 Consider the program

$N \geq 0$
{$i := 1; \; t := 1;$
while $i \leq N$ **do**
begin $t := t * x; \; i := i + 1$ **end**}
$t = x^N$

In Section 5.3.7.2, a suitable invariant for the loop in this program was determined to be

$$1 \leq i \leq n + 1 \land t = x^{i-1}.$$

What changes in the invariant and the **while** test would be required to preserve the given postcondition for each of the following changes to the program?

a) The order of the two statements in the loop body is reversed.
b) The initialization of i is changed to start i at 0.
c) The initialization of t is changed to start t at x.
d) The initialization of both i and t is changed to start i at 0 and t at x.
e) The order of the statements in the loop body is reversed, and i is started at 0.
f) The order of the statements in the loop body is reversed, and t is started at x.
g) The order of the statements in the loop body is reversed, and the initialization of both i and t is changed to start i at 0 and t at x.

5.18 The rule for computing the Fibonacci numbers (denoted F_0, $F_1, \cdots$) is

$$F_0 = F_1 = 1$$
$$F_n = F_{n-1} + F_{n-2}, \; \text{for } n > 1.$$

Write and verify two programs to compute the Fibonacci numbers. In one of your programs use a **while** statement; in the other use a **for** statement.

Comment on the relative difficulty of the proofs.

5.19 Write formal specifications and verifications of the programs of the following exercises:

   a) Exercise 2.1
   b) Exercise 2.14
   c) Exercise 2.15
   d) Exercise 2.16
   e) Exercise 3.15

5.20 Write formal specifications for the two polynomial evaluation programs of Section 6.1.1, and verify them. (Show strong correctness, including termination. A single specification should do for both programs.)

5.21 Write formal specifications for the two exponentiation programs of Section 6.5.5, and verify them. (Show strong correctness, including termination. A single specification should do for both programs.)

5.22 Specify, write, and verify a program for performing binary search in a sorted array.

5.23 Consider the following sorting program:

true {**for** $i := 1$ **to** n **do**
        Reorder $A[i]..A[n]$ so that the smallest element is in $A[i]$}
    $A[1]$ to $A[n]$ are sorted.

Of course, the English sentence in the program must be replaced by some appropriate PASCAL statement, but for the purposes of this exercise, just leave it as it is. An appropriate invariant for this loop is "INV$(i) \equiv A[1]$ to $A[i]$ are sorted and are all less than or equal to each of the values $A[i+1]$ through $A[n]$." Prove the program using this invariant. What must you assume about A?

5.24 Specify, write, and verify a program that performs integer addition using only "add 1" and "subtract 1" as primitives. You may assume that the program is restricted to nonnegative integers.

5.25 We can prove statements about a simple form of input and output by first modeling the input and output streams as variables that contain sequences of inputs and outputs, as does PASCAL. Roughly, if F is a variable of type *file* and contains the sequence of values

$$(v_1, v_2, \cdots, v_n),$$

then the statement

    $Read(F, x)$

sets x to v_1 and F to the new sequence

    $(v_2, \cdots, v_n)$.

However, this is not a complete description.

    a) What is missing in the description? What special cases have we ignored?

    b) What is the following?

        $wp(Read(F, x), Q(F, x))$

    c) What is the following?

        $wp(Write(F, x), Q(F))$

        Why didn't we write $Q(F, x)$?

5.26 Prove Eq. (5.10), using induction on k. If we take array bounds into account, how does Eq. (5.10) change?

5.27 Prove formally that $(U - L)$ in figure 5.3 strictly decreases and remains nonnegative, and, hence, that the binary search terminates.

5.28 Derive the weakest precondition of the PASCAL **repeat** statement. That is, compute

    $wp(\textbf{repeat } S \textbf{ until } B, Q)$.

5.29 Derive the weakest precondition of the **if–then** statement (that is, with no **else**). Compute

    $wp(\textbf{if } B \textbf{ then } S, Q)$.

5.30 Represent the **case** statement in terms of **if** statements and derive a weakest precondition rule for it. That is, compute

    $wp(\textbf{case } E \textbf{ of } E_1: \ S_1; \ E_2: \ S_2; \ \cdots; \ E_n: \ S_n; \textbf{ otherwise } S_*, Q)$

5.31 Finding the wp of the case statement in Exercise 5.30 is actually made more difficult by the explicit expansion into **if** statements. The difficulty arises because the expansion imposes an order on the tests. Suppose instead that the informal, intuitive semantics of **case** statements are as follows:

Find all the i's such that $E = E_i$. If there are no such values of i, execute S_*. If there is at least one such i, pick any one of them and execute only that S_i.

Express the wp for this version of the case statement.

5.32 Starting from the solution to either Exercise 5.30 or 5.31, derive the wp for the **numberedcase** statement.

5.33 We have formulated the weakest precondition for the **if** statement as

$$wp(\textbf{if } C \textbf{ then } S_1 \textbf{ else } S_2, Q) \equiv$$
$$(C \supset wp(S_1, Q)) \wedge (\sim C \supset wp(S_2, Q)).$$

It is sometimes more convenient to work with the following alternative form:

$$wp(\textbf{if } C \textbf{ then } S_1 \textbf{ else } S_2, Q) \equiv$$
$$(C \wedge wp(S_1, Q)) \vee (\sim C \wedge wp(S_2, Q))$$

Show that the two formulations are equivalent.

5.34 Use the FCL/2 expansions of the **for** statements in Exercise 3.8 to derive their wp's.

5.35 Show that

$$wp(S, B) \wedge wp(S; S, Q \wedge \sim B) \equiv wp(S, B \wedge wp(S, Q \wedge \sim B)).$$

5.36 Formidable though the formal procedure proof rule (Eq. (5.18)) seems, it is actually no more than one might expect intuitively. This exercise explores Eq. (5.18) as it applies to the procedure *Update* defined as

```
const n = 100;
var i: integer;
A: array [1..n] of integer;
y: integer;
procedure Update;
    pre 1 ≤ i ≤ n;
    post (i = i' + 1) ∧ (y = A[i']);
    begin y := A[i]; i := i + 1 end;
```

a) What list of variables from *Update* corresponds to x in Eq. (5.18)?

b) If you were given only the preconditions and postconditions and the procedure header of *Update*, could you determine x? Why, or why not?

c) Is the body of *Update* correct with respect to its preconditions and postconditions? Prove it.

d) Consider the logical assertion $(1 \leq i \leq n) \wedge (q = 6)$. After calling *Update*, you would expect q to be unchanged. That is, you would expect

   $$(1 \leq i \leq n) \wedge (q = 6) \; \{Update\} \; q = 6.$$

   Show that applying Eq. (5.18) gives just this result. If z replaces q. Show that the formula still works.

e) Show that

   $$(\forall j \; A[j] = j) \wedge (1 \leq i \leq n) \; \{Update\} \; y = i - 1$$

   using Eq. (5.18).

5.37 Section 5.3.11 mentions that Eq. (5.19) doesn't work if one of the variables in the list, **x**, of variables changed by f is contained in the list of actuals. Give an example. What implicit assumption has been violated?

5.38 Discuss the problems involved in developing a proof rule for value-returning procedures (that is, for "functions" that alter one or more parameters).

5.39 The following statements can be arranged to make a program fragment that computes the perimeter of a triangle whose coordinates are given in the variables $<x_1, y_1>$, $<x_2, y_2>$, $<x_3, y_3>$. If the statements are executed in the order given, they won't work correctly. The questions below deal with the orders that *will* work.

A: $Side1 := SQRT\,((x_2 - x_1)^2 + (y_2 - y_1)^2$

B: $Perimeter := Side1 + Side2 + Side3$

C: $Side3 := SQRT\,((x_1 - x_3)^2 + (y_1 - y_3)^2)$

D: $Side2 := SQRT\,((x_3 - x_2)^2 + (y_3 - y_2)^2)$

Let $\square$ be the relation "must not be executed after" for program statements. In other words, if $P \square Q$, P must not follow Q in the program listing, and if the statements are arranged in such a way that $\square$ holds between all pairs of statements, the program will work correctly.

a) For statements A, B, C, D, give the definition of $\square$ that allows the statements to be used in all the different orders that work, but in no others. List the elements (pairs) of the relationships, give the connection matrix, and draw the graph.

b) Which of the following properties does the relation □ have?

transitivity   reflexivity

symmetry   antisymmetry

c) Is □ a partial ordering? If not, what pairs must be added to the relation (what arcs must be added to the graph) to make it one? Show three different topological sorts of the program statements under the partial ordering.

5.40 As you have probably noticed, elementary program verification as presented here, even for simple problems, is a tedious and often unrevealing process. What do you think is necessary to make it practical for wide-spread use?   There are programs that generate verification conditions for a program, its specifications, and its loop invariants. There are also programs that can prove some of the resulting verification conditions. Do you think it would be sufficient to perfect these? Why? What do you think of the prospect of taking a large existing program, writing specifications for it, and verifying it? Justify your answers.

5.41 Write a program that computes the weakest precondition for a sequence of assignment statements and a postcondition. This involves processing the assignment statements in the proper (reverse) order, substituting the right-hand side expression for the left-hand variable whenever it appears in the assertion. The substitution must be done in such a way that the validity of the assertion is preserved; this may require the insertion of parentheses.

You may make the following simplifying assumptions:

- The input consists of a sequence of assignments followed by a single assertion, that is enclosed in braces. They appear one per line and contain no spaces. Variable names are single characters (there are no assignments to elements of arrays).

- It is not necessary to simplify the assertions as you make the substitutions.

Execute your program for the following cases:

a) $x := y;$
   $\{x = 3\}$

b) $w := v + 1;$
   $x := z * w;$
   $\{y = x\}$

c) $z := 5;$
$y := z;$
$x := y;$
$\{x >= y \wedge y >= z\}$

d) $c := b;$
$d := c;$
$e := d;$
$f := e;$
$\{f = a \wedge d = h\}$

e) $c := b;$
$d := c;$
$e := d;$
$f := e;$
$\{f = a \wedge d = h\}$

f) $a := x + 1;$
$b := a - 1;$
$x := 4 - b;$
$a := x;$
$\{a = 0\}$

5.42 Extend your program of Exercise 5.41 to include the **if–then–else** statement.

Chapter 6

Determining Efficiency of Computations

Just as Chapter 5 was concerned with methods of reasoning about *what* a program does, this chapter concerns similar methods of reasoning about *how well* a program does it. Although the first requirement of a "good" program is that it be correct (that is why we presented the material on verification in the preceding chapter), correctness is not enough. A good program must also be modifiable, easy to use, efficient, and so on. A good programmer must learn to evaluate his programs critically for all these qualities. Of all these qualities, efficiency has the largest collection of quantitative techniques. In this chapter, we will concentrate on techniques for evaluating program efficiency. To some people, the "efficiency" of a program means *only* execution speed. However, we use the word in a broader sense: A program may consume resources other then execution time, and so we measure efficiency in many different ways. The primary differences involve

- *Resources.* The resource might include time, space, number of components, bandwidth, response, etc.
- *Units Measured.* Time may be measured by counting seconds, steps, statements, or the number of times some given operation is performed; space may be measured in bits, words, or records; bandwidth can be measured in bits-per-second or as a fraction of the capacity of a channel.

In this chapter we continue our discussion of control by examining ways to measure and compare the execution speeds of programs—that is, their efficient use of the resource *time.* Later, in Chapter 12, we deal with

145

storage requirements; in Chapter 18 we deal with the special problems of analyzing recursive programs.

Just as we can reason about the correctness of a program using either analytic or experimental techniques, we can reason about the efficiency of a program either by examining the program text or by observing the program in action. Although we emphasize analytic techniques that can be applied directly to the program text, we also suggest some simple experimental techniques for collecting data on actual program performance.

6.1 Two Easy Examples

Consider two programs that solve the same problem. We compare their efficiency by estimating the number of operations or commands that each must execute. Frequently, estimates accurate enough for comparison can be obtained by focusing on one or two operations that are critical to the computation and by counting the number of times they are performed. If these operations are selected carefully, the estimates can be quite accurate. We do this in the following two sections in order to illustrate the technique.

6.1.1 POLYNOMIAL EVALUATION

As a first example, consider the problem of evaluating a polynomial

$$P(x) = a_0x^n + a_1x^{n-1} + \cdots + a_{n-1}x + a_n$$
$$= \Sigma_{i=0}^{n}\ a_ix^{n-i}$$

At their first attempt to devise an algorithm that evaluates polynomials, many people begin by computing the powers of x, then multiplying the powers by the corresponding coefficients and adding things up. They produce programs that look something like:

```
y[0] := 1;
for i := 1 to n do
    y[i] := y[i − 1] * x;
P := 0;
for i := 0 to n do
    P := P + a[n − i] * y[i];
```

This program fragment requires n multiplications for the first loop and $n + 1$ multiplications for the second; the total number of multiplications required is therefore $2n + 1$. It requires $n + 1$ additions involving x in the second loop. We ignore the subtraction in the subscripts because operations on small integer indices are often more efficient than operations involving general variables.

There are several ways to improve this program. For example, you can eliminate a loop and replace the vector y with a simple variable, and doing so will not change the computation in any essential way. However, consider the following equivalent expression for the polynomial:

$$P(x) = (\cdots ((a_0x + a_1)x + \cdots + a_{n-1})x + a_n$$

This factorization leads to a simpler way of computing the result, one that *halves* the number of multiplications required:

$P := a[0];$
for $i := 1$ **to** n **do**
$\quad P := P * x + a[i];$

The algorithm expressed in this program, known as Horner's Rule, was discovered by Newton in the early 18th century.* It requires n multiplications and n additions. It also has been proved to be the *best* possible way to evaluate general polynomials: That is, it requires the smallest possible number of multiplications and additions, and is, in short, more elegant.

To see that the two algorithms are really different, compare their loop invariants. For the first program, the invariant for the loop that actually evaluates the polynomial holds at the beginning of the loop:

$$P = \Sigma_{j=0}^{i-1} \ a_{n-j}x^j$$

However, for the faster program, the loop invariant is

$$P = \Sigma_{j=0}^{i-1} \ a_j x^{i-j-1}$$

In the former, each coefficient is associated immediately with the proper power of x; in the latter, the powers of x build gradually, in parallel. It is this fundamentally different view of how to do the computation that makes the speedup possible.

In this particular example, where the cost of multiplication is small compared to typical loop overheads, the cost of the loop itself may be substantially larger than that of the multiplications. Thus, halving the number of multiplications does not halve the total execution time—indeed the improvement may be quite small. However, if x were a matrix or polynomial,[†] the multiplications on their data structures would be very

---

* The simple form was discussed by Newton in 1711. A generalization was given by Horner in 1879. Horner's name stuck.

† Although the most familiar polynomials are functions on scalars, polynomials operating on other types (for example, matrices or other polynomials) are well defined and, indeed, quite useful.

expensive, and halving the number of multiplications would result in a substantial improvement.

6.1.2 SERIES EVALUATION

Let's turn now to another problem—that of evaluating a series that approximates a trigonometric function. The following series approximates $\sin^{-1}x$ to any desired accuracy.

$$\sin^{-1}x = x + \frac{x^3}{2\cdot 3} + \frac{1\cdot 3\cdot x^5}{2\cdot 4\cdot 5} + \frac{1\cdot 3\cdot 5\cdot x^7}{2\cdot 4\cdot 6\cdot 7} + \cdots$$

$$= \Sigma_{i=0}^{\infty}\ \frac{x^{2i+1}\Pi_{j=1}^{i}\ (2j-1)}{(2i+1)\ \Pi_{j=1}^{i}\ 2j}$$

Suppose we write a program to evaluate the first N terms of this series. If we write the program directly from the description of the series, it will probably resemble

```
S := 0;
for i := 0 to N do
    begin
    T := x ** (2 * i + 1) / (2 * i + 1);
    for j := 1 to i do
        T := T * (2 * j - 1) / (2 * j);
    S := S + T
    end
```

This program uses the operations addition ($+$), multiplication ($*$), division ($/$), and exponentiation ($**$) on reals, together with the operations addition ($+$), subtraction ($-$), and multiplication ($*$) on integers.* However, the most important thing to notice about the program is that it contains a nested **for** loop. The body of the inner loop performs multiplication and division on reals, and it is executed i times on the ith iteration of the outer loop. Since we know that

$$\Sigma_{k=0}^{N}\ k = (N\cdot (N+1)\ /\ 2)$$

---

* PASCAL, for reasons known best to its designers, does not have an exponentiation operation. Here, we have remedied that omission—in print, at least.

we know that the inner loop is executed $(N^2 + N) / 2$ times. In this case we settle for the following crude estimate: For large values of N, the execution time is proportional to $N^2 / 2$. The reason this estimate is good enough becomes clear as we examine another program for evaluating the same series.

Closer examination of the series reveals that each term is similar to its predecessor. If we take advantage of this by retaining the common portion of the expression from one iteration to the next, we can obtain the following improved program:

```
x2 := x * x;
S := x;  T := x;
for i := 2 step 2 until 2 * N do
    begin
    T := T * x2 * (i − 1) / i;
    S := S + T / (i + 1)
    end
```

Now this program, like the preceding one, involves operations on reals and integers. However, this program has no nested loop. The number of arithmetic operations in the loop in the second program is about the same as in the first program, but for large values of N, the cost of the first program is dominated by the execution of the inner loop. Eliminating this inner loop makes a significant improvement in the speed of the program; the execution time of the second program is proportional to N. Thus for even moderate values of N, the second program is far superior.

6.1.3 IMPROVEMENTS COME IN DIFFERENT FLAVORS

The two examples in this section show two ways to improve the efficiency of a program. In the first example, the performance was improved by devising an entirely new algorithm. In the second, the improvement arose from careful programming—from avoiding recomputation of a value already available.

These examples also show two essentially different kinds of improvements In the first example, rewriting reduced the execution cost by a constant factor. In the second example, the cost reduction depended on N—the size of the input (specifically, on the degree of the polynomial). In this latter instance, the savings grow as the size of the problem grows. Both changes are improvements, of course, but the second is much more significant.

The remainder of this chapter presents some techniques for analyzing the efficiency of programs. Section 6.2 is an excursion into mathematics presenting a technique for approximate arithmetic that we will rely on in the remainder of the chaper. Section 6.3 discusses strategies for deciding when it is appropriate to try to improve program performance. Section 6.4 presents experimental methods for efficiency analysis, and Section 6.5 presents analytic methods.

6.2 Order Arithmetic

When comparing programs or algorithms, the first thing you should pay attention to is gross differences in cost (for example, the amount of time or space consumed). For a few programs, the cost is fixed and can be calculated by examining the program text. More frequently, however, cost depends on the input or some other factor that varies from one use of the program to the next. Thus cost may be treated as a function of

- Length of input
- Number of digits of accuracy required in the answer
- Number of terms in an equation
- Size of problem (for example, row length of a matrix)
- Size of stored data being referenced by a program

or any of a large number of other things. When we make gross comparisons of programs whose cost is such a function, we often refer to the "order–of–magnitude" of the cost. The notation used is sometimes called "Big–Oh," and is always of the form $O(f(n))$, where $f(n)$ is some function of the positive integer, n. The Big–Oh notation simply means that the cost function is bounded by (is less than) some multiple of the function $f(n)$. For example, if we say

$$P = n^3 + O(n^2)$$

we mean that that P equals n^3, plus some terms that are "are on the order of n^2"—that is, they do not grow faster than $k \cdot n^2$, where k is some constant term.

More precisely,

Definition 6.1. A function $g(n)$ is said to be $O(f(n))$, written

$$g(n) = O(f(n))$$

if there is a positive constant M with

$$|g(n)| \leq M f(n)$$

for all sufficiently large values of n. As part of an expression, $O(f(n))$ stands for some function satisfying this property. □

We do not need to say just what the constant M is; indeed, it may be *different* for each appearance of O. We do need to know that M is a constant and does not depend on n. For example, recall that

$$1 + 2 + 3 + \cdots + n = n(n + 1) / 2 = n^2 / 2 + n / 2.$$

We can use order arithmetic to say that

$$1 + 2 + 3 + \cdots + n = n^2 / 2 + O(n),$$

or, with less precision, that

$$1 + 2 + 3 + \cdots + n = O(n^2).$$

This notation is a big help when we want to make gross comparisons of algorithms, because it allows us to avoid worrying about detail and allows us to concentrate on the things that really increase computation cost.

There are a number of rules for operating with order-of-magnitude quantities. The most important is that "equality" is *not* symmetric. By convention, we always write one–way equalities so that the right–hand side does not give away more information than the left–hand side. That is, we always write

$$n^2 / 2 + n / 2 = O(n^2),$$

and we never write

$$O(n^2) = n^2 / 2 + n / 2.$$

This helps keep us from treating this special kind of approximate equality like real equality. For example, although

$$5n^2 + 3n + 2 = O(n^2),$$

we must restrain ourselves from applying the transitive law and "deducing" that

$$5n^2 + 3n + 2 = n^2 / 2 + n / 2,$$

which is, of course, false.

Here are some other equivalences that allow you to manipulate equations involving order–of–magnitude quantities:

$$f(n) = O(f(n)),$$

$$K \cdot O(f(n)) = O(f(n)),$$

$$O(f(n)) + O(f(n)) = O(f(n)),$$

$$O(f(n)) \cdot O(g(n)) = O(f(n) \cdot g(n))$$

The differences that you can detect by considering only the order of the cost are illustrated in Figs. 6.1 and 6.2. Figure 6.1 plots 2^N, N^2, N $\log_2 N$, N, and $\log_2 N$ on the same axes for $N = 1$ to 100. (To appreciate how much faster 2^N grows than N^2, note that $2^{48} = 2.81475 \times 10^{14}$, but that $48^2 = 2.304 \times 10^3$ —a difference of a factor of about 10^{11}. Figure 6.2 shows why the order is more important than the constant multiplier. The curves kN and $k \log_2 N$ are plotted for $k = .25, .5, 1, 2,$ and 4. You can see that in all cases $k_i N$ eventually dominates $k_j \log_2 N$. Note, by the way, that the base to which logarithms are computed does not affect the order of magnitude of the result because changing the base of the logarithms from 2 to c changes the value by a constant factor of $\log_2 c$.

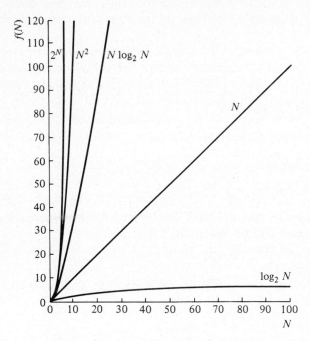

Fig. 6.1 Growth of several cost functions

If you know that the costs of two programs are roughly the same order, you may want to make finer comparisons by examining the constant multipliers of the main term and some smaller-order terms. Sometimes a factor of two will make a significant difference, but you should make sure there is not a drastically better way to accomplish the same improvement before you invest the time.

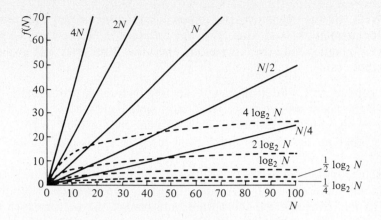

Fig. 6.2 Growth of kN and $k \log_2 N$

6.3 Gathering and Using Information about Efficiency

In the past, emphasis has sometimes been placed on the performance, or efficiency, of programs at the expense of other desirable properties, such as a user's ability to understand and modify them. The resulting backlash against single–minded emphasis on efficiency has made it fashionable in some circles to totally ignore efficiency. Indeed, correctness must not be sacrificed for performance, and when you're first trying to solve a problem its better to concentrate on getting it right.* However, totally ignoring efficiency is just as silly as ignoring correctness; efficiency is worth *reasonable* attention.

It is important to understand when it makes sense to try to improve the performance of a program and when it does not. Three factors must be considered before a program is modified to gain efficiency:

- How much will the change improve the overall performance of the program? If the faster section is only executed once, it probably is not justified.
- Will an improved program be executed often enough to pay back the costs of installing and debugging the change? If the program isn't used frequently, the changes may not be worthwhile.

---

* As one wag observed, a program can be made *arbitrarily* fast if it doesn't have to be right.

- What will the change cost in terms of your time, debugging, updating documentation, and long-term maintenance? If the new version is very complicated, the savings in execution time may not justify the other costs.

As a general rule, order-of-magnitude improvements are usually worthwhile if the faster programs are reasonably simple and the order constant is small. Also, constant factors of improvement are justified in the sections of the program that are executed most often (for instance, in the "inner loops").

Before modifying a program with the intent of improving it, you need to get estimates of the program's present performance, identify the sections that consume the most time, and predict the performance of the improved version. Often the most time-consuming portions of a program are quite small; several researchers have reported that, in typical programs, 3%–5% of the program text gives rise to 80%–95% of the execution time. Finding *which* 5% of the text is the critical code is not so easy, however; most programmers are often very poor guessers when they try to pick out the sections of their programs that consume the most time. You must, therefore, carefully direct your efforts to discovering the real bottlenecks in your programs.

Program speed can be determined in two essentially different ways. First, you can execute the program with some sample data and measure the amount of time it takes. Second, you can analyze the underlying algorithm to discover the relationship between the input and the number of steps required. Both techniques have their uses, and they are discussed in the next two sections.

6.4 Experimental Determination of Performance

An important class of techniques for determining program performance involves measuring the amount of time a program takes to process representative input data. These techniques can be applied without regard to the complexity of the program text itself, except to the extent that the complexity of the computation depends on the selection of input data. As a result, they are often useful for large programs such as compilers and operating systems, and the results of such measurements describe exactly the program measured.

Unfortunately, direct measurement has some disadvantages. First, the results describe only the program measured; these techniques are not capable of distinguishing between the amount of time inherently required by the algorithm, and the time that might be saved by making relatively minor changes in the program. Second, the results may depend heavily on the choice of test data. These data are useful only to the extent that they are truly representative; this may be very difficult to determine in a complex program. Third and last, direct measurements of programs, like all experimental measurements, are subject to error. The program measured will probably not require the same amount of time every time it is executed, even with the same input. Errors arise from interference by other programs using the same computer, imprecision in the timing tools, and other factors outside the control of the programmer. These problems are usually dealt with by statistical techniques that provide for such anomalies.

Two schemes for determining program performance are easily adapted to evaluating simple programs. The first scheme involves counting the number of times certain statements are executed; it can be applied to any program. The second consists of measuring the actual execution time of the program. To implement this scheme, you must have access to a routine that provides information about execution time. Both techniques are described in the following subsections. Other, more sophisticated methods of measuring performance are available, but they require special-purpose monitoring programs that are not available for all computers.

Note that the two techniques presented here (counting and timing) measure the same things. Further, if we could *count* all the operations, multiply the number of each one by the *time* it requires, and add all the products, we would obtain an estimate of the actual running time. Although optimizing compilers can make this estimate too conservative, it is a good first approximation.

6.4.1 COUNTING OPERATIONS

The first program in this chapter was one for evaluating a polynomial; we reasoned informally about it in order to illustrate the kinds of improvements that can result by changing the algorithm used to solve a problem. Let's return to that program to illustrate another point, namely the technique of operation counting. We add counters for those multiplications and additions that involve x, and we add statements to increment those counters whenever the corresponding operations are performed.

```
MultCnt := 0;
AddCnt := 0;
y[0] := 1;
for i := 1 to n do
    begin
```

```
      y[i] := y[i - 1] * x;
      MultCnt := MultCnt + 1;
    end;
  P := 0;
  for i := 0 to n do
    begin
      P := P + a[n - i] * y[i];
      AddCnt := AddCnt + 1;
      MultCnt := MultCnt + 1;
    end;
```

Executing this program with polynomials of degrees 0, 1, 2, 5, and 10 gives the following results:

Degree	Additions	Multiplications
0	1	1
1	2	3
2	3	5
5	6	11
10	11	21

These results correspond exactly to the prediction of $2n + 1$ multiplications and $n + 1$ additions for a polynomial of degree n. Note, however, that this does not prove the claim about the number of operations. It only shows that the statement is correct for five particular values of n.

6.4.2 MEASURING TIME

Programming systems often provide a routine that returns the amount of computer time your program has used. We usually assume that such a routine, called *RunTime*, is provided, and that the result is in arbitrary units (we sometimes provide an approximate conversion to actual time.) A naive approach is to invoke the timing routine before and after the program to be timed, as in

```
    t := RunTime;
    y[0] := 1;
    for i := 1 to n do
      y[i] := y[i - 1] * x;
    P := 0;
    for i := 0 to n do
      P := P + a[n - i] * y[i];
    time := RunTime - t;
```

Unfortunately, this approach may not be quite adequate. First, it fails to account for the time used to invoke the function *RunTime* itself. Second, and more important, the time measurements provided by the *RunTime* program may not be precise enough to provide worthwhile information. For example, if the clock used to provide time estimates is accurate to the nearest 0.01 second, but the program segment being tested runs in a few milliseconds (0.001 second), the "clock ticks" will come too far apart to provide useful information. This can be remedied by measuring the time required to execute the program segment N times and dividing the result by N. However, the value of N must be large enough to furnish accuracy but not so large as to waste execution time.

In general, you will use a loop to execute your program segment repeatedly. You may also need to reinitialize some of your program variables before each execution. Be sure to determine the cost of these overhead operations and subtract that cost from your result. Our example now becomes

```
{Determine overhead time}
LoopCost := RunTime;
for k := 1 to Many do;
t := RunTime;
LoopCost := t − LoopCost;


{Perform test}
t := RunTime;
for k := 1 to Many do
   begin
   y[0] := 1;
   for i := 1 to n do
      y[i] := y[i − 1] * x;
   P := 0;
   for i := 0 to n do
      P := P + a[n − i] * y[i];
   end;
time := RunTime;
time := (time − t − LoopCost) / Many;
```

6.5 Analytic Determination of Performance

The time required to execute a program can also be predicted by analyzing the program itself. Techniques that allow you to analyze program behavior in detail are easily simplified, so you can obtain estimates of time requirements with very little effort.

The time required by a program usually depends on the input values. For example, a program that searches for a value in an array, or one that iterates until it converges, depends on specific input values. Frequently, however, execution time depends on the size, or amount, of the input rather than on its specific value. This is true, for example, for matrix multiplication. The analysis techniques developed here are particularly useful for the latter class of programs.

Since an analysis based solely on the program text is done without reference to a specific machine or compiler, we need some unit other than clock time (seconds) with which to measure costs. We might assume that the cost of each kind of statement in the programming language is constant and count the number of times each is used. That is, we might assume that all assignmment statements take the same amount of time, all **if**'s take the same amount of time, and so on. This simplification ignores variations in the cost of evaluating expressions. Alternatively, we might assign different costs to different kinds of statements, making the ratios of these costs roughly proportional to the time required by the statements. More frequently, we will choose a few operations that are central to the algorithm and count the number of times *they* are used.

The particular choice of operations depends on what kind of algorithm is being analyzed. For example, for numerical algorithms we count arithmetic operations, and for sorting algorithms we count comparisons. This process clearly does not account for all the costs. However, as the total running time of the program tends to be proportional to the number of the abstract operations required, the number of abstract operations offers a fair estimate of the running time. Moreover, the resulting estimates are easier to obtain, and they are independent of programming details.

6.5.1 VERIFYING STATEMENTS ABOUT OPERATION COUNTS

It is not actually necessary to execute the program of Section 6.4.1 in order to determine the number of additions and multiplications required. We can instead deduce the number of such operations we expect using any techniques at our disposal, then verify a statement about the values that *MultCnt* and *AddCnt* will have when the (augmented) program finishes.

At this point we are interested only in the number of operations required, not the "answer" (that is, the value of *P*). Therefore we write an assertion that states only the fact we are trying to prove:

$$MultCnt = 2n + 1 \ \wedge \ AddCnt = n + 1.$$

We can now select loop invariants and prove the assertion about costs. Recall that invariants must hold just before the loop body is executed. An invariant for the first loop is

$$MultCnt = i - 1 \ \wedge \ AddCnt = 0 \ \wedge \ n > 0$$

For the second loop, a suitable invariant is

$$MultCnt = n + i \ \wedge \ AddCnt = i \ \wedge \ n > 0$$

Thus we need to verify the following program as annotated with assertions:

```
assert n > 0;
MultCnt := 0;
AddCnt := 0;
y[0] := 1;
for i := 1 to n do
    begin
    assert MultCnt = i − 1 ∧ AddCnt = 0 ∧ n > 0;
    y[i] := y[i − 1] * x;
    MultCnt := MultCnt + 1;
    end;
P := 0;
for i := 0 to n do
    begin
    assert MultCnt = n + i ∧ AddCnt = i ∧ n > 0;
    P := P + a[n − i] * y[i];
    AddCnt := AddCnt + 1;
    MultCnt := MultCnt + 1;
    end;
assert MultCnt = 2n + 1 ∧ AddCnt = n + 1;
```

We won't bother to carry out all of the steps of the proof in detail; however, working back from the end of the program, we can check the invariant for the second loop. This involves showing that the invariant after the loop implies the postcondition and that the invariant is preserved by the body. Showing that the invariant implies the postcondition is trivial:

$$(MultCnt = n + n + 1 \ \wedge \ AddCnt = n + 1 \ \wedge \ n > 0)$$

$$\supset \ (MultCnt = 2n + 1 \ \wedge \ AddCnt = n + 1).$$

(We have made a slight simplification here; since $n > 0$, $max(n + 1, 1)$ is $n + 1$.) To show that the invariant is preserved, we must show that

$$MultCnt = n + i \land AddCnt = i \land n > 0 \land 0 \leq i \leq n$$
$$\{P := P + a[n - i] * y[i];$$
$$AddCnt := AddCnt + 1;$$
$$MultCnt := MultCnt + 1\}$$
$$MultCnt = n + i + 1 \land AddCnt = i + 1 \land n > 0.$$

By computing weakest preconditions (substituting for the assignment statements), we establish the validity of the loop invariant. Assuming that n is nonnegative (otherwise the loop won't execute at all), we find that the weakest precondition for the loop is the invariant with 1 substituted for i:

$$MultCnt = n \land AddCnt = 0 \land n > 0.$$

This assertion does not involve P, so it is also the postcondition of the first loop. The verification of the first loop proceeds just like that of the second loop; the weakest precondition for the loop is

$$MultCnt = 0 \land AddCnt = 0 \land n > 0,$$

and after accounting for the initializing assignment statements, we find that the precondition for the entire program is, indeed, *true*. Note that we have done something much stronger here than we did in Section 6.4.1. There we argued informally about the efficiency of this program; here we *proved* that the program uses $2n + 1$ multiplications and n additions (assuming, of course, that the counting statements were inserted correctly.) In the previous case we determined only the number of operations required for a few given values of n.

6.5.2 A FORMAL TECHNIQUE FOR THE ANALYSIS OF EXECUTION TIME

In this section we formalize this intuitive notion of "proving an assertion about the number of operations." Instead of keeping a separate integer counter for each different kind of operation, we keep a single counter and add—symbolically—a constant amount for each operation the program performs. We do this for the costs of program statements (**if**, **for**, etc.), as well as for arithmetic operations. More specifically, we analyze the execution time of a program by attributing a cost to each expression or statement, then by estimating the number of times each will be evaluated. We state the rules so that costs can be associated with all the statements and expressions in a program.

The resulting analysis is much more detailed than we normally need, so we then simplify it in two ways: First, we concentrate on only the important operation, as we did in the examples in the previous section. Second, we use order-of-magnitude arithmetic instead of exact arithmetic.

We approach the analysis of efficiency the same way we did correctness, by giving a rule for each construct of the language. Just as for correctness, we associate logical predicates with the program. These logical predicates use the pseudovariable π to accumulate costs. For example, the final assertion will include a term that expresses the notion, "the cost of the program was K"—and that term will be $\pi = K$.

For deducing things about π, we introduce the notion of a *cost precondition*, which is much the same as the weakest precondition introduced in Chapter 5, but with additional expressions for time accounting. Cost preconditions use symbolic operations that increment π by appropriate amounts to formalize the intuitive notion of counting operations.

Definition 6.2. If S is a statement, then the *cost precondition* of S with respect to a logical statement Q is defined

$$cp(S, Q) \equiv wp(\pi := \pi + K_S; S, Q),$$

where K_S is an expression whose value is the cost of executing S. The cp is false when S does not terminate. If E is an expression with no side-effects on the program variables, then

$$cp(E, Q) \equiv wp(\pi := \pi + K_E, Q),$$

where K_E is the cost of evaluating E. Again, the cp is false if the evaluation of E does not terminate. □

Thus, cp's are simply wp's that also take into account the effects that statements and expressions have on the total time used by a program.

The following cp rules express the costs of executing some of the constructs in our programs. The symbolic constants σ_i, denote the costs of individual operations; the constants are keyed to statement types by their subscripts. For example, $\sigma_{:=}$ is the cost of an assignment. Figure 6.3 in Section 6.5.3 gives descriptions and sample values of these constants. The Q's are logical predicates in which expressions involving π refer to the program cost.

$$cp(\textbf{skip}, Q) \equiv Q$$

$$cp(S1; S2, Q(\pi)) \equiv cp(S1, cp(S2, Q(\pi + \sigma_;)))$$

$$cp(x := E, Q(x, \pi)) \equiv cp(E, Q(E, \pi + \sigma_{:=}))$$

$$cp(V[E1] := E2, Q(\pi)) \equiv$$
$$cp(V[E1], cp(E2, Q(\pi + \sigma_{:=}))),$$

provided that V does not appear in Q.

$cp(\textbf{assert } R, Q) \equiv R \wedge Q$

$cp(\textbf{if } B \textbf{ then } S1 \textbf{ else } S2, Q(\pi)) \equiv$
$\quad B \wedge cp(S, cp(B, Q(\pi + \sigma_{ift})))$
$\quad \vee \sim B \wedge cp(S2, cp(B, Q(\pi + \sigma_{ife})))$

$cp(\textbf{while } B \textbf{ do } S, Q(\pi)) \subset cp(B, I(\pi + \sigma_{wst})),$
$\quad\quad\quad \text{provided } B \wedge I(\pi) \supset cp(S, cp(B, I(\pi + \sigma_{wh})))$
$\quad\quad\quad \text{and } (\sim B \wedge I(\pi) \supset Q(\pi))$

$cp(\textbf{for } i := E1 \textbf{ to } E2 \textbf{ do } S, Q(\pi)) \subset$
$\quad\quad cp(E1, cp(E2, I(E1, \pi + \sigma_{fst}))),$
$\quad\quad\quad\quad \text{provided } I(i, \pi) \wedge E1 \leq i \leq E2 \supset cp(S, I(i + 1, \pi + \sigma_{for})$
$\quad\quad\quad\quad \text{and } I(max(E1, E2 + 1), \pi) \supset Q(\pi)$

$cp(E1 \textbf{ op } E2, Q(\pi)) \equiv cp(E1, cp(E2, Q(\pi + \sigma_{op})))$
$\quad\quad \text{where "op" is a binary operator}$

$cp(\textbf{op } E, Q(\pi)) \equiv cp(E, Q(\pi + \sigma_{op}))$
$\quad\quad \text{where op is a unary operator}$

$cp(x, Q(\pi)) \equiv Q(\pi + \sigma_{get})$

$cp(c, Q(\pi)) \equiv Q(\pi + \sigma_{const})$

$cp(V[E], Q(\pi)) \equiv cp(E, Q(\pi + \sigma_{ss}))$

$\quad\quad \text{where } S_i \text{ denotes a statement}$
$\quad\quad\quad\quad\quad E_i \text{ denotes an expression}$
$\quad\quad\quad\quad\quad x \text{ denotes a variable}$
$\quad\quad\quad\quad\quad B \text{ denotes a boolean expression}$
$\quad\quad\quad\quad\quad V \text{ denotes a vector}$
$\quad\quad\quad\quad\quad c \text{ denotes a constant}$

You should be fairly familiar with predicate transformers such as these, but let's look at a few to be sure that their meaning is clear. For example, the semicolon rule

$$cp(S1; S2, Q(\pi)) \equiv cp(S1, cp(S2, Q(\pi + \sigma_i)))$$

is identical to the corresponding *wp* except that, in addition, we have indicated that $\pi + \sigma_i$ must be substituted for all occurrences of π in Q.

The rule says that we may find the cost of executing a sequence of statements by computing the individual costs of the two statements ("pushing Q" back through $S2$ and then $S1$) and adding in the cost, $\sigma_{;}$, of any code executed between statements. This constant often has the value 0, but some compilers insert certain debugging information between statements, for example.

Consider another example, the cp for binary operators,

$$cp(E1 \text{ op } E2, Q(\pi)) \equiv cp(E1, cp(E2, Q(\pi + \sigma_{op})))$$

Operations are assumed to be free of side effects; thus the only change in Q as the result of such an operator is that time passes—hence the substitution for π.

To see how these rules work, we prove that the cost of the statement "$i := i + 1$" is $\sigma_{get} + \sigma_{const} + \sigma_+ + \sigma_{:=}$, or the cost of fetching one constant and one variable, executing one assignment statement, and performing one addition. The program, with assertions, is

$$\pi = 0 \; \{i := i + 1\} \; \pi = \sigma_{get} + \sigma_{const} + \sigma_+ + \sigma_{:=}$$

These assertions say that if the cost is 0 at the beginning, it will be $\sigma_{get} + \sigma_{const} + \sigma_+ + \sigma_{:=}$ at the end. In other words, we are proving an assertion about the cost of executing the program, namely that the statement accesses one constant and one variable, and it performs one addition and one assignment. The assertion states what is intuitively clear: we estimate the cost of executing the program by counting the individual operations involved. The rules given above do no more than provide a precise, systematic way to do this counting.

The proof proceeds, in gory detail, as follows. First we determine the cost predicate that must hold at the beginning of the statement:

$$cp(i := i + 1, \pi = \sigma_{get} + \sigma_{const} + \sigma_+ + \sigma_{:=})$$

$$\equiv cp(i + 1, \pi + \sigma_{:=} = \sigma_{get} + \sigma_{const} + \sigma_+ + \sigma_{:=})$$

$$\equiv cp(i, cp(1, \pi + \sigma_+ + \sigma_{:=} = \sigma_{get} + \sigma_{const} + \sigma_+ + \sigma_{:=}))$$

$$\equiv cp(i, (\pi + \sigma_{const} + \sigma_+ + \sigma_{:=} = \\ \sigma_{get} + \sigma_{const} + \sigma_+ + \sigma_{:=}))$$

$$\equiv (\pi + \sigma_{get} + \sigma_{const} + \sigma_+ + \sigma_{:=} = \\ \sigma_{get} + \sigma_{const} + \sigma_+ + \sigma_{:=})$$

$$\equiv \pi + \sigma_{get} + \sigma_{const} + \sigma_+ + \sigma_{:=} = \sigma_{get} + \sigma_{const} + \sigma_+ + \sigma_{:=}$$

$$\equiv \pi = 0$$

Now we can show that the given precondition implies the computed predicate:

$$\pi = 0 \supset \pi = 0.$$

The cost precondition rules, like the wp's, can all be applied mechanically—except when loops are involved. All loops require the verifier to invent an invariant, and the invariants we need here must have a term to describe the contribution of each execution of the loop body to the total cost. For example, consider the loop that increments a counter ten times. We showed that the cost of the simple increment ($i := i + 1$) is $\sigma_{get} + \sigma_{const} + \sigma_+ + \sigma_{:=}$. We'll use this statement as a loop body to see where the loop overhead costs enter the whole cost expression. Note that since the cost of the loop body does not depend on the loop counter, we can simplify the description of the loop by introducing a constant to represent the cost of the loop body:

$$\sigma_{incr} = \sigma_{get} + \sigma_{const} + \sigma_+ + \sigma_{:=} .$$

We now deal with the program fragment

assert $\pi = 0$;
for $j := 1$ **to** 10 **do**
    begin
    assert $\pi + \Sigma_j^{10} (\sigma_{for} + \sigma_{incr}) = \sigma_{fst} + 10(\sigma_{for} + \sigma_{incr})$;
    $i := i + 1$;
    end;
assert $\pi = \sigma_{fst} + 10(\sigma_{for} + \sigma_{incr})$;

The postcondition clearly expresses the fact that the body is executed ten times, the cost of proceeding to the next iteration is incurred ten times, and the cost of starting the loop is incurred once. The summation term in the invariant expresses the number of executions of the body that remain; the loop invariant asserts that the cost already incurred (π) and the cost that will be incurred during the remaining iterations (the sigma term) add up to the total cost of the loop. This example shows an important shortcut we will often use when the loop cost is the same every time: the cost of such a loop is just the startup cost (σ_{fst}) plus the number of times the loop is executed multiplied by the sum of the body cost and σ_{for}.

6.5.3 SAMPLE VALUES FOR CONSTANTS

In order to make the analyses in this chapter concrete and simple, we sometimes assign values to the constants σ. The values given Fig. 6.3 were measured empirically for a typical compiler, using techniques such as those described in Section 6.4. Although these values are specific to a particular computer and compiler, they are sufficiently representative to illustrate the analytic techniques.

Symbol	Value	Description
$\sigma_;$	0.0	Sequential execution
$\sigma_{:=}$	1.1	Assignment
σ_{ift}	0.8	Conditional, **then** branch
σ_{ife}	0.8	Conditional, **else** branch
σ_{wst}	1.5	**While** loop, terminal test
σ_{wh}	0.8	**While** loop, normal iterations
σ_{fst}	2.0	**For** loop, terminal test
σ_{for}	1.8	**For** loop, normal iterations
σ_{call}	5.6	Function call (parameterless)
σ_{val}	1.3	Value parameter
σ_{var}	1.1	Var parameter
$\sigma_>$	0.5	relation $>$
$\sigma_\geq$	0.5	relation $\geq$
$\sigma_=$	0.5	relation $=$
$\sigma_{\neq}$	0.5	relation $\neq$
σ_+	0.5	integer $+$
σ_-	0.7	integer $-$
σ_*	2.5	integer $*$
σ_{div}	5.5	integer div
σ_{mod}	5.6	integer mod
σ_{r+}	2.0	real $+$
σ_{r-}	2.3	real $-$
σ_{r*}	2.8	real $*$
σ_{get}	0.1	simple variable fetch
σ_{const}	0.0	integer constant fetch
σ_{ss}	2.2	single subscript

Fig. 6.3 Typical operation times (in microseconds).

In the remainder of this chapter we will assume that $\sigma_. = \sigma_{const} = 0$ and that $\sigma_{ift} = \sigma_{ife} = \sigma_{if}$. The assumption that $\sigma_. = 0$ is particularly important, since it allows us to find costs for individual statements and sum those costs.

6.5.4 EXAMPLES REVISITED

Let's consider the polynomial evaluation programs of Section 6.1.1. We will give the costs that the cp rules add to π for individual statements in a parallel column next to the listing of the program. The derivations of these values can be completed as in the preceding section.

```
assert π = 0;
y[0] := 1;                      σ:= + σss
for i := 1 to n do              σfst + σget + n(σfor + σ:= +
    y[i] := y[i−1] * x;              σr* + σ_ + 2σss + 3σget)
P := 0;                         σ:=
for i := 0 to n do              σfst + σget + (n+1)(σfor + σ:= +
    P := P + a[n−i] * y[i];          σr+ + σr* + σ_ + 2σss + 4σget)
assert π = final cost
```

Adding components and combining terms, we obtain the result that "final cost" must be

$$n(2\sigma_{for} + 2\sigma_{:=} + \sigma_{r+} + 2\sigma_{r*} + 2\sigma_- + 4\sigma_{ss} + 7\sigma_{get})$$
$$+ 2\sigma_{fst} + \sigma_{for} + 3\sigma_{:=} + \sigma_{r+} + \sigma_{r*} + \sigma_- + 3\sigma_{ss} + 6\sigma_{get}.$$

There are three things we might do to simplify this rather formidable expression.

1. Assign values to the σ_i and reduce the expression to a single value.
2. Select one or two operations as the important operations of the computation; set all other σ_i to 0; but maintain the counts of the selected operations symbolically.
3. Use order-of-magnitude arithmetic to relate the cost to the size of the input.

In order to do the first of these we need values of the σ_i for a specific machine and compiler. Even then, the values will be "average-case" estimates, and results will differ from actual times. Furthermore, optimizing compilers can often rearrange code, reducing the number of redundant fetches and duplicate computations. Even though the results are not exact, they may be useful. If we use the values in Fig. 6.3, the expression for the execution cost reduces to

$$24.3\, n + 21.8.$$

The second simplification is very common. We observed that additions and multiplications involving x and the coefficients are the important elements, and that these operations *must* be performed in order that the polynomial can be evaluated. We could, therefore, change the loop organization or even eliminate the loops entirely—using a different program for each value of n—to change many of the costs, but the operations on x and the coefficients would remain. If we ignore all costs except for the essential arithmetic, the cost estimate for the program is found by

$$(2n + 1)\sigma_{r*} + (n + 1)\sigma_{r+},$$

which is surely much simpler than the original expression. Moreover, it is independent of any implementation, and it still captures the essence of the cost.

Finally, using order-or-magnitude arithmetic, we can note that the cost expression has the form

$$An + B$$

where A and B are constant. We can therefore say the cost of the program is

$$O(n)$$

as a function of its input.

Now let's perform the same analysis for the program that implements Horner's rule:

```
assert  π = 0;
P := a[0];                          σ:= + σss
for i := 1 to n do                  σfst + σget +
    P := P * x + a[i];              n(σfor + σ:=
                                        + σr+ + σr* + σss + 3σget)

assert  π = final cost
```

We find that final cost must be

$$n\,(\sigma_{for} + \sigma_{:=} + \sigma_{r+} + \sigma_{r*} + \sigma_{ss} + 3\sigma_{get})$$
$$+ (\sigma_{fst} + \sigma_{get} + \sigma_{:=} + \sigma_{ss}).$$

Using the values for σ_i given above, this becomes

$$10.2n + 5.4.$$

If we consider only arithmetic operations on the coefficients and on x, the cost is, as expected,

$$n\sigma_{r*} + n\sigma_{r+}.$$

This program, like the slower one, has order $O(n)$.

6.5.5 ANOTHER EXAMPLE

To integrate these three techniques and show some of their strengths and weaknesses, let's consider the problem of computing x^n, where n is a positive integer. One way to do this is to multiply x by itself n times:

```
assert  π = 0 ∧ n ≥ 0;
prod := x;                        σ:= + σget
for i := 2 to n do                σfst + σget +
     prod := prod * x;            (n − 1)(σfor + σ:= + σ* + 2σget)
assert π = final cost
```

The invariant for the loop is simple:

$$prod = x^{i-1} \; \land \; n \geq 0 \; \land \; \pi + \Sigma_i^n (\text{loop cost}) = \text{final cost}.$$

The program clearly computes x^n. The consolidated expression for the final cost is

$$n(\sigma_{for} + \sigma_{:=} + \sigma_* + 2\sigma_{get}) + (\sigma_{fst} - \sigma_{for} - \sigma_*).$$

Substituting values for the constants σ_i yields

$$5.6n - 2.3.$$

Alternatively, focusing on arithmetic operations leads to the observation that the algorithm above requires one multiplication each time through the loop, or a total of $n - 1$ multiplications.

A second way to compute x^n is to compute x^2 from x, then x^4 by squaring x^2, then x^8 by squaring x^4, and so forth, multiplying some of these values together to get the desired product.

```
assert  π = 0 ∧ n ≥ 0;
t := x;  m := n                   2σ:= + 2σget
prod := 1;                        σ:= + σget + σ>= +
while m ≥ 1 do                    σwst + ⌈log₂(n+1)⌉(σwh + σget + σ>=)
   begin                          + ⌈log₂(n+1)⌉ [
   if m mod 2 = 1                      σif + σ= + σmod + σget
      then prod := prod * t;           + (if m odd then
                                              (2σget + σ:= + σ*))
   t := t * t;                     + σ:= + σ* + 2σget
   m := m div 2                    + σ:= + σdiv + σget
   end                            ]
assert  π = total cost
```

The invariant for the preceding loop is more complex, for it must establish a value for t as well as a value for *prod*. In addition, the value of n is changed in the loop:

$$x^n = prod \cdot t^m \wedge m \geq 0$$

$$\wedge \; \pi + \Sigma_{j=1}^{\lceil \log_2(m+1) \rceil} \; (\text{cost of the } j^{\text{th}} \text{ remaining loop}) = \text{total cost.}$$

Analysis of the loop shows it will be executed a total of $\log_2 n+1$ times. The computation to the right of the program has taken advantage of this to state directly the number of times the loop will be executed. We let K_{odd} denote the number of "one" bits in the input n. The cost of this more complex algorithm is thus

$$18.9 \lceil \log_2 (n+1) \rceil + 3.8 \, K_{odd} + 5.6.$$

The exact number of multiplications and divisions per loop is either 2 or 3, depending on K_{odd}. If $n = 2^k$ for some k, only the last iteration will perform the multiplication $prod := prod * t$. If $n = 2^k - 1$ for some k, then *every* iteration will perform it. (To understand why this is so, and indeed to understand what the algorithm does and why it works, examine the binary representation of n. The multiplication involving *prod* is done once for each 1-bit.) We can either evaluate the expression above for particular n, counting the bits in n in order to evaluate the expression, or estimate the number of bits that will be 1s in a large number of program uses. For the latter, we need to know the probability that a bit in a binary number is 1. This is slightly greater than .5, and the term involving n will have a value between $20 \lceil \log_2(n+1) \rceil$ and $21 \lceil \log_2(n+1) \rceil$.

The important thing about the algorithm expressed by this program is that the number of times the loop is executed is

$$\lceil \log_2 n+1 \rceil,$$

and the order of the algorithm is

$$O(\log_2 n).$$

Figure 6.2 displays the values of the order-of-magnitude costs of these two programs. If you are more concerned with execution time than programming time, it probably seems clear that the second algorithm is much better than the first. Remember, however, that the analysis that led to these graphs ignored some of the costs in a real program. The table in Fig. 6.4 shows measured execuion times for the two algorithms on a typical modern computer (the same computer for which constants were given in Fig. 6.3. Because of the overhead associated with the conditional statement and the temporary variables, the fast algorithm isn't faster for small values of *n*. Indeed, it doesn't become faster until $n = 18$, as the table shows.

N	Execution Time (microseconds)	
	"slow"	"fast"
1	9.56	28.32
2	15.56	47.61
3	21.67	51.19
4	27.80	67.13
5	34.06	70.68
6	40.07	70.36
7	46.15	74.14
8	52.30	86.81
9	58.36	90.27
10	64.65	89.92
11	70.74	93.58
12	76.85	89.94
13	83.05	93.74
14	89.18	93.52
15	95.07	97.07
16	101.78	106.63
17	107.49	109.94
18	113.63	109.68
19	119.59	113.29
20	125.74	109.54
21	131.83	113.26
22	138.03	112.79
23	144.12	116.62
24	150.19	109.60
25	156.27	113.21
26	162.37	112.90
27	168.55	116.62
28	175.20	113.47
29	181.02	116.67
30	187.05	116.45
31	193.13	119.96

Fig. 6.4 Comparative execution times of two algorithms

However, a *really* interesting result appears when powers of 3 are evaluated: The program fails when you attempt to evaluate 3^{16}; the last time the loop body is executed, it attempts to compute 3^{32}, which is too large to fit into an integer on this computer! For the same reason, the largest power of 4 that can actually be computed with this program on this machine is 4^{15}; the largest power of 5 is 5^7. Thus, for computing powers of integers on our particular computer, the "slow" program is, for all practical purposes, superior to the "fast" program.

The predicted and observed values are compared graphically in Figure 6.5.

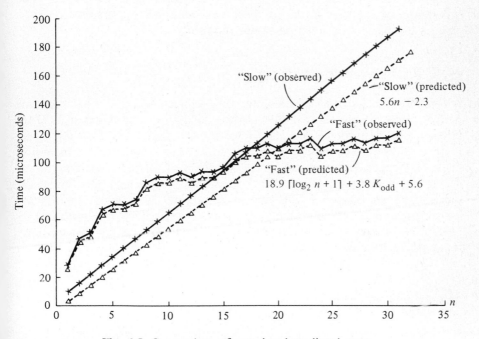

Fig. 6.5 Comparison of actual and predicted costs

It is tempting to conclude that the fast algorithm is useless. On the contrary, the results of the example we use simply indicates that an algorithm, like all tools, should be used with caution. Consider, for example, the problem of raising *complex* numbers or multiple-precision integers to integer powers. In these cases both the available range of exponents and the costs of single operations (for example, multiplication and assignment) are larger than in the examples above. Moreover, if such data types are not provided directly by the language, a function call may be needed for each operation. Suppose, for the sake of illustration, that assignment and multiplication have costs of 10 and 15 microseconds, respectively. Then re-evaluating the costs to reflect the increased costs for the complex or multiprecision operations (but not the operations on integer exponents) would yield the following cost functions

"slow": $27.0n - 14.8$

"fast": $49.2 \log_2 \lceil (n+1) \rceil$

$\qquad + 25.2 K_{odd} + 23.4$

Here the fast algorithm is faster on average when n is about 9 or 10. The advantage of the fast algorithm becomes even more pronounced when powers of matrices or polynomials are required.

6.6 Further Reading

Several textbooks are devoted to the subject of analyzing program performance. See, for example, Aho, Hopcroft, and Ullman (1974); Borodin and Munro (1975), and Horowitz and Sahni (1978). The basic mathematical techniques used in this chapter, including order-of-magnitude arithmetic, as well as analyses of some data structures, are to be found in Knuth, vol. 1 (rev. 1973). Algorithms involving polynomials are treated extensively in Knuth, vol. 2 (1969). Algorithms for sorting and searching problems are studied extensively in Knuth, vol. 3 (1973). In addition, a good survey of sorting techniques and suggestions for selecting the ones you need is included in Martin (1971).

Exercises

6.1 Add counters to our fast polynomial evaluation program to determine the number of additions and multiplications it requires.

6.2 Does the analysis of the polynomial evaluation programs depend on the particular values used as coefficients? Why or why not?

6.3 Insert calls on a routine such as *RunTime* into the fast polynomial evaluation program and tabulate results for polynomials of degree 0, 1, 2, 5, and 10.

6.4 Insert calls on a routine such as *RunTime* into the slow polynomial evaluation program and tabulate results for polynomials of degree 0, 1, 2, 5, and 10.

6.5 Perform experimental execution cost analyses for the programs for evaluating the $\sin^{-1}x$ series.

6.6 Perform analytic execution cost analyses for the programs for evaluating the $\sin^{-1}x$ series.

6.7 Write "crude" and "clever" programs (in the sense of Section 6.1.2) for evaluating the series

$$(1 + x)^{1/4} \doteq 1 + \frac{1}{4}x - \frac{1 \cdot 3}{4 \cdot 8}x^2 + \frac{1 \cdot 3 \cdot 7}{4 \cdot 8 \cdot 12}x^3$$

$$- \frac{1 \cdot 3 \cdot 7 \cdot 11}{4 \cdot 8 \cdot 12 \cdot 16}x^4 + \cdots$$

for $x^2 \leq 1$.

Compare the performances of your programs by

a) Inserting counters
b) Inserting calls on a routine such as *RunTime*
c) Using analytic techniques

6.8 Give an "order" for each of the following expressions—that is, for each expression, e, find the simplest expression that you can, e', such that $e = O(e')$. State your reasoning for each answer.

a) $x^3 + x^2$
b) $\log(x^2)$
c) $x^2 + \log(x)$
d) $\log 2^x$

6.9 Which of the following "equations" are true? Which are true in the reverse direction? (Remember that by convention, the symbol $=$, in this context, is not the same as ordinary equality.) Give your reasoning for each answer.

a) $O(1) = O(10^{53})$
b) $O(k \cdot f(n)) = O(f(n))$, where k is a positive constant.
c) $O(k \cdot f(n)) = O(f(n))$, where k is a negative constant.
d) $O(x^3) = O(15x^3 - 25x + 6)$
e) $O(n) = O(n^2)$
f) $O(x^3) / O(x) = O(x^2)$. Note: in the reverse direction, this claims that any function that is $O(x^2)$ is bounded by the

quotient of some two functions, one of which is $O(x^3)$ and the other of which is $O(x)$.

g) $O(f(n)) / O(g(n)) = O(f(n) / g(n))$

h) $O(f(n)) / n = O(f(n) / n)$

i) $O(f(n)) + O(g(n)) = O(\max(f(n), g(n)))$, where the value of max is the largest of its two arguments.

6.10 Fill in the details of the proof of Section 6.5.1.

6.11 Prove formally that

assert $\pi = 0$;
for $j := 1$ **to** 10 **do**
  begin
  assert $\pi + \Sigma_j^{10} (\sigma_{for} + \sigma_{incr}) = \sigma_{fst} + 10(\sigma_{for} + \sigma_{incr})$;
  $i := i + 1$;
  end;
assert $\pi = \sigma_{fst} + 10(\sigma_{for} + \sigma_{incr})$;

Assume that σ_{incr} accounts for the entire cost of incrementing an integer.

6.12 Exercise 3.15 asks about the relative efficiencies of iterative and recursive routines for computing Fibonacci numbers. Write both versions of these routines and instrument them using the techniques of section 6.4.

6.13 Write versions of both exponentiation routines with calls on a routine such as *RunTime* and collect data on actual performance. Compare these execution times to the times listed in the table in Fig. 6.4. Explain any discrepancies you note.

6.14 Obtain values for the cost coefficients σ_i for your computer. Use them to redo the analyses of this chapter.

6.15 Run the programs of this chapter on your computer. Compare the relative costs of your run to the relative costs given in this chapter.

6.16 Compare the results of the previous two exercises. Explain any discrepancies.

6.17 It is not always necessary to carry out an analysis with as much precision as the complete set of σ's provide. The computations can be simplified by grouping σ's of similar magnitude and keeping track of only a few constants. Reduce the table of σ's to four major groups; then repeat the analyses of the exponentiation and polynomial evaluation programs.

6.18 Insert counters in the programs of the following exercises to determine their performances. In each case, consider carefully what operations should be counted.

- a) Exercise 2.1
- b) Exercise 2.14
- c) Exercise 2.15
- d) Exercise 2.16
- e) Exercise 3.15

6.19 Insert calls on a routine such as *RunTime* in the programs listed in Exercise 6.18 to determine their performances.

6.20 Determine the execution time requirements of the programs listed in Exercise 6.18 using analytic techniques.

6.21 Derive the cost precondition for each of the following statements:

- a) **case**
- b) **numberedcase** (Hint: Look back at the SMAL representation.)
- c) **repeat**
- d) each of the variations of the **for** statement given in Exercise 3.8.

6.22 The "Big-Oh" notation, as in $O(g(n))$, is not entirely satisfactory for some purposes. For example, if a function is $O(n^2)$ does this mean that it is not $O(n)$? What does this say about the limitations of the notation?

6.23 The notation $\Omega(g(n))$ is analogous to $O(g(n))$. We say that $f(n)$ is $\Omega(g(n))$ or $f(n) = \Omega(g(n))$ iff there is a positive constant, K, such that $|f(n)| \geq K g(n)$ for all sufficiently large n. For example, $3n + 17$ is $\Omega(n)$ (choose $K = 3$) and also $\Omega(\log n)$ (choose any K).

- a) Show that $6n^2 + 35n + 7 = \Omega(n^2)$. Show that it is not $\Omega(n^3)$.
- b) Show that $1 + (1 / n)$ is $\Omega(1)$.
- c) Show that $\sum_{i=1}^{n} 1 / i = \Omega(\log n)$.
- d) Show that $n \log n = \Omega(n)$. Show that it is not $\Omega(n^k)$, where $k > 1$.
- e) If $f(n) = O(n^2)$ and is never 0, show that $n^3 / f(n)$ is $\Omega(n)$. Is it also $O(n)$?
- f) Consider a program that does a linear search through an array. That is, it starts at the first element of the array and continues checking each successive element until it finds one it is looking for (or comes to the end of the array.) Is the time required by this program $O(n)$? Is it $\Omega(n)$? Why?

6.24 The notation $\Theta(e)$ is analogous to $O(e)$ and $\Omega(e)$ (Exercise 6.23). We say that $f(n)$ is $\Theta(g(n))$ (or $f(n) = \Theta(g(n))$) iff $f(n)$ is both $O(g(n))$ and $\Omega(g(n))$. Thus, $3n + 17$ is $\Theta(n)$, but not $\Theta(n^2)$ or $\Theta(1)$.

a) Show that $6n^2 + 35n + 7 = \Theta(n^2)$. Show that it is $\Theta(n^2 + n)$.

b) Show that $n \log n$ is not $\Theta(n^k)$ for any k.

c) Show that $2 + \sin x$ is $\Theta(1)$.

d) Show that if $f(n)$ is $\Omega(n^2)$, then $3n + (n^2\log n) / f(n) = \Theta(n)$.

e) Consider the linear search program in Exercise 6.23 (f). Either find function $f(n)$ such that the time required for search is $\Theta(f(n))$ or say why there isn't one.

6.25 Which of the following are $O(n^2)$? $\Omega(n^2)$? $\Theta(n^2)$? (See Exercises 6.23 and 6.24).

a) $n^2 \log n$

b) $(f(n) + 1)^2$, where $f(n) = O(n)$.

c) $1 / f(n)$, where $f(n) = O(1 / n^2)$

6.26 Consider why we added the phrase "for sufficiently large n" to the definition of $O(g(n))$. Give an example of a function that would *not* be $O(n^2)$ had the definition stated only "for all n," but which *is* $O(n^2)$ as the definition stands.

6.27 One of the great advantages of order notation is that it often allows us to avoid a great deal of detailed analysis. For example, consider the loop

for $i := 1$ **to** n **do** S;

Often, we can make statements about the time required by this loop without looking at S more than superficially. What assumptions do you need about S to conclude that the entire loop runs in time $O(n)$? $\Omega(n)$? $\Theta(n)$? What must you do to check these assumptions? Under what assumptions about S is the loop $O(n^2)$? $\Omega(n^2)$? $\Theta(n^2)$?

6.28 In Section 5.3.7.3, we discussed the use of well-founded sets to show that a **while** loop terminates. Suppose that we have a program such as

while B **do** S

and also suppose that the expression $E(n)$ is the constantly decreasing quantity used in the proof of termination. If S requires $O(1)$ time (that is, no more than some fixed constant,) what can you say in general about the total time required by the loop? When can you give a more optimistic

estimate? If S is $\Omega(1)$, what is the best (that is, the most accurate) Ω
bound you can give in general? (The more information it gives, the better
a bound is. A bound of $\Omega(n)$ is better than one of $\Omega(1)$, because the
former gives more information about the function.)

6.29 Repeat exercise 6.9 with Ω in place of O.

6.30 Repeat exercise 6.9 with Θ in place of O.

6.31 Another way to think of the notation $O(g(n))$ is that it denotes the
set of functions

$$\{f \mid \text{for some } K, |f(n)| \leq K\, g(n) \text{ for all sufficiently large } n\}.$$

Using this interpretation, decide what $f(n) = O(g(n))$ means. What does
$O(f(n)) + O(g(n))$ mean?

Part Two

Fundamental Structures of Data

In Part One we discussed control, and said almost nothing about data. In Part Two we'll reverse the situation, discussing data and ignoring control. Later, in Part Three, we'll discuss how the two interact in real programs.

We begin this part with an introduction to the mathematical bases relating to data—namely, the concepts of *memory*, *variable*, *name*, *value*, and *type*. Of particular importance are the methods used to precisely define the properties of a data type. We use these methods to define a rich collection of data types—both ones that appear in practical programming languages and ones that programmers are usually required to synthesize for themselves. After several data types have been defined, we demonstrate a variety of ways each can be represented. As we'll find no single *best* representation for any type, we show how to analyze the context in which a type will be used in order to discover the best representation for its particular situation. We then show how to give a formal proof that a particular implementation of a type satisfies its definition. Finally, we discuss how to analyze the space efficiency of a given representation.

Chapter 7

Mathematical Models of Data

7.1 Introduction

In this chapter we discuss five related concepts: *memory*, *variable*, *name*, *value*, and *type*. Some of these may already be familiar to you from an introductory programming course. Here we'll try to treat them more precisely than such courses usually do. In particular, we want to develop a mathematical basis for these concepts so that they can be used in careful analyses of real programs.

Memory is often described as a collection of "boxes," or variables, each of which can contain a value. You are permitted to examine the contents of a box (a variable) or replace the value in it. The concept of type is explained, if at all, as a property of a value that determines the nature of the operations that can be performed on that value. We can be a bit more precise in our "intuitive" descriptions that follow:

- A *value* is an element of some set associated with a *type*. For example, 3 is a value from the set $\{0, +1, -1, +2, -2, \cdots\}$ associated with the type *integer*. Similarly, *true* is a value from the set $\{true, false\}$ associated with the type *Boolean*.

- A *variable* is a component of a *memory* and has both a *name* and a *type* associated with it; a variable "holds" or "contains" a value of this type.

- A *memory* consists of a collection of variables and has just two operations—fetch and store. Given a variable name and the current contents of memory, the operation *fetch* retrieves the value stored in the named variable. Given a name, a value, and the current contents of memory, the *store* operation returns the new contents of memory. The new contents differs from the old *only* in the named variable, and the new value of this variable will be that specified in the store operation.

As you can see, these notions are intertwined—each description depending to some extent on an understanding of the others. And, although the descriptions are more precise than those usually given in introductory textbooks, we have not yet reduced them to mathematical terms. We do that in the following sections. For simplicity, we divide our discussion into two parts, first treating all concepts *except* type in one main section and then, in a section by itself, treating type in considerable detail.

7.2 Memory, Variables, Names, and Values

In this section we explore two descriptions of memory. One characterizes memory as a pair of functions; the other uses the "assignment axiom," as discussed in Chapter 5. The two descriptions are equivalent. Each, however, is useful under certain circumstances, and understanding both reinforces our understanding of memory–related issues.

7.2.1 A FUNCTIONAL DESCRIPTION OF MEMORY

Perhaps one of the most obvious ways to describe memory is as a function—a mapping from the names of individual memory cells to the values stored in them.

> *memory*: *name* → *value*

By using this description, we can think of every access to a variable named *x*, as an invocation of the function *memory*(*x*).

We are using the word of "name" rather informally here, but your intuitions should carry you through the discussion—although a more complete treatment is given in Section 8.2. You should note, however, that "name" does not simply mean the identifier used in the declaration of a variable. A variable may be referred to by many identifiers during the course of a program execution. For example, the identifier *x* may refer to the first parameter of a subroutine, while a given call to that subroutine may pass a variable with identifier *y*. While the subroutine executes, therefore, the identifiers *x* and *y* both refer to the same variable. In the present context the "name" of a variable is unique; each variable has only one name. Thus a name, in our usage, is roughly an address of a variable in computer memory. However, we can't quite equate names with addresses. For example, the address of a variable in computer memory gives no indication of how big the variable is—how many words it spans. But to fetch the value of that variable, we certainly need to know this additional information. Here, such technicalities would only serve to muddy the waters, and we will continue to treat names and addresses as practically synonymous while keeping the terms distinct.

The view of memory as a function is entirely compatible with the intuitive description of memory as a collection of "boxes," each box containing a value. As you recall, a mathematical function is defined as a set of ordered pairs: the first element of each pair is an element of the *domain*; the second element is an element of the *range*. Consider some function, $f(a_i) = v_i$. Mathematically, the function f is defined by

$$f = \{<a_1, v_1,>, <a_2, v_2>, \cdots, <a_n, v_n>\}.$$

In particular, consider the memory function as defined by

$$memory = \{<a_1, v_1>, <a_2, v_2>, \cdots, <a_i, v_i>, \cdots, <a_n, v_n>\}.$$

You can think of each ordered pair as one of the "boxes" in the intuitive description. The first element of each pair is the name, or address, of the box. The second element is the value contained in it. Invoking the memory function is nothing but retrieving, or "fetching," the value stored in a particular box. This description is both simple and intuitively appealing; and so long as we assume that the contents of memory remain unchanged, it is accurate. An assignment statement, however, can change the contents of memory.

Intuitively, the net effect of an assignment statement is to change the content (or value) of one of the boxes—that is, to change the second element in one of the ordered pairs. If at some time

$$memory = \{<a_1, v_1>, <a_2, v_2>, \cdots, <a_i, v_i>, \cdots, <a_n, v_n>\},$$

and then we perform the assignment,

$$a_i := x,$$

then *memory* changes to

$$memory = \{<a_1, v_1>, <a_2, v_2>, \cdots, <a_i, x>, \cdots, <a_n, v_n>\}.$$

That is, we must change $<a_i, v_i>$ to $<a_i, x>$. Technically, this is a completely new set—and, in particular, $<a_i, v_i>$ and $<a_i, x>$ are completely different pairs. Still, the intuition that the pairs are "boxes" leads to the same net effect.

Note that the assignment operation, by replacing one of the ordered pairs has actually produced a new function. In mathematics, a function that produces a function as a result is called a *functional.* Thus assignment, as described, is a functional because

assignment: (*name* → *value*) × *name* × *value* → (*name* → *value*).

That is, the assignment functional takes a function (*name* → *value*), a *name*, and a *value* as arguments, and it produces a new function (again, *name* → *value*), as a result.

7.2.2 AN AXIOMATIC DESCRIPTION OF MEMORY

In Chapter 5, we discussed proof rules for various language constructs. We also noted that these proof rules could be interpreted as definitions of the semantics (or meaning) of the constructs. In particular, the assignment axiom,

$$wp(x := E, \ Q(x)) \ \equiv \ E \text{ is defined } \wedge \ Q(E), \tag{7.1}$$

can be viewed as a mathematical definition of assignment semantics—and, more broadly, as a mathematical definition of memory. The Eq. (7.1) assignment axiom, however, inadequately describes memory. Why it's inadequate, and what an adequate formulation would be is the topic of the rest of this section.

Let's recap briefly what was said in Chapter 5. The reasoning behind Eq. (7.1) is as follows:

> If the predicate Q is to be true after executing the assignment statement, then it must be true for the particular value of x at that point. Only the value of x is changed by the assignment; all other variables still have the same values they had prior to the assignment. The only effect that the assignment can have on the truth value of Q is through its effect on the variable x. Thus, if we substitute (into Q) the value that x has *after* the assignment, this new predicate must also be true *before* the assignment.

Note that this recap captures one of the essential properties of computer memory: variables retain their values unless they are explicitly altered.

As we noted, the Eq. (7.1) version of the assignment axiom is not an adequate description of memory. To see why, let's compare it with the functional description of memory in Section 7.2.1. Aside from the notation form, one of the most striking differences is that the functional description characterizes an entire memory, while the axiomatic description characterizes only the effect of assignment on an individual variable within a memory. And that's where the Eq. (7.1) axiomatic description fails: It's too narrow. Suppose we execute the statement

$$a[i] := 3,$$

and the assertion $a[i] = a[j]$ is supposed to be the postcondition. If we substitute, $a[i]$ for x in Eq. (7.1), so that

$$Q(a[i]) \equiv a[i] = a[j],$$

and blindly apply the axiom, we obtain

$$wp(a[i] := 3, a[i] = a[j]) \equiv Q(3) \equiv 3 = a[j].$$

But what if $i = j$ prior to executing the assignment? If that's so, then assigning to $a[i]$ also assigns to $a[j]$. Furthermore, if $i = j$, then $3 = a[j]$ is not the proper precondition. The value of $a[j]$ does not have to be $3-$ in fact, if $i = j$, then prior to the assignment $a[j]$ can have *any* value. The problem, of course, is that the assignment axiom, Eq. (7.1) fails to account for the fact that there may be two or more ways to denote the same variable. These several identifications for the same variable are called *aliases*. To improve the assignment axiom we must account for all possible aliases for the variable that is being changed. This is fairly easy for arrays of variables: All we need do is introduce a way of identifying the new state of the array after assignment.

Definition 7.1. If A is defined to be an

array $[Q]$ of P,

i is a value of type Q, and e is a value of type P; then

$$\alpha(A, i, e)$$

is the value of A after the assignment of the value e to $A[i]$. □

Note that the α-function produces an entirely new array—an array that has all of the values except one of old array A. Another way of characterizing the α-function is in terms of the following theorem, which can be derived from the Definition 7.1.

Theorem 7.1. $\alpha(A, i, e)[j] = $ if $i = j$ then e else $A[j]$

That is, if we select the jth element of the array produced by $\alpha(A, i, e)$ we obtain either e or the previous value of $A[j]$, depending on whether $i = j$.

Now we can extend the assignment axiom to cover assignments to array components (the old version still works for simple variables):

$$wp(A[i] := E, Q(A)) \equiv E \text{ is defined} \wedge Q(\alpha(A, i, E)). \tag{7.2}$$

This equation means that to obtain the weakest precondition, $wp(A[i] := E, Q(A))$, we simply substitute $\alpha(A, i, E)$ for all occurrences of A in Q. Note, again, that Eq. (7.2) says that we, in essence, are assigning a new value to the *entire array*; it just so happens that the new value is just like the old one except possibly in the ith position.

For example, to compute

$$wp(a[i] := 3, \quad a[i] = a[j]),$$

we note that

$$Q(a) \equiv a[i] = a[j].$$

Hence,

$$wp(a[i] := 3, \quad a[i] = a[j]) \equiv b[i] = b[j], \text{ where } b = \alpha(a, i, 3).$$

Now, from theorem 7.1, we have

$$
\begin{aligned}
b[i] = b[j] &\equiv \alpha(a, i, 3)[i] = \alpha(a, i, 3)[j] \\
&\equiv (\text{if } i = i \text{ then } 3 \text{ else } a[i]) = \\
&\quad (\text{if } i = j \text{ then } 3 \text{ else } a[j]) \\
&\equiv 3 = (\text{if } i = j \text{ then } 3 \text{ else } a[j]) \\
&\equiv i = j \supset 3 = 3 \ \wedge \ i \neq j \supset 3 = a[j] \\
&\equiv i = j \ \vee \ 3 = a[j].
\end{aligned}
$$

In other words, in order for $a[i]$ to equal $a[j]$ after the assignment, either i and j must have the same value or $a[j]$ must originally have been 3.

7.2.3 RELATING THE FUNCTIONAL AND AXIOMATIC DESCRIPTIONS

Let's think of the collection of all the variables in a program to be in one huge array, M. Further, imagine that the strings corresponding to the identifiers in the program are used as indices into this array. That is, if the identifier x is declared in a program, we will access the corresponding variable as $M['x']$. A program fragment such as

> **var** x, y, z: *integer*;
> **begin** $\cdots$ $x := y + 1$; $\cdots$ **end**

will become

> **var** M: **array** [*string*] **of** *integer*;
> **begin** $\cdots$ $M['x'] := M['y'] + 1$; $\cdots$ **end**

Then we can apply the array form of the assignment axiom and, for example, obtain

$$
\begin{aligned}
wp(M['x'] := M['y'] + 1, \ Q(M)) \\
\equiv Q(\alpha(M, 'x', M['y'] + 1)).
\end{aligned}
\tag{7.3}
$$

Compare this equation with the functional description of memory in Section 7.2.1. In the functional description, memory was characterized as a set of <*name, value*> pairs, and assignment was defined as an operation that produces a new <*name, value*> pair. In Eq. (7.3), the array M is the set of <*name, value*> pairs; it is the set $\{<'x', v_x>, <'y', v_y>, \cdots\}$. Further, the α-function is also the assignment functional. That is, the α-function produces a new array—a new set of <*name, value*> pairs—after an assignment. Thus, although we haven't proved anything formally (and although the array M is a hypothetical construction, being illegal in most programming languages,) you should be able to see that the two descriptions of memory are equivalent, each producing a new <*name, value*> set from an old one.

7.3 Type

So far in this textbook we've used the word *type* frequently. We have said, for example, that a data value has a particular type and that a variable is allowed to contain (or "take on") values of that type. Look at the following program fragment.

```
var x, y:  integer;
begin
x := 3;
y := x + 2;
end
```

We say that *integer* is a program language type and that the variables x and y are of type *integer*, or that they can contain values of type *integer*. The numeral "3" denotes a type *integer* value that can be stored in (or assigned to) an integer variable. The expression "$x + 2$" evaluates to an *integer* value. Because type *integer* variables x and y contain mathematical integers, certain mathematical operations involving them, such as addition, are permitted.

These properties, namely having values and operations on these values, are not peculiar to the type *integer*. The other common types—including *real, boolean, bits, complex*, **array of** *integer*—also have values and operations. In general,

> **Definition 7.2.** A *data type* (or *type*) consists of a set of values (its *domain*), and a set of operations. □

Note that to have a type you need both values and operations. The set $\{0, +1, -1, +2, -2, \cdots\}$ would *not* be the type *integer* if the appropriate operations $(+, -)$ were not included.

It is possible to invent lots of useful data types that are not normally provided by most programming languages. A *point-in-2-space* type, for example, might be defined as

- *Domain*: The set of ordered pairs representing the Cartesian coordinates of the points.
- *Operations*: Performing vector addition and subtraction, finding distance between two points, etc.

Another type might be *calendar-date*, defined as

- *Domain*: Ordered triples of integers. The first integer in each triple represents a month, the second represents a day, and the third represents a year.
- *Operations*: Determining the day of a particular date, printing the date in any of several formats (for example, 1/23/79 *or* 23 Jan 79 *or* Tuesday, January 23, 1979), computing the phase of the moon for a given date, etc.

Now let's consider a few simple syntactic issues. Generally we invoke the operations of a type with the familiar functional notation—for example, *sin*(x), *cos*(a), *ln*(z), etc.   In both mathematical and programming languages, however, we write certain common operators in the customary infix form (with the operator between its two operands)—for example, $a + b$. This is merely a matter of convention and convenience; formally we may think of the infix form as simply a shorthand for the functional form, that is, in this example, a call on the addition operation identical to $+(a, b)$

When constant values are written directly in program text we call them *literals*. Thus, for example, 1 and 23 are literals that denote specific constant integer values; *true* is a literal of type *boolean*; and, in a suitable language, $<2, 3>$ and 1-23-79 would be literals of types *point-in-2-space* and *calender date* respectively.

Definition 7.3. A *literal* is a constant whose value is given by its external, or written, form. □

When we discuss types it is convenient to classify them as either *scalar* or *structured.*

Definition 7.4. A *scalar* (or unstructured) *type* is a type whose domain consists of indivisible entities, or scalars. A scalar has no component parts that can be accessed or manipulated independently. □

Familiar scalar types in programming languages include *integer*, *boolean*, and *real*.

> **Definition 7.5.** A *structured type* has a domain consisting of structured values constructed from simpler component values, sometimes called *components* or *elements*. □

The most familiar structured types in programming languages are record types and array types.

7.4 Describing Types

In the following chapters we define a number of important data types: some of these types will be found in most modern programming languages, others won't. Before we can define the types, however, we need to discuss how one goes about constructing a definition, whether it is to be written in mathematical symbols or in English.

According to Definition 7.2, a type consists of a domain and a set of operations. This immediately suggests that to define a particular type, we must identify its domain and then describe the effects of each of its operations.

Identifying and describing the domain is generally straightforward. If a domain is finite (and small), it can be enumerated; in example, the domain of the type *boolean* is just {*true*, *false*}. In other cases it may be defined as a subset of a well-known domain—for example,

The set of *positive* integers,

The set of complex numbers *with modulus less than one*, or

The set of XYZ Corporation's *married* employees.

Sometimes, however, a domain is defined "constructively"—that is, by enumerating a few basic members of the domain and then providing rules that generate (or construct) the remaining members from those enumerated. For example, the domain of strings of letters might be defined as follows:

1. All single letters are strings.
2. Any string followed by a single letter is a string.
3. There are no other strings.

Describing the operations of a type requires giving both a syntactic specification and a semantic specification. A syntactic specification merely indicates the names of the operations and the types of operands that each operation requires.   A semantic specification gives the meanings of syntactically correct expressions—the values or effects they produce.

7.4.1 SYNTACTIC SPECIFICATION

The syntactic specification of a type is quite straightforward. Perhaps the most concise presentation is a list of procedure headers—one for each operation.   For example, suppose that we wish to give a rigorous specification of the type *integer*; we start with its syntactic specification:

Operations:
   function $'+'(a,\ b$: *integer*): *integer*;
   function $'-'(a,\ b$: *integer*): *integer*;
      likewise for $'*'$, **'div'**, **'mod'**, $'\uparrow'$
   function $'='(a,\ b$: *integer*): *boolean*;
      likewise for $'\neq'$, $'<'$, $'>'$, $'<='$, $'>='$
   function abs(a: *integer*): *integer*;

Literals:
   maxint, minint, and
   all lexemes described by the regular expression
      $(0+1+2+\cdots+7+8+9)\ (0+1+2+\cdots+7+8+9)^{*}$

Note that we are defining such operations $+$, $*$, etc., as if they were ordinary functions or procedures, although they are usually written as infix operators in program text. Also note that we have defined two extra literal values—*maxint* and *minint.* These correspond to the largest and smallest integers representable on a particular computer.

These syntactic specifications, together with the given syntax of our programming language, tell us that if x, z, q, and r are declared type *integer*, then

$$(x + z) * (q - r + 5)$$

is a legal expression. These specifications do not, however, tell us what the expression "means" or "computes"; we need a semantic specification for that.

Another common notation for the syntactic specification of operations—taken from mathematical notation—is simply a listing of the types of the values in each operation's domain and range, as follows:

   $+$:   *integer* $\times$ *integer* $\rightarrow$ *integer*

   $-$:   *integer* $\times$ *integer* $\rightarrow$ *integer*

>:   *integer* × *integer* → *boolean*

maxint: → *integer*

minint: → *integer*

The use of the cartesian product symbol (×) in the specification indicates
that + takes a pair of values as input (the cartesian product of two sets
yields a set of pairs). Note that we show the literals *maxint* and *minint* as
parameterless functions. Technically, all literals can be described this way,
as functions that always return the same value and take no arguments.
Thus, for example, "*true*" and "3" are such functions.

7.4.2 SEMANTIC SPECIFICATION

Once we know syntactically valid expressions, our next concern is their
meaning. There are several schools of thought on how best to proceed.
Rather than advocate any one approach, we take a pragmatic point of view
and suggest that the approach chosen is largely a matter of taste and
circumstances. We discuss three approaches: The use of (1) natural
languages, such as English; (2) algebraic specifications; and (3) abstract
models. In later chapters we will use natural language where rigorous
specifications are not needed; we tend to use abstract models when we
wish to be precise.

7.4.2.1 Natural Language Specification

There is much to be said for the use of natural languages such as English
as an informal specification tool. The evidence is clear: English has
served to specify the semantics of every major programming language
(hence, the semantics of their data types), the properties of every
operating system, and the machine language of every existing computer.
It can often serve to give an informal specification on which to base a
more formal one.

The trouble with a natural language specification occurs when
informality has been taken as license for vagueness. For example, saying
"**div** divides one integer by another" is less than helpful; you probably
assumed as much on the basis of past experience. A good specification
must identify all the possible actions that can occur. It would have been
more helpful to say,

> The value of "x **div** y" is x divided by y, with any remainder discarded. The
> value is not defined if y is 0.

Some people say that, when precision is required, we must use
mathematical notation and programming languages because "natural
language is ambiguous." This is true to an extent, but it would be more

accurate to say, "Natural language is only as precise as we make it." On
the other hand, precise semantic specifications in a natural language tend
to be bulky; it often takes a considerable number of words to achieve a
desired precision. This bulkiness has its own disadvantages since the longer
an explanation, the harder it is to understand.

7.4.2.2 Algebraic Specification

Mathematicians typically describe objects such as integers by giving a set of
axioms about them—a set of true statements from which all interesting
properties may be deduced.   Axiomatic definitions can, in principle, be
used to define any data type, but they are most useful for those basic types
that are subsequently used to define more elaborate types. We will use one
of these basic types, *sequence*, to illustrate axiomatic definition.

The type *sequence* is a structured type, built of simpler components.
That means that we can't really define a single type, *sequence*, but rather
must define a whole collection of types, one for each possible component
type.   Therefore, in what follows, the type name *sequence* is short for
sequence of D, where *D* is any particular component type.

First, we give a syntactic specification for the operations on type
sequence. There is one constant (parameterless function) of type *sequence*,
the *null sequence*:

$<>: \rightarrow sequence$

There is a unary operator for making simple one-element sequences out of
values of type *D*:

$<*>: D \rightarrow sequence$

Here, the asterisk shows where the operand would go—the expression
$<5>$, for example, indicates the sequence containing the single element 5.
This peculiar form of operator—which surrounds its operands rather than
preceding or separating them—is called a *matchfix* operator.

The syntax of the other operators on *sequence* is as follows:

~: $sequence \times sequence \rightarrow sequence$
last: $sequence \rightarrow D$
leader: $sequence \rightarrow sequence$
first: $sequence \rightarrow D$
trailer: $sequence \rightarrow sequence$
length: $sequence \rightarrow integer$

To specify the semantics of type *sequence* operations, we provide the
following set of axioms, in which *d* stands for any value of type *D* and *x*, *y*,
and *z* are values of type *sequence*.

a) $last(x \frown <d>) = d$

b) $leader(x \frown <d>) = x$

c) $x \frown (y \frown z) = (x \frown y) \frown z$

d) $first(<d> \frown x) = d$

e) $trailer(<d> \frown x) = x$

f) $length(<>) = 0$

g) $length(x \frown <d>) = 1 + length(x)$

h) $<> \frown x = x \frown <> = x$

i) If $x \neq <>$, $y \neq <>$, $first(x) = first(y)$, and $trailer(x) = trailer(y)$, then $x = y$

Note that these axioms do not indicate what happens when *first*, *last*, *leader*, and *trailer* are applied to the empty sequence. Such applications are intended to be erroneous, and it is advisable to make that intention explicit. For example, we can add the following to our axioms:

$$first(<>) = last(<>) = leader(<>) = trailer(<>) = ERROR.$$

We have said nothing about the domain for type *sequence*. In fact, it is characteristic of axiomatic definitions that we do not have to do so; the information we need to compute the results of any particular computation involving sequences is implicit in the axioms.

Finally, since writing out long sequences as a string of concatenations can be rather tedious, we define the following shorthand:

Definition 7.6. The notation $<d_1, d_2, \cdots, d_n>$ is an abbreviation for

$$<d_1> \frown <d_2> \frown \cdots \frown <d_n>. \quad \Box$$

7.4.2.3 Abstract Models

A type consists of a domain of values and a set of operations. In some circumstances we can describe the domain of one type in terms of another type—and then use the operations of the second type to describe the operations of the type we are defining.

For example, suppose we want to define the type *integer* as it is implemented in a particular programming language. In most ways, the behavior of the programming language type is just like that of mathematical integers—the only differences arising from the finite size of integers possible in computers. That is, the only difference is the *domain* of type *integer* and mathematical integers. So, to define the programming language type *integer*, we might very well describe its behavior in terms of the behavior of mathematical integers. In doing so, we would be careful to point out the differences between them, but we would benefit from relying on an established mathematical definition.

Comparative definitions of this kind are called *abstract models* because they model the domain and operations of a type using the domain and operations of some previously defined type or types. In the literature, you will also find the term *operational specification* used for this kind of definition. We use these abstract models (or operational specifications) extensively in the following chapters.

Suppose, for example, that we want to define the type *string* as it appears in many programming languages. As you know, strings are sequences of characters, so it would be logical to define the type *string* in terms of the type *sequence* defined in the previous section. Thus, we might choose as the domain of the type *string* the set of sequences of characters with length less than or equal to N, where N is the maximum length of a string:

$$\{a \mid a = \,<a_1, \cdots, a_n>, \; a_i \text{ is a character}, \; n \le N\}.$$

Now we can give the syntax and semantics of the operations on type *string* as a set of annotated functions headers:

> **function** *concat(a, b: string): string*
>   **pre** *length(a)* + *length(b)* $\le N$
>   **post** *RESULT* = $<a_1, \cdots, a_n, b_1, \cdots, b_m>$,
>   where $n = $ *length(a)* and $m = $ *length(b)*

> **function** *substr(a: string, f, t: integer): string*
>   **pre** $f \ge 1 \wedge t \le$ *length(a)* $\wedge f \le t$
>   **post** *RESULT* = $<a_f, \cdots, a_t>$

The syntactic specification here is given by the two function headers, while the semantic specification is given by the pre- and postconditions that follow. The interpretation of **pre** and **post** is as in section 5.3.9: if the precondition is true before the operation is performed, then you are assured that the postcondition will be true afterward. These **pre** and **post** clauses show how values of the type *sequence* model values of type *string* and how operations on objects of type *string* can be modeled as operations (or combinations of operations) on corresponding objects of type *sequence*.

7.5 Further Reading

Most of the current ideas on data structure and type derive from the pioneering work of C. A. R. Hoare. His essay, "Notes on Data

Structuring," found in Dahl, Dijkstra, and Hoare (1972), contains a thorough treatment of the structures we describe in this and the next chapter. That work, however, takes a somewhat narrower view of the concept type than we do and, equates "type" with "domain of values." The idea that a structure and its access functions together form a complete program unit (which we call a type) grew out of experience with the programming language SIMULA 67—Dahl, Dijkstra, and Hoare (1972): In striving for a somewhat different objective, the authors discovered and subsequently introduced features suitable for constructing types.

The two approaches to the formal definition of data types—algebraic specification and the use of abstract models—are the principal techniques currently under investigation. Algebraic specification was first introduced by Guttag (1975); Liskov and Zilles (1975) and Guttag (1977) provide less detailed presentations.

Exercises

7.1 Prove Theorem 7.1.

7.2 State and prove a theorem to the effect that the functional and axiomatic formulations of memory are equivalent.

7.3 Consider the following program:

```
var a, b, c: integer;
begin
a := 1;
b := 2;
c := 3;
 {Point 1}
a := b + 1;
b := a + c;
c := a + b;
end.
```

    a) Establish a correspondence between program identifiers and names. Show the value of the memory function at Point 1 by listing the pairs of the function.

    b) Show the value of the memory function after each assignment in the remainder of the program.

7.4 Consider the following program:

```
var V: array[1..5] of integer;
```

```
begin
  V[1] := 13;
  V[2] := 15;
  V[1] := 17;
  V[4] := 19;
  {Point 1}
  V[7] := 25;
end.
```

a) Let V_0 be the value of V immediately after the declaration. Use the α-function in writing the value of V after each of the assignments.

b) Using the rule given in Theorem 7.1 and the value of V at Point 1, show the computations that yield the values of
   - i) $V[1]$
   - ii) $V[2]$
   - iii) $V[4]$

c) Each time there is a new assignment, the expression describing the resulting array is larger. Can you think of any improvements?

7.5 Consider the following program:

```
var V: array [1:5] of integer;
begin
    i, j, k: integer;
    i := 1; j := 2; k := 3;
    V[i] := j; V[j] := k; V[k] := i;
    {Point 1}
    j := j + 1;
    V[i] := j; V[j] := k;
    {Point 2}
    k := k + 1;
    V[j] := k; V[k] := i;
    {Point 3}
end.
```

a) What are the values of i, j, k, $V[i]$, $V[j]$, and $V[k]$ at Point 1, Point 2, and Point 3?

b) Write a functional description (see Section 7.2.1) of V at Point 1, Point 2, and Point 3.

c) Use the α-function (see Section 7.2.2) to describe V at Point 1, Point 2, and Point 3. Assume that the initial value of V is $V0$.

7.6 Write an axiomatic description of the relationship between the *fetch* and *store* operations of memory. Use Eq. (7.1) as a guide. If $M0$ is the initial state of memory, and the program fragment

$$x := 3; \, y := x + 2; \, x := x * y$$

is executed, what is the final memory state, M, in terms of *fetch* and *store* operations on $M0$? Use your axiom to prove that $fetch(M, y) = 5$ and that $fetch(M, x) = 15$.

7.7 Give domain, sample operations, and typical literals for types

    a) *real*

    b) *complex*

    c) **array of** *integer*

    d) *telephone number*

    e) *set of integer*

7.8 Indicate which of the following might be scalar types and which might be structured types. For each type, provide an appropriate set of operations and, for structured types, provide the elements.

    a) *complex number*

    b) *point-in-2-space*

    c) *calendar date*

    d) *telephone number*

    e) *set*

    f) *rational number*

7.9 Any structured type that allows direct assignments to its elements may be recast as a type that does not. For example, to get the effect of the assignment $A[i] := y$, we can define a procedure called *SetEle* (for example) that takes as arguments an array variable, A, an index value, i, and a value, y, and that also has the same effect on A as the assignment. Now the operation of assigning to an element of the array looks just like an ordinary operation of a scalar type. Define the structured type **array** [lower..upper] **of** T so that it becomes an ordinary scalar type. What operations other than assignment to an array element must be recast? Write a complete specification, using *sequences* of T's as the abstract model.

7.10 Many scalar types can be recast as structured types. A structured type *integer*, for example, could be described as having two components—a sign and a magnitude. What components, if any, might each of the following, typically scalar, types have were they also recast?

 a) *reals*
 b) *booleans*
 c) *complex*
 d) *references*

7.11 Complete the syntactic specification for the type integer by filling in the following operations.

$$
\begin{array}{ccc}
* & \textbf{div} & \textbf{mod} \\
** & \neq & < \\
> & \leq & \geq
\end{array}
$$

7.12 Rewrite the syntactic specification for the type *integer* using functional types (for example, "$f: D \rightarrow R$") instead of procedure headers.

7.13 The following syntactic specifications should suggest the operations they are meant to denote. Provide precise semantic specifications. In some cases, you might be able to think of several different and reasonable approaches. When you do, choose one.

 a) +: *array* $\times$ *array* $\rightarrow$ *array* (component-by-component)
 b) =: *array* $\times$ *array* $\rightarrow$ *boolean*
 c) =: *array* $\times$ *scalar* $\rightarrow$ *boolean*
 d) *: *scalar* $\times$ *array* $\rightarrow$ *array*
 e) *: *array* $\times$ *array* $\rightarrow$ *array*   (matrix multiplication)
 f) *: *array* $\times$ *array* $\rightarrow$ *array*   (component-by-component)
 g) **: *array* $\times$ *integer* $\rightarrow$ *array*

7.14 Is "$\sim$" on sequences associative? Commutative? Why?

7.15 Let E, F, and G be sequences with the values

$$E = <>$$
$$F = <u, v, w>$$
$$G = <y, z>$$

What are the values of the following expressions?

 a) $<x> \sim G$ b) *first*(F)
 c) $G \sim <x>$ d) *trailer*(F)
 e) $F \sim G$ f) *first*$($*last*$(F))$
 g) $E \sim F$ h) *first*$($*trailer*$(G))$

i) *last(F)*   j) *length(leader(G))*
k) *leader(F)*   l) *trailer(trailer(trailer(G)))*

7.16 Devise an abstract-model specification for *sequence* (see Section 7.4.2.2). Use sets for the abstract model.

7.17 Consider the type *boolean* with literals *true* and *false*, and operations ∧, ∨, ⊃, ≡, and ~. For type *boolean*, construct the following:

   a) A syntactic specification.

   b) An informal English definition.

   c) An algebraic specification.

   d) An abstract model.

7.18 Suppose type *complex* (for complex numbers) has components *real* and *imag*, literals 1 and *i*, and operators +, −, *, :=, and =. For type *complex*, do the following:

   a) Describe its domain precisely.

   b) Provide a syntactic specification.

   c) Write an informal English specification.

   d) Give an algebraic specification.

   e) Give an abstract model; base this model on *sequences of two reals*. Do you think this is an appropriate model? Why?

7.19 Suppose type *rational* has components *num* (for numerator) and *denom* (for denominator), literals formed by two integers and a slash (such as 1/2, 4/12, 79/54, etc.) and operators +, −, *, /, :=, =, and <. For type *rational*, do the following:

   a) Describe its domain and literals precisely.

   b) Write a syntactic specification.

   c) Write an informal English specification.

   d) Give an algebraic specification.

   e) Construct an abstract model; base this model on *sequences of two integers*. Do you think this is an appropriate model? Why?

7.20 Suppose type *point-in-2-space* has components *x* and *y*, literals <1, 0>, <3.5, 4.7>, etc., and operations +, * (multiply point by constant), distance between two points, :=, and < (meaning "closer to origin"). For type *point-in-2-space*

   a) Describe its domain and literals precisely.

   b) Write a syntactic specification.

c) Write an informal English specification.

d) Write an algebraic specification.

e) Construct an abstract model; base this model on any type you wish.

7.21 Suppose type *calendar date* has components month, day, and year, literals such as "23 Jan 79", and operations + (between integers and *calendar-dates*), −, <, :=, =, and *DayofWeek*. For type *calendar date*, provide the following:

a) Precise description of its domain and literals.

b) A syntactic specification.

c) An informal English specification.

d) An algebraic specification.

e) An abstract model.

7.22 PASCAL text files (type *text*) are sequences of characters. The procedure *Read*, when provided a file argument and an argument of type *char*, removes a character from the front of the sequence and places it in the second argument. The routine *Write*, when given a file argument and an argument of type *char*, places that character at the end of the file.

a) Formally specify the PASCAL type *text* and the operations *Read* and *Write* as described above.

b) Add the *eof* predicate to your specification: (*eof*(*F*) is true iff the file, *F*, contains no more characters.)

c) In general, a given file can be either read or written, but not both. One of the routines *RESET* or *REWRITE* must be called to make a file "readable" or "writable." Read the descriptions of these routines in the PASCAL Report. List the difficulties you see in trying to specify them. Write an abstract-model specification that includes the requirement that a file must be reset before it is read, and rewritten before it is written.

7.23 Consider the problem of describing an object of which there can only be one instance in an entire program. For example, if we had a program that involved looking up words in a dictionary, we might have exactly one dictionary used by the program. In that case, it might be more convenient to give specifications that assume there is only one dictionary, rather than to introduce *Dictionary* as a new data type. For instance, we might write the following specifications:

function *DefinitionOf*(*Word*: *String*): *String*;

procedure *Initialize*;
  post For any word, *W*, *DefinitionOf*(*W*) = '* *Undefined** ';

procedure *AddWord*(*Word, Definition*: *String*);
  pre *DefinitionOf*(*Word*) = '* *Undefined** ';
  post *DefinitionOf*(*Word*) = *Definition* ∧
      For all words, *W*, such that *W* ≠ *Word*,
      DefinitionOf(*W*) = *DefinitionOf'*(*W*);

The notation "*DefinitionOf'*(*W*)" means "the value of *DefinitionOf*(*W*) before the execution of *AddWord*." Assume that *String* is a string of characters (for the purposes of this exercise, ignore the difficulties in properly implementing type *Dictionary* in standard PASCAL). Note that we have defined the effects of all procedures in terms of their effects on the values of the function *DefinitionOf*, rather than in terms of their effects on variables. Clearly, there must be some global variables at work holding the set of words currently defined (along with their definitions), but these are hidden and can be accessed only through the routines listed above. (These routines are called Parnas modules, after their inventor, David Parnas. The syntax of his modules differs considerably from our routines, but the concept behind them is the same.)

a) Specify a new procedure, *DeleteWord*, that deletes a definition from the dictionary. Use the same style as for *AddWord* and *Initialize* above.

b) Rewrite (a) so it is a definition of type Dictionary, each of its routines accepting an argument of type *dictionary*. Exclude hidden global variables.

Chapter 8

Data in Programming Languages

The previous chapter discussed the basic concepts *variable*, *memory*, and *type*. In this chapter we re-examine these concepts as they occur in programming languages. No programming language supports all data structures and types. Some types will be missing from any language you encounter, and others will be restricted—sometimes severely. Even when the same concept appears in several languages, both its syntax and its detailed semantics will probably be expressed a bit differently in each one. This and the following sections describe the ways these basic data types appear in many programming languages, and we place special emphasis on

- some less common data types and data structures, and
- typical variations and concessions to the limitations of contemporary computers.

8.1 Scalar Types

Programming languages generally provide an assortment of scalar data types, complete with operations and literals. Frequently, these scalar types correspond to types directly supported by the computer hardware—such as *integer*, *real*, *character*. In addition, programming languages usually provide certain types—*complex*, for example—because they are useful, even though they are not directly supported by the hardware.

The semantics of scalar types tend to vary according to limitations of

or differences among computers. The most obvious variations involve the
finiteness of *real* and *integer* numbers. The amount of accuracy (number
of digits of precision) associated with type *real* varies from one computer to
another, and can sometimes be specified by the programmer, as in PL/1.
The maximum size of an integer also depends on the underlying computer
hardware. Another source of variation is the hardware's built-in sorting
order for characters; this order affects the results of comparisons between
character strings.

Literal values are usually provided for scalar types. The exact syntax
varies from language to language, but we can always examine a literal in a
program and determine its type. The syntax of literals is usually (but
unfortunately not always) the same as the syntax for input data of the
same type. In well-designed languages, the output routines print values in
a format acceptable to input routines. This is a major convenience when
we want to use one program to prepare data for another.

A few languages allow us to declare scalars that are initialized to a
value we provide and may never be altered after that. In such languages,
a declaration such as

constant *PI* = 3.14159;

allows us to write *PI* throughout the program and know that its value will
never be altered accidentally.

8.1.1 ENUMERATED TYPES

When we examined control, we looked at the simplest possible system that
allowed interesting computation, the FSM. Perhaps the simplest imag-
inable data types are those having only a few constants, comparison and
assignment operations, and nothing else. Such types are called *enumerated
types*, because in defining them it is sufficient to simply list, or enumerate,
their literals. It is then assumed that the literals represent distinct, unequal
values and that the equality and assignment operators are valid and have
the standard specifications:

- *Equality* (=). The predicate is *true* iff the left and right arguments
  have the same values.
- *Assignment* (:=). The operator replaces the value of the variable on
  the left-hand side with the value of the expression on the right-hand
  side.

For example, if we wished to have a type day taking on values
Monday, Wednesday, etc., we might give the definition:

Literals: *Sunday, Monday, Tuesday, Wednesday, Thursday, Friday,*
   Saturday

Operations: = , := (as defined)

We can define variables of type *day*:

   var *x, y: day*;

and can assign and test:

   x := *Wednesday*;
   y := *x*;
   if *x* = *Friday* **then** · · ·

However, the operations will never permit us to create any value of type
day other than the ones denoted by the seven literals from the definition.
These literals stand for all the values of this type that can possibly exist.

We can add a slight amount of power by allowing the programmer to
impose a linear ordering on the elements. When the programmer does
this, the result is called an ordered enumerated type. These are just like
enumerated types, except for three additional operations. If T is an
enumerated type, we add the operations ' < ', *succ* (for successor), and
pred (for predecessor) with the definitions:

   function ' < '(*a, b: T*): *boolean*;
      post *RESULT* = *true* iff *a* appears before *b* in the enumeration;

   function *succ*(*a: T*): *T*;
      pre *a* is not the last element of the enumeration
      post *RESULT* = the element that immediately follows *a* in the
         enumeration;

   function *pred*(*a: T*): *T*;
      pre *a* is not the first element of the enumeration
      post *RESULT* = the element that immediately precedes *a* in the
         enumeration;

The specifications of these functions can, of course, be made more formal
but the English specifications should suffice. These are *standard* definitions
that are to be assumed automatically when an enumerated type is also to
be an ordered type; they are not defined anew explicitly for each
enumerated type. Naturally, the predicates ≤, ≥, >, and ≠ may be
synthesized from <, =, and **not**. (See also Exercise 8.1.)

For example, we can define an ordered version of the type *day* as:

Ordered literals: *Sunday, Monday, Tuesday, Wednesday,*

Thursday, Friday, Saturday

Operations: $=$, $:=$, *succ, pred* (as defined.)

Since *day* is an ordered type, the operations $<$, *pred*, and *succ* are implicitly defined. Thus the following statements are true:

succ(Monday) $=$ *Tuesday.*

pred(Thursday) $=$ *Wednesday.*

succ(Saturday) is undefined.

succ(pred(Thursday)) $=$ *Thursday.*

succ(pred(Sunday)) is undefined. (See Exercise 8.2.)

Monday $<$ Tuesday.

Wednesday $<$ Friday.

As one might guess, enumerated types are used to represent data that function as *n*-position switches, where *n* is the number of literals of the type.   For example, suppose we define an enumerated type that corresponds to the suits in a deck of cards; we could then declare a variable, *x*, of this type:

type *suit* $=$ (*club, diamond, heart, spade*);

var *x*: *suit*;

We can use *x* to hold information about the suit of a particular card—or possibly which suit is currently trump. However, we can only use this value to discriminate among a collection of possible *actions*; we cannot use it in some more general computation (it makes no sense, for example, to multiply *clubs* by *diamonds*). (See Exercise 8.3.)

PASCAL is one of the few languages that allows programmers to define new enumerated types.   PASCAL defines the operations of assignment ($:=$), successor (*succ*), predecessor (*pred*), and the ordering relations ($=$, $<>$, $<=$, $>=$, $<$, $>$) automatically for each enumerated type.

8.2 Names and References

Earlier, in discussing memory, we used the term "name." We did not, however, provide a careful definition. In this section we are more precise.

In particular, we want to accomplish the following: First, to lay the groundwork for a discussion of the reference data type; second, to take another look at the meaning of assignment (which will also lead into a discussion of procedure parameter-passing mechanisms).

Before starting, we would like to remind you of a quotation from Lewis Carroll's *Through the Looking Glass.*

> "The name of the song is called '*Haddocks' Eyes.*'"
>
> "Oh, that's the name of the song, is it?" Alice said, trying to feel interested.
>
> "No, you don't understand," the Knight said, looking a little vexed. "That's what the name is *called.* The name really *is* '*The Aged, Aged Man.*'"
>
> "Then I ought to have said, 'That's what the *song* is called'?" Alice corrected herself.
>
> "No, you oughtn't; that's quite another thing! The *song* is called '*Ways and Means*': but that's only what it's *called,* you know!"
>
> "Well, what *is* the song, then?" said Alice, who was by this time completely bewildered.
>
> "I was coming to that," the Knight said "The song really *is* '*A-sitting on a Gate*': and the tune's my own invention."
>
> So saying he stopped his horse and let the reins fall on its neck; then, slowly beating time with one hand, and with a faint smile lighting up his gentle, foolish face, as if he enjoyed the music of his song, he began.

The distinctions made by the White Knight are both real and important. They are the differences between a thing, the name (or names) of that thing, and the name of the name(s).

Let's consider an example—the number 4, which is a mathematical property possessed by, for example, the collection of X's on the following line:

X X X X.

The character string "four" is a name for this concept; indeed, there are many names for the same concept—4, IV, quatre, IIII, and so on. In English, if we want to talk about a thing, we simply use its name. But, if we want to talk about the name itself, we must indicate explicitly that we are doing so. The phrase

I saw four birds on a fence.

refers to the mathematical concept four, but the phrase

I saw the word "four" painted on the fence.

refers to a specific name of four, namely "four." We could go on to get names of names of things:

I saw the phrase "the word 'four'" painted on a fence.

In the context of our discussion of types, the distinction we draw becomes crucial when we realize that *a name may itself be a legitimate value.* Thus, we can treat *name* itself as a type. When dealing with *name* as a type in a programming language, it is customary to use the terms *reference* or *pointer.* Naturally, if *reference* is a type there must be reference variables, variables that contain references, and reference-valued functions, functions that return references as their values. There must also be reference constants, of which variable names are the prime examples.

For each type, T, we assume that there exists a type *reference-to-T*, which we denote $\uparrow T$. A variable of type $\uparrow T$ will contain values that are names of, or references to, variables that, in turn, contain values of type T. So, for example, we may have a variable of type $\uparrow integer$; the value of this variable will be the name of some *integer* variable.

As with all types, references have associated operations. In addition to the equality and assignment operators, the principal one is the *dereferencing operator* that accepts a reference value and produces the value stored at the location designated by the input reference.

References are simultaneously one of the most important, powerful features of any programming language and one of the most subtle and dangerous. References allow the programmer to construct arbitrary data structures. Since good, problem-specific data structures are essential to the design of well-structured programs, references are an important tool in the construction of good programs. Unfortunately, references can also be used to construct unnecessarily complex, "fragile" data structures that are hard to understand. As with all powerful tools, references must be used with care and restraint.

In the subsequent sections we say much more about references. In these we become very explicit, and so will need two pieces of notation. An up-arrow ($\uparrow$), denotes the postfix dereferencing operation, as in PASCAL. Thus,

Definition 8.1. If X is a value of type $\uparrow T$, then $X\uparrow$ is the object of type T that X references. $\square$

We use a beta, β, in a prefix position to denote creation of a reference value. Thus,

Definition 8.2. If Y is a variable of type T, then βY is a reference to Y—that is, a value of type $\uparrow T$. $\square$

Although reference types do not appear explicitly in all languages, the concept of reference is extremely important. It underlies the meaning of the assignment statement and of parameter passing in procedures, as we see in the next two sections.

8.2.1 USE OF REFERENCES IN ASSIGNMENTS

Recall the intuitive meaning of the assignment operator:

> The expression on the right-hand side of := is evaluated, and its value replaces the value of the variable that is named on the left-hand side.

This same definition still holds when we consider assignments involving reference types, but there are some subtleties. There are two cases to consider: first, assignment between ordinary variables that are not themselves of reference types, and second, assignments involving explicit reference types.

First let's consider the simple case—assignment when no explicit reference types are involved. The assignment operator requires a value of some type, say T, and a reference to a storage location of the same type, $\uparrow T$. The operator does not yield a value, but does have an effect on the location named by the first, the $\uparrow T$, parameter.

This works for assignment to scalars, to vector or array elements, and to fields of records. In the latter two cases, structure access acts as an expression that produces a reference value. For example, in the assignment statement

$$A[5] := 34;$$

we evaluate the access function $A[5]$ to obtain the address of the fifth element of vector A, then store the value 34 at that address.

Now let's consider a second, slightly more complicated case—one in which the assignment statement involves explicit references. Such assignment statements can be a bit tricky.

Suppose w and v are variables of type T, r and s are variables of type $\uparrow T$, that r currently contains the name of v, and s currently contains the name of w. Their are several assignments involving w and r that might make sense. For example, suppose that T is *real* and that we start with the situation at the top of Fig. 8.1. We can assign the name of w to r as in the middle of the figure. Alternatively, we can assign the value of w to the variable that r names, or "refers to" as shown at the bottom of the figure.

You can probably imagine reasonable uses for both of these assignments. Certainly we need to be able to distinguish them. In order to do so, we use the operators $\uparrow$ and β, denoting the assignments shown in Fig. 8.1 with $r := \beta w$ and $r\uparrow := w$.

Figure 8.2 illustrates some assignment statements involving more than one reference variable. Again, the top of the figure shows the initial configuration. The statement $r := s$, where both r and s are of type $\uparrow real$,

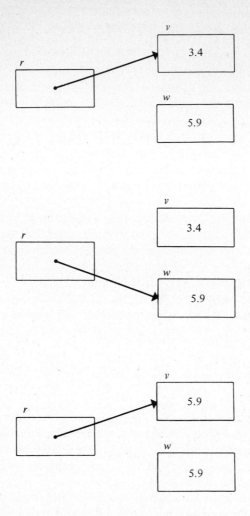

Fig. 8.1 Top: initial situation.   Middle: after $r := \beta w$.   Bottom: initial situation after $r\uparrow := w$

denotes, in most languages, *pointer copy*: "Change r so that r and s contain the same name." Its effect on the initial configuration is shown in the middle of Fig. 8.2.   The assignment $r\uparrow := s\uparrow$, on the other hand, means, "Assign the value of the variable that s points at to the value of the variable that r points at." Its effect on the initial configuration is illustrated at the bottom of Fig. 8.2.

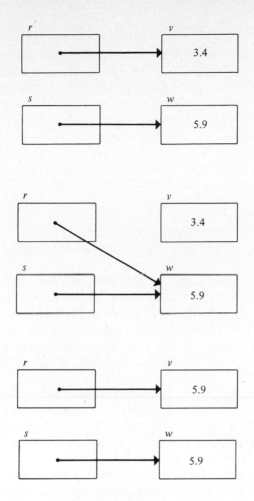

Fig. 8.2 Top: initial situation. Middle: after $r := s$. Bottom: initial situation after $r\uparrow := s\uparrow$

8.2.2 USE OF REFERENCES FOR PARAMETER PASSING

Another common use of references is for passing parameters to procedures. A language needs a technique for connecting the formal parameter names in the procedure declaration to the actual parameters provided by the caller. In section 15.3 we describe several ways to do this. The one that is of interest here is that of passing a reference as an actual parameter. This technique is used even in languages without explicit references. For example, FORTRAN compilers sometimes pass references even though the programmer is not allowed to use them explicitly.

8.2.3 EXPLICIT REFERENCE TYPES

There are as many different treatments of explicit reference types as there are programming languages. The most common languages, FORTRAN and COBOL, do not have reference types. PASCAL provides a reference type for any previously defined type. That is, if T is a type, then there is a type $\uparrow T$, whose variables contain names of things of type T. However, PASCAL provides only a dereferencing operator $(x\uparrow)$; reference creation (βx) is not provided.

Generally, languages that provide explicit references also provide operations of equality, assignment, and dereferencing—the general nature of which was discussed in Section 8.2. Some languages do not need an explicit dereferencing operator, since we can always determine from context (that is, from the way a name is used) whether dereferencing must be performed. The only cases where a question of the need for dereferencing might occur are similar to the one described in Fig. 8.2.

Some languages, such as ALGOL W and PASCAL, require the programmer to declare what types of variables a reference variable can point to. Other languages, such as PL/1, BLISS, and most machine languages, do not require such a declaration; a pointer can point at anything. The absence of this restriction might seem to make a language more powerful. Unfortunately, the lack of knowledge about what a variable may be pointing at makes the program using it difficult to understand. We will return to this problem in Section 8.4.3.

Obviously, reference variables are useless without something to reference. Most languages supporting references also have operations for creating new objects and returning references to them. In PASCAL, if P is a reference variable to a type T, then the statement $new(P)$ creates a new object of type T and places a pointer to it in P. Other languages provide similar operations with a different syntax.

It is useful to be able to describe a reference that does not point at anything. We use the constant value *nil* for the *null reference*. This value is frequently used to mark the end of a data structure built from references.

8.3 Structured Types

The principal characteristic of objects of a structured type is that they are built of simpler components. Therefore, the principal operations on structured types are those that access the components—called *access functions* or *selectors*—and those that, given the individual components' values, construct a structured object—called *constructors*. In addition, some languages provide assignment and equality comparison on structured types.

The most familiar structured objects are arrays and records, which we discuss in the following two sections.

8.3.1 ARRAYS

Definition 8.3. An *array* is a collection of objects, all of the same type—called the *element type*—and indexed by a set of values—called the *index set.* The members of the index set are all of the same type, called the *index type.* A type whose domain consists of arrays and whose operations include a *subscripting operation*—that is, an operation that selects the object of an array having a given index—is called an *array type.* □

By far, the most common index set is a finite interval of *integer* numbers. The notation "$-2..2$" in PASCAL, for example, indicates a type whose domain is the set

$$\{-2, -1, 0, 1, 2\}.$$

The numbers -2 and 2 in this example, are the limits of the interval—its lower and upper bounds. The declaration

V: **array** $[-2..2]$ **of** *integer*;

declares V to be of an array type—namely, **array** $[-2..2]$ **of** *integer*—and says that the values of V are collections of five *integer* values that have indices -2, -1, etc., that are accessible with the subscripting operator as

$V[-2]$, $V[-1]$, $V[0]$, $V[1]$, $V[2]$.

Definition 8.3 indicates that a subscripting operation is characteristic of an array type. We did not include the converse operations, *array constructors*, in the definition because for some reason, most commonly used languages, including FORTRAN, PASCAL, PL/1, and COBOL, do not provide full-fledged array constructors, although FORTRAN and PL/1 do provide a limited form of constructor for initializing arrays. For arrays with ordered index sets, such as intervals of the integers, we will use the ALGOL 68 notation for array constructors. This notation is essentially the same as that for finite sequences. For example, after executing the statement

$V := (43, 29, 62, 7, 0),$

the array V will have a value such that

$V[-2] = 43$, $V[-1] = 29$, $V[0] = 62$, $V[1] = 7$, $V[2] = 0$.

So far, we have restricted ourselves to arrays whose elements are scalars. This is not an essential restriction; indeed, it is quite common to declare arrays that have arrays or other non-scalars as elements. For example, after the declaration

var A: **array** [1..10] **of array** [1..5] **of** *real*;

the elements $A[1]-A[10]$ are all arrays with *real* elements. Getting at the individual elements of type *real* requires two subscripting operations— $A[1][1]$, $A[1][5]$, $A[10][5]$, etc.

Another way to get the same effect is to have the index set consist of ordered pairs of numbers, such as

$$\{<1, 1>, <1, 2>, \cdots, <10, 5>\}.$$

A typical declaration for such an array looks like this:

var B: **array** [1..10, 1..5] **of** *real*;

The index set of B is the Cartesian product of the intervals 1..10 and 1..5— that is, it is the set of all pairs whose first member is between 1 and 10 and whose second member is between 1 and 5. Subscripting operations on B take pairs of values, as in

$$B[1, 1], B[1, 2], \cdots, B[10, 5].$$

Because the elements of the index set of B are pairs, we say that B is two-dimensional. We can similarly construct three-dimensional arrays— having triples in the index set—or, indeed, arrays with any number of dimensions. The one-dimensional case is sufficiently common to warrant a special name:

Definition 8.4. A one-dimensional array—that is, an array whose index set is an interval of the integers or a contiguous sequence of values from an enumerated type—is called a *vector*.

Finite intervals of integers or tuples of integers are by no means the only possible index sets. The PASCAL language, for example, also allows the use of enumerated types as index types. For example, if we define the enumerated type *GreatLakes* to have the values *Erie*, *Ontario*, *Huron*, *Michigan*, and *Superior* (ordered here by increasing volume of water,) then we can define

area: **array** [*GreatLakes*] **of** *integer*,
elevation: **array** [*GreatLakes*] **of** *real*;

and give them values

> *area* := (9910, 7340, 23000, 22300, 31700);
> *elevation* := (570.38, 244.77, 578.68, 578.68, 600.38);

Now we can access individual elements to find, for example, that

> *area*[*Erie*] = 9910
> *area*[*Superior*] = 31700
> *elevation*[*Huron*] = *elevation*[*Michigan*] = 578.68

The operations of assignment and equality on whole vectors (as opposed to individual elements of other vectors) have obvious meanings, but are not always allowed in programming languages. Some languages, such as PL/1 and APL, allow us to write the equivalent of, for example,

> *A* := *B*,

where *A* and *B* are vectors, and the statement mean "Copy all the elements of *B* into *A* in the corresponding positions." They may also allow

> *A* := 0,

which means "Set every element of *A* to zero." This statement is an example of a *coercion* (see Section 8.4.2), in which a scalar value (0) is converted, or coerced, to a vector value.

Some languages provide additional operations on arrays. In PL/1, for example, certain infix scalar operations can also operate on arrays element by element. For example, if *A*, *B*, and *C* are arrays,

- "**if** *A* = 0 **then** · · ·" means "if *all* the elements of *A* are zero, do the statement after **then** [in some versions of PL/1]."
- "*A* := *B* * *C*" Where *A*, *B*, and *C* are arrays with the same dimensions, means "Set each element of *A* to the product of the corresponding elements of *B* and *C*." (Note that this is *not* ordinary matrix multiplication, which should suggest that this method of defining operations is not necessarily good.)

Another language, APL, provides a very rich assortment of operations on arrays, including matrix inversion and solution of linear equations.

Most languages do not have array literals (although APL and ALGOL 68 do). Thus, we usually build particular arrays by explicitly assigning values to individual elements. If, for example, we were working with square matrices and wanted the identity matrix, we could construct it with code something like the following:

> **var** *Identity*: **array** [1..*N*, 1..*N*] **of** *real*;
>   · · ·

```
for I := 1 to N do
  begin
  for J := 1 to N do
    Identity[I, J] := 0.0;
    Identity[I, I] := 1.0;
  end;
```

Another set of operations, collectively known as *slices*, select a subset of vector elements and treat them like new vectors. First, the operation known as *cross-sectioning* selects an entire row or column of a multiple-dimensioned array. If a procedure, F, expects a one-dimensional vector, and A is two-dimensional, then the calls,

$$F(A[*, j]); \quad \text{and} \quad F(A[i, *]);$$

respectively, pass to F the vectors, or cross sections, consisting of the jth column and the ith row of A.

We can also name a "subvector" of a vector. Some languages, such as APL and ALGOL 68, for example, allow statements of the form $F(V[2..5])$ that pass to F the vector consisting of the elements of V with indices 2, 3, 4, and 5. The new vector has index set $\{1, 2, 3, 4\}$.

8.3.2 RECORDS

Definition 8.5. A *record type* is a structured type whose components (called *fields*) are accessed by a set of explicitly named selectors called *field accessors*, *field names*, or *field selectors*. The components of a record type may be of different types. □

A *record-type* declaration describes how records of a given type are to be constructed. We may think of this kind of declaration as a template, or die, used to manufacture individual records. We may define, for example,

```
type person =
  record
    age: integer;
    name: array [1..44] of char;
    height, weight: real;
    sex: (Male, Female);
    married: boolean
  end
```

This declaration defines *person* as a type whose values have six components. The first part of any value of type *person* is an integer, the second part is a string, and so forth. If *Ralph* is a variable of type *person*, then we might refer to its fields as *age(Ralph)*, *name(Ralph)*, etc., suggesting the functional nature of the selectors. A second notation is often used for selecting elements: a *qualified name* is constructed by appending a dot and the field name to the variable name—*Ralph.name*, *Ralph.age*, etc.

Our language makes it difficult to initialize records (or arrays, for that matter) easily. It would certainly be more convenient to be able to write, for example,

$x := PersonConstructor(23, 'Larry \cdots', 1.8, 70, Male, false);$
$y := PersonConstructor(22, 'Evelyn \cdots', 1.6, 55, Female, true);$

instead of twelve assignment statements. Here, we intend "PersonConstructor" to be a *constructor*—taking as many arguments as there are fields in a *person* record type and returning a record initialized with those field values. In our language, we must write such contructors explicitly. Other languages, such as ALGOL 68 and ALGOL W, provide these constructors automatically.

A number of languages permit the programmer to define new record types. Historically, modern facilities for record definition were derived from the facilities provided by such languages as COBOL for defining the structure of records in a file. Our use of the term "record," in fact, comes from business data-processing terminology.

8.4 Other Data-Related Issues that Arise in Programming Languages

The following sections are a *potpourri* of programming-language issues that relate to data.

8.4.1 UNIONS

It is sometimes convenient to allow a variable to contain values of different types at different times. Some programming languages allow the declaration of a variable to specify that legal values are the union of two or more types; we call these union types.

For example, suppose that we have defined the enumerated types

type *MasculineName* = (*Tom, Dick, Harry*);
type *FeminineName* = (*Mary, Diane, Anastasia*);

and now wish to define type *Name* so that the variables of that type can have any of the values in the set {*Tom,* $\cdots$ *, Anastasia*}. We might consider a syntax such as

type *Name* = **union**(*MasculineName, FeminineName*);

For variables declared

    var x, y: *Name*

It would then be valid to write

    $x :=$ *Tom*; $y :=$ *Mary*; $x := y$

But it would not necessarily be valid to write

    if $x = y$ **then** $\cdots$

because we haven't defined what it means to compare values from potentially different types.

In general, if we had defined

    procedure $f(z$: *MasculineName*$)$ $\cdots$

then the statement $f(x)$ might or might not make sense, depending on whether x had last been assigned a value from type *MasculineName*. This suggests the need for an operator to determine whether x is currently acting as type *MasculineName* or *FeminineName*. The ALGOL W language, for example, has an operator, **is**, that allows the programmer to determine whether the value of a certain variable is of a given type. For example, the expression,

    x **is** *MasculineName*

yields *true* whenever x contains a value of type *MasculineName*, and allows a program to take different action—to invoke different procedures, for example—depending on the type of value contained in x. (Actually, this example is a slight generalization of ALGOL W, which only allows unions of reference types.) Students of current events will quickly realize that this is discrimination, and indeed, union types for which there is an operator for determining type are known as *discriminated unions*.

In place of the **is** operator, our language—being a dialect of PASCAL—uses *variant records* to get discriminated unions. To define the type *Name*, we would write

    type   *Gender* = (*Masculine, Feminine*);
               Name =   **record**
                     case *kind*: *Gender* **of**
                       Masculine: (*He*: *MasculineName*);
                       Feminine: (*She*: *FeminineName*)
          end

Name is a record type having fields "*kind*," "*He*," and "*She*." The "*He*" field is valid if *kind* = *Masculine*, and the "*She*" field is valid if *kind* = *Feminine.* Thus, the call on f (in $f(x)$), would be written $f(x.He)$. The **is** operator is replaced by tests on the *tag field*, "*kind*":

    x.kind = *Masculine*

The part of the **record** declaration beginning with "**case**" is called the *variant* part. In PASCAL, it may be preceded by ordinary field definitions, although we haven't done so in the example above.

8.4.2 CONVERSIONS BETWEEN TYPES: COERCIONS

It occasionally happens that two distinct types represent objects so similar that there is a natural correspondence between the values of the two. In such cases, it is often convenient to be able to treat values of one type as if they were values of the other. The most common examples are found in types *integer* and *real.* If i is type *integer* variable, and x a type variable *real,* then $i + x$ makes perfect sense and is allowed in most programming languages. We *might* say that what is happening is that the operator for addition used here is simply a new operator, that happens to work between *integer*s and *real*s, instead of between *integer*s and *integer*s or *real*s and *real*s. It is more usual, however, to think that what's going on is a conversion of the *integer* value of i to a *real* value, followed by ordinary addition of type *real* numbers. Such an implied, automatic conversion is called a *coercion.*

    Coercion is used only in certain situations. For example, FORTRAN allows a type *real* value to be assigned to a type *integer* variable (FORTRAN coerces it to an *integer* before making the assignment), but it does not allow a type *real* expressions to be used in array subscripts where integers are expected.

    At least one language, PL/1, carries this idea to extremes, providing coercions from nearly any type to any other type. Most languages content themselves with a few convenient coercions (as between type *integer* and *real* values) and, perhaps, also provide explicit conversion functions (called *transfer functions*), such as PASCAL's *trunc, round, ord,* and *chr,* for other useful conversions.

8.4.3 VARIABLE DECLARATIONS

You are probably familiar with the variable declaration simply as a programming language construct that "defines variables." Programming language designers, however, distinguish several functions of a variable declaration:

    1. It introduces vocabulary—that is, it provides identifiers for objects and determines the scope of that vocabulary.

2. It controls the allocation of objects and the *extents* (lifetimes) of those objects.

3. It associates *type* information with each name it introduces.

We'll consider these functions in detail in Chapter 16. The following three sections give a brief introduction.

8.4.3.1 Scope

The *scope* of an identifier declared in any given declaration is the section of program text in which that declaration applies—that is, in which the identifier has the meaning established by the declaration. Thus, the scope of *x* in

```
begin
var x: integer;
    .
    .
    .
end
```

is the program text between the **begin** and the **end**. (Note that in this example, and in the rest of the section, we do not use the standard PASCAL syntax—in which the **var** declarations precede the **begin** and the remaining program text—so that we can better explain the nesting of scopes.) Although for convenience we often refer to the "scope of a variable name," it is technically more accurate to refer to the scope of a *declaration*. This is because we can have situations such as

```
begin
var x: integer;
    . . .
    begin
    var x: real;
        . . .
    end
    . . .
end
```

in which the identifier *x* is valid everywhere, but has a different meaning in the inner block.

Note, by the way, that although *x* has a different meaning in the inner block, the object created by the declaration in the outer block still exists while the inner block executes. That is, the *scope* of a declaration does not necessarily determine the *extent* of the objects it creates. The scope of a declaration in languages, such as PASCAL, ALGOL 60, or FORTRAN, is a *static* property; it does not change with time and does not depend upon the particular events of a given execution of a program.

8.4.3.2 Extent

The *extent* of a variable is its lifetime—the period of program execution during which the variable exists, occupies storage, and is accessible. As we noted, a variable may exist during the execution of program text outside the scope of its declaration; unlike the *scope* of a declaration, the *extent* of a variable may depend on program execution. In particular, we distinguish several categories of extent:

1. *Static*: Created (allocated storage) at the beginning of program execution that is not deallocated until program termination.
2. *Local*: Created upon each entry to a block containing a declaration of the variable, and destroyed upon exit from that block.
3. *Dynamic*: Created and possibly destroyed by explicit statements in the programming language.

As an example of each category of variable extent, consider:

1. *Static*: All variables in most FORTRAN implementations are static. If a subroutine G is declared

   SUBROUTINE G

   INTEGER I, J

   Space for I and J is usually allocated by the compiler before the beginning of program execution. Consequently the same locations are used for each invocation of G, and the values of I and J are therefore preserved between calls. (ALGOL 60 allows specific declaration of such variables.)
2. *Local*: In PASCAL, ALGOL, and many similar languages, variables declared within procedures are created on entry and destroyed upon return from the procedure.
3. *Dynamic*: The statement *new*(*P*) in PASCAL creates a new object that will stay around as long as anyone is using that object, regardless of what procedures are entered or exited.

The rules governing the scope and extent of objects can have a profound effect on the complexity of the machine code that implements them. Introducing an ability to handle variables with dynamic extent, in particular, considerably complicates the "run time support system"—the collection of subprograms that every program written in the language needs in order to run.

8.4.3.3 Nontyped Languages

In the standard algorithmic languages, variables hold values from only one type. Even languages that support union types have this property, since we consider

> **union**(*real, integer*)

to be but one, in this case, structured type. We say that languages in which the type of each variable is fixed by a declaration are *typed*.

There are *nontyped* languages as well. In APL, for example, we may write the sequence of statements

> $x \leftarrow 3$
>
> $x \leftarrow 7\ 9\ 10$

The first statement sets x to the *integer* 3; the second sets the same variable, x, to the vector of *integers* $<7, 9, 10>$. In nontyped languages, there *are* types, but they are implied, and variables are not constrained to hold values from any particular type.

Nontyped languages offer much power and flexibility, but do so at considerable expense. First, compilers cannot easily generate efficient machine code for manipulating nontyped variables. Second, we lose a certain amount of safety and clarity when using nontyped variables. For example, in a nontyped PASCAL, you might have

> $y := 6;$
> **if** $x > 0$ **then** $y := 'ABC'$;
> $z := y + 3;$

This program is probably in error; if $x > 0$, we will end up trying to add 3 to $'ABC'$. In a typed language, the compiler would catch this error. In a nontyped language, the error will be discovered during execution—and then only at the expense of constant checking.

Certain typed languages—or more precisely, certain *compilers*—display an "illegitimate" form of nontyping caused by their failure to check the validity of some statements. For example, in most FORTRAN compilers, the program

> INTEGER I
>
> CALL G(I)

will be accepted even if G is a subroutine taking an argument of types REAL or LOGICAL (*boolean*). The effects of such a call vary from compiler to compiler. Unfortunately, the effect is often predictable for any given compiler, and this effect can be exploited by "clever" programmers.

There is no better way to write a purposely confusing program than to take advantage of such type mismatches; it is a practice to avoid.

8.4.4 STORAGE MANAGEMENT AND ALLOCATION

When we write

    var P: $\uparrow T$;

    ·

    ·

    ·

    $new(P)$; $P := nil$;

(where T is a record class), we create an object that can no longer be referenced. Such an object can be thrown away or reused without anyone's noticing. Our example is a trivial one, but we often have use for data objects during only one phase of a program, after which, the space they occupy can be reused. Some language implementations, such as PL/1, allow one to destroy objects. Other implementations, such as those for LISP, automatically destroy objects when they discover that the program can no longer reach the objects, an operation known as *garbage collection*. The combined topic of the allocation and deallocation of storage is called *storage management*, and will be discussed more fully in Chapter 16.

8.5 Further Reading

C.A.R. Hoare's essay in Dahl, Dijkstra, and Hoare (1972) is useful. Unfortunately, no *single* book contains a satisfactorily detailed comparison of the ways various languages treat data types.

Since 1972, researchers in programming language design have paid considerable attention to facilities that allow programmers to define new data types. See, for example, the descriptions of the experimental languages CLU, EUCLID, and Alphard in Liskov, *et al.* (1977), Lampson, *et al.* (1977), and Wulf, *et al.* (1976).

Exercises

8.1 Write the predicates $<$ and $>$ on an enumerated type in terms of *succ*, $=$, and the constant *MaxElt*, defined to be the last value in the enumeration.

8.1 Why is *succ*(*pred*(*Sunday*)) undefined?

8.2 Define ordered enumerated types for

a) Card suits.

b) Card ranks (How do you decide whether the ace is high or low?).

c) Departments in a university.

d) College classes (freshman, sophomore, etc.).

e) Gender.

f) Marital status.

g) Colors of the rainbow.

h) Months of the year.

8.3 The normal definition of ordered enumerated types as described in Section 8.1.1 provides for functions *succ* and *pred*, but not for constants *MaxElt* and *MinElt* (the first and last values in the enumeration). Is this an advantage or a disadvantage, and why?

8.4 Which of the following is the larger number, and why?

$$\mathbf{5}_{8}$$

8.5 Assume that we have the following declarations:

```
type SList = record
   N : Integer;
   Next : ↑SList
   end;
var q,r,s : ↑SList;
   T : SList;
```

Assume that the situation diagramed in Fig. 8.3 exists before *each part* of this problem. Boxes with two slots are *SLists*. The top slot is the *N* field; the bottom is the *Next* field. Reference values are arrows; *nil* appears as a ground symbol (=). Undefined values are blank. For named variables, each name appears above the upper left corner.

Diagram the result of each of the following sequences of assignments (a–p). Assume that the situation of Fig. 8.3 is restored *before each fragment*. Three of the fragments are illegal; tell why for each.

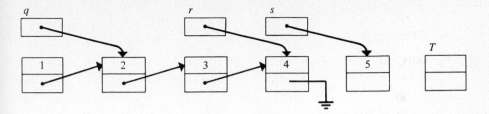

Fig. 8.3 Initial list structure for Exercise 8.6

a) $q := q\uparrow.next$

b) $q\uparrow := q\uparrow.next\uparrow$

c) $q\uparrow.next := q\uparrow.next\uparrow.next$

d) $q := r\uparrow.next$

e) $q\uparrow := r\uparrow.next\uparrow$

f) $s\uparrow.next := q\uparrow.next;$
   $q\uparrow.next := s$

g) $s\uparrow.next := s; \ T := q\uparrow;$
   $q\uparrow := s\uparrow; \ s\uparrow := T$

h) $q := T$

i) $r\uparrow.next := q;$
   $q := q\uparrow.next\uparrow.next\uparrow.next$

j) $s := r$

k) $r := \beta T$

l) $q\uparrow.next := \beta T$

m) $r\uparrow.next := \beta r$

n) $s\uparrow.next := \beta(q\uparrow.next\uparrow)$

o) $r\uparrow.N := G$

p) $q\uparrow.next\uparrow.N := S\uparrow.N$

8.7 Assume the declarations and initial conditions of Exercise 8.6. Assume that q, r, s, and T are the *only* named program variables. Is it possible to write a program fragment that achieves the effect shown in Fig. 8.4 below? If so, write that fragment. If not, give your reasoning.

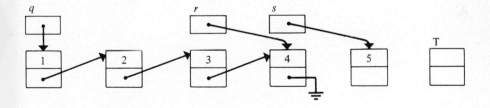

Figure 8.4

8.8 Assume the initial conditions and declarations given in Exercise 8.6. Write sequences of assignments that convert the situation diagramed in Fig. 8.3 to each of the configurations (a–g) in Fig. 8.5.

8.9 Suppose in some program there is an unnamed record that has the property that no pointer variable in the program could ever be made to

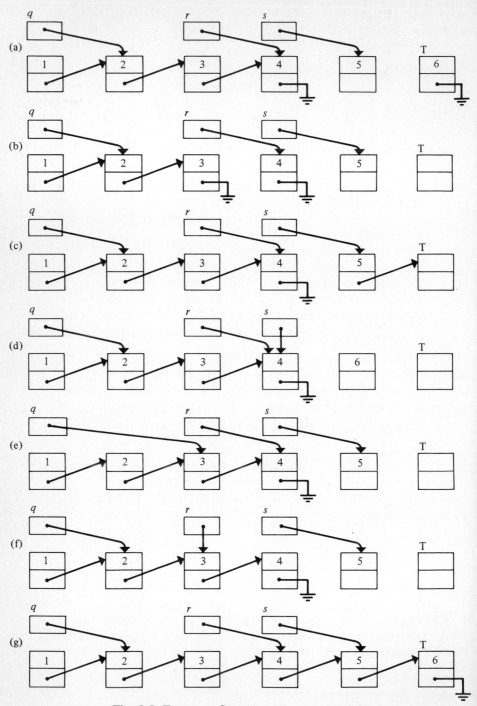

Fig. 8.5 Target configurations for Exercise 8.8

point at it. If the values of the fields of that record were to be changed somehow (say by magic), would there be any effect on the program's output? Why or why not?

8.10 Write statements to change the configuration at the top of Fig. 8.6 to each of the configurations (a), (b), and (c) in that figure, without using any constant real numbers nor the variables *v*, *w*, *x*, or *y* on the left side of an assignment statement.

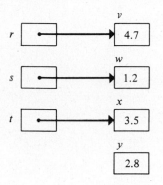

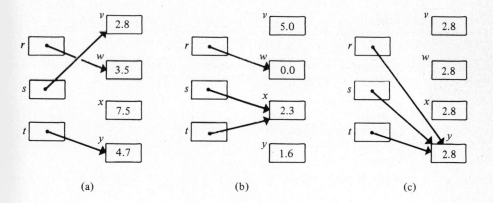

(a)   (b)   (c)

Fig. 8.6 Exercise 8.10

8.11 Suppose your programming language does not supply assignment or equality operators for structured types. Write a procedure *Assign* and a function *Equal* for each of the following types:

    a) 3*DPoint* = **array** [1..3] **of** *real*
    b) *Grid* = **array** [−2..2] **of array** [-2..2] **of** *integer*

c) *Debtor* = **record** *Name*: *string*; *Amount*: *real* **end**

d) *Sample* = **record** *Size*: *integer*, *Mean*: *real* **end**

e) *Inventory* = **array** [Color] **of** *integer*

8.12 Consider the following program:

```
type
      {List element contains data and link}
      ListElt =
          record
              Data: integer;
              Link: ↑ListElt;
          end;

var
          I: integer;
          Head: ↑ListElt;
          Temp: ↑ListElt;

begin
      Head := nil;
      {Point 1}
      for I := 1 to 5 do
          begin
          new(Temp);
          Temp↑.Data := I;
          Temp↑.Link := Head;
          Head := Temp;
              {Point 2}
          end;
      {List the list just created}
      Temp := Head;
      {Point 3}
      while Temp ≠ nil do
          begin
          Writeln(Temp↑.Data);
          Temp := Temp↑.Link;
      {Point 4}
      end;
end.
```

Trace the execution of this program by drawing the data structures as they
exist at Points 1 through 4.

8.13 In Section 8.3.1, we stored the areas and elevations of the Great
Lakes in vectors. Restate these definitions in order to store the same
information in a structure that is

  a) A vector of records.
  b) A record of vectors.
  c) An array indexed by type *GreatLakes* and by another enu-
     merated type.

8.14 Exercise 7.13 asks you to specify a number of procedures or
functions that allow you to operate on entire arrays. Implement those
operations, using only operations that act on the scalar components.

8.15 Declare vector or array types to satisfy the following informal
descriptions. Also include declarations of any index types you find
appropriate:

  a) An element for each month of the year.
  b) An element for each integral temperature between $-10^{\circ}C$ and
     $+100^{\circ}C$.
  c) An element for each course offered by the mathematics
     department.
  d) The cost of beginner, intermediate, and professional Frisbees,
     which come in red, yellow, blue, or chartreuse. (The price may
     vary by color; chartreuse Frisbees are in great demand in some
     places.)
  e) The cards in a card deck, so that a card may be selected by suit
     and rank.

8.16 Think of a way to initialize an identity matrix without making two
assignments to the elements on the diagonal. Is it more or less efficient
than the one given in the text, and why?

8.17 Assume you are using a language that permits the passing of both
cross sections and subvectors as parameters. Write a procedure

   Zerotriangle(*A*: **array** [1..*, 1..*] **of** *real, Upper: boolean*)

that sets the upper (lower) triangle of *A* to zeroes if *Upper* is true or false.
The asterisks indicate that the corresponding array bound may be any
positive integer (note that this is modified PASCAL.) The only operations
that *Zerotriangle* may perform on *A* are slicing and calls on the procedure
declared

   Zero(*A*: **array** [1..*] **of** *real*)

that sets each element of its argument to 0. Assume the existence of appropriate functions for finding the upper and lower bounds of an array.

8.18 Declare records to satisfy the following informal descriptions:

   a) The semester grade record of a student taking five courses.
   b) Information about cars to which parking permits have been assigned.
   c) Information about books in the library.
   d) The personnel record of an employee of a nuclear power station.

For each of these, discuss the nature of the fields you used, why you included them and—possibly—excluded others, and how you might expect each field to be used.

8.19 The registrar's office of a certain university processes student records with a computer. One program needs the following information about students:

For all students: Name, identification number, gender.

For sophomores, juniors, and seniors: Major Department.

For freshmen: High school.

For seniors: Employer or graduate school when relevant.

Define a variant record type that is suitable for storing this information. You may find is helpful to first define an auxiliary enumerated type.

8.20 In a certain set of simple bibliographic references, all entries have an author and a title. In addition, books have publishers and dates; articles are published in magazines or journals, and have page numbers and dates; and unpublished manuscripts have no bibliographic information.

   a) Write a definition for a variant type suitable for storing and processing bibliographic references. You may find it useful to first define an auxiliary enumerated type.
   b) Write a program to print out a vector of bibliographic references.

8.21 Coercion from type t_1 to type t_2 is called *widening* if every value of t_1 has a corresponding value of t_2. If not, the coercion is called *narrowing*. Widening is always safe. Narrowing, if allowed, is dangerous because loss of information occurs from mapping of more than one value of t_1 to the same value of t_2.

Define the type *rational*. A value of type *rational* is any value expressible as the ratio of two integers (that is, p/q for integer p, q). Write routines to add and multiply values of *rational* variables. Write routines to coerce values of type *integer* to values of type *rational* (and vice versa), and to coerce values of type *rational* to values of type *real* (and vice versa). For each new coercion, indicate with a comment whether it is a widening or narrowing operation. Indicate the places where narrowing is dangerous or ambiguous, and justify your handling of these situations.

8.22 Consider the program shown in Fig. 8.7.

```
begin
var i, j, k, l, m: integer;
i := 1; j := 2; k := 3;
Print(i, j, k, l, m, n);
    begin
    var i, j, n;
    i := 4; m := 5; n := 6;
    Print(i, j, k, l, m, n);
    end;
Print(i, j, k, l, m, n);
end.
```

Fig. 8.7 Figure for Exercise 8.22

Assume that a *Print* command prints $ when asked to print the value of a variable that has not been initialized and # when asked to print the value of a variable that has not been declared (assume the compilation error for this was suppressed). What does this program print?

8.23 Consider the following program:

```
begin
var a, b, c, d: integer;
procedure P(var x,y: integer);
    begin
    var z: integer;
    z := 5; x := z + y;
    end;
a := 1; b := 2; c := 3; d := 4;
P(a, b);
P(c, d);
end.
```

How many variables are declared? Which program identifiers can name each variable at some point during execution?

8.24 For each programming language you know, list the extents that are possible—and how to specify syntactically those you want.

8.25 The interaction of references with scope rules can cause some interesting problems. For example, in the program

begin var p: ↑ T; **begin var** x: T; $p := \beta x$ **end** · · ·

we presume that the storage associated with x is released when the inner block is exited, yet p, being declared in an outer block, still points to that storage. We call this p a "dangling reference." Why is this a serious problem? Suggest some simple rules to prevent dangling references from occurring.

8.26 The following declarations define two record types and a vector of each type.

```
type
    PersonalData =
        record
            Name: packed array[1..8] of char;
            SSNo: integer;
            Wage: real
        end;
    PhoneList =
        record
            Name: packed array[1..8] of char;
            Phone: integer;
        end;


var
    PersonalFile: array [1..20] of PersonalData;
    PhoneFile: array [1..20] of PhoneList;
```

Write a program that reads data from a file to fill the arrays and prints the following:

a) *PersonalFile* sorted on *Name*, *SSNo*, and *Wage*.

b) *PhoneFile* sorted on *Name* and *Phone*.

c) The elements of *PersonalFile* and *PhoneFile* that share the same *Name*.

Chapter 9

Nonelementary Data Structures

Any program has a subject matter—a set of mathematical or physical objects about which it is supposed to compute things. The data structures of a program represent these objects, or features of them such as their heights, quantities, names, or relationships. The variety of subject matter requires a corresponding richness in the possible data structures at the programmer's disposal. In this chapter, we introduce several useful structured types which form the basis for most of the data organization schemes used in practice.

We define the abstract properties—the behavior—of these types here, using the abstract modeling technique of Section 7.4.2.3. In keeping with our philosophy of separating an abstraction from its implementation, we wait until Chapter 10 to discuss how these types are represented and how their operations are implemented. Bear in mind, however, that in some sense, this chapter, too, deals with representation, since the data structures we discuss here are used in turn to represent the subject matter of programs.

9.1 Abstract Structured Types: Linear Structures

In many cases the components of a structured type have a natural, linear order—that is, we can talk about one element coming *before* (or *after*)

233

another, we can talk about the *first, second,* and n^{th} element, and so on. The most common example of such a structure, which is built into virtually all programming languages, is the vector (section 8.3.1). In this section we will consider some examples of other linear structures; these structures differ from vectors in either (or both) of two important respects:

1. All of the operations defined on the type are restricted to access only the *first* or *last* element of the structure. Thus, unlike vectors, these structures do not allow the value of an arbitrary element to be accessed or altered.

2. The length of the structure–that is, the number of elements it contains, is not fixed. It is possible to add new elements, or delete existing ones, dynamically as the program executes.

We now consider four examples of such structures: deques (pronounced "decks"), queues, stacks, and linear lists. Queues and stacks are special deques.

9.1.1 DEQUES

Imagine a deck of cards such as in Fig. 9.1, and a card game in which you may either draw from or replace cards into the deck, provided each draw or replacement is made at either the top or bottom of the deck. In such a game the deck of cards is behaving as a deque. The cards are the elements of the structure and their order in the deck is linear. The fact that insertions and deletions may be made at either end (but only at the ends) is the distinguishing characteristic of a deque.

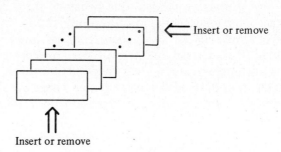

← Insert or remove

Insert or remove

Fig. 9.1 A deck/deque

The major operation on deques (containing objects of type *T*) are

Insertfront: *deque* × *T* → *deque*
Insertback: *deque* × *T* → *deque*

Removefront:	*deque* → *deque* × *T*
Removeback:	*deque* → *deque* × *T*
Empty:	*deque* → *boolean*
Full:	*deque* → *boolean*
Clear:	*deque* → *deque*

To be more precise we can use the abstract modeling technique introduced in section 7.4.2.3. Specifically, consider the modeling domain to be finite sequences, denoted

$$<x_1, x_2, \cdots, x_n>,$$

of objects of type T (that is, x_i is of type T). Sequences and their properties were defined in section 7.4.2.2.

Using sequences, we can define the operations on the type *deque* as in Fig. 9.2. Assume that *maxdeq* is some positive integer constant denoting the maximum size of a *deque*.

Remember, in specifications like these, the postcondition may need to state a relation between the value of an argument (here, D) at the beginning of the procedure and its value after the procedure is executed. To accomplish this, we let the variable name itself (D) refer to the value *after* the procedure; we "prime" the name (D') to refer to the value *before* the procedure started.

For example, we might have the following sequence of calls and resulting values for a *deque* variable F, assuming that the type T is *integer* and *maxdeq* equals 3.

Clear(F);	{Now $F = <>$}
Insertfront(F, 3);	{$F = <3> \sim F' = <3> \sim <> = <3>$}
Insertback(F, 7);	{$F = F' \sim <7> = <3> \sim <7> = <3, 7>$}
Insertfront(F, 6);	{$F = <6> \sim F'$
	$\quad = <6> \sim <3,7> = <6,3,7>$}
Removeback(F);	{returns $last(F) = 7$;
	$\quad F = leader(<6, 3, 7>) = <6,3>$}
Insertfront(F, 19);	{$F = <19> \sim F'$
	$\quad = <19> \sim <6,3> = <19,6,3>$}
Insertback(F, 10);	{? $full(F)$, so precondition is violated}

Note, again, that we have defined only operations which insert and remove elements from the front and back of the structure; we have not

procedure *Clear*(**var** *D*: *deque*);
  post $D = <>$;

procedure *Insertfront* (**var** *D*: *deque*; *x*: *T*);
  pre $\sim full(D)$;
  post $D = <x> \sim D'$;

procedure *Insertback* (**var** *D*: *deque*; *x*: *T*);
  pre $\sim full(D)$;
  post $D = D' \sim <x>$;

function *Removefront* (**var** *D*: *deque*): *T*;
  pre $\sim empty(D)$;
  post $RESULT = first(D') \wedge D = trailer(D')$;

function *Removeback* (**var** *D*: *deque*): *T*;
  pre $\sim empty$ (*D*);
  post $RESULT = last(D') \wedge D = leader(D')$;

function *Empty*(*D*: *deque*): *boolean*;
  post $RESULT \equiv (D = <>)$;

function *Full*(*D*: *deque*): *boolean*;
  post $RESULT \equiv (length(D) \geqq maxdeq)$;

Fig. 9.2 Specifications for deques

provided any way to access or modify an element while it is in the middle of a deque.

9.1.2 QUEUES

A queue is a restricted form of deque: insertions are permitted only at the back and removals only from the front.

Queues are quite common in the "real"—that is, noncomputer—world. A common example is the supermarket checkout line, in which customers get in line at the back and the cashier processes people from the front. We call this a first-in-first-out discipline, or FIFO; elements in the queue are removed in the same order as they were put in. Strictly speaking, there are several kinds of queues. FIFO queues are one kind; stacks, discussed below, are another. FIFO queues are the most familiar however, and so hereafter the word "queue" is used to mean "FIFO queue."

There are two major queue operations. They are "add something at the back end," called *Enq* (for "enqueue"), and "remove something from the front end," called *Deq* (for "dequeue"):

Enq: *queue* × *element* → *queue*

Deq: *queue* → *queue* × *element*

The specifications for *queue* are derived from those of *deque* by

1. Removing the operations *Insertfront* and *Removeback*, and
2. Renaming *Insertback*, *enq*; and *Removefront*, *deq*.

9.1.3 STACKS

Another restriction of a deque permits only insertions and deletions from the same end; such a structure is called either a stack or a LIFO queue (LIFO stands for last-in-first-out).

The *stack* gets its name from its similarity to the plate dispensers sometimes found in cafeterias or to the piles of boxes you see in warehouses. In both cases, new plates (or boxes) get added and taken away at the top. (You don't lift the top 5000 pounds of widget crates to get the widget crate at the bottom of the pile; you take one off the top).

The two most important operations on stacks are "put something on the top," called *Push*, and "take something off the top," called *Pop*. Again, we can specify the abstract properties by naming the functions and specifying their behavior. The abstract model for a stack, which is given in Fig. 9.3 is very similar to that for deques and queues, and, in particular, is just another restriction of a deque.

You should contrast these three specifications, the ones for deques, queues, and stacks, to be sure that you understand both the structures and the specification technique. All three of these structures are used in real programs, but queues and stacks are especially important and you should pay particular attention to their specifications.

9.1.4 LINEAR LISTS

Informally, a linear list is a sequence in which one can insert, delete, or access items at any point (the middle as well as the ends).

9.1.4.1 Simple Linear Lists
The usual definition is recursive.

procedure *Push*(**var** *S*: *stack*; *x*: *T*);
    pre ~*full*(*S*);
    post $S = \,<x> \sim S'$;

function *Pop*(**var** *S*: *stack*): *T*;
    pre ~*empty*(*S*);
    post $S' = \,<RESULT> \sim S$;

function *Top*(*S*: *stack*): *T*;
    pre ~*empty*(*S*);
    post $RESULT = first(S)$;

procedure *Clear*(**var** *S*: *stack*);
    post $S = \,<>$;

function *Empty*(*S*: *stack*): *boolean*;
    post $RESULT \equiv (S = \,<>)$;

function *Full*(*S*: *stack*): *boolean*;
    post $RESULT \equiv (length(S) = maxstack)$;
      {where *maxstack* is some positive integer constant.}

Fig. 9.3 Specifications for stacks

Definition 9.1. A linear list either

- is empty, or
- consists of a first data item (of some type *T*), and the list of all other data items after the first (the *tail* of the list).

For a nonempty linear list *L*, we define

- L_{first} = the first item on *L*, and
- L_{tail} = the tail of *L*.

We will use "ε" to denote the empty list. □

Figure 9.4 shows a typical linear list, displayed as suggested by the definition. Figure 9.5 shows the same list in more standard notation—as a sequence of boxes containing data items (that is, *T*s) with each box pointing at its tail. This latter notation comes from the standard representation of lists, which we will see in section 10.1.4.1. We prefer it, mostly because it is easier to read and write.

(*first*: 1,
  tail: (*first*: 2,
     tail: (*first*: 3,
        tail: ϵ)))

Fig. 9.4 A typical linear list

$$[1] \rightarrow [2] \rightarrow [3] \rightarrow \epsilon$$

Fig. 9.5   The same list in a different format

Operations on lists generally take and return references to their list arguments and results, rather than the lists themselves. This turns out to be convenient for implementation, and sometimes necessary. Figure 9.6 shows one representative set of specifications for the type *listofT*. In particular notice that these specifications use the β-operator defined in section 8.2. The β-operator, you will recall, creates a reference to an object. It is used here to explicitly indicate where references are created; note, for example, that operations such as *Tail* return references. As always, bear in mind that many variations are possible.

For example, suppose that the type, *T*, of the data items is *integer* and that *La* and *Lb* point to the lists

  La: $[3] \rightarrow \epsilon$

  Lb: $[7] \rightarrow [10] \rightarrow [0] \rightarrow \epsilon$.

Then

FirstItem(*La*)	= 3
IsEmpty(*La*)	$\equiv$ *false*
IsEmpty(*Tail*(*La*))	$\equiv$ *true*
Tail(*Lb*)	*points to* $[10] \rightarrow [0] \rightarrow \epsilon$
EmptyList	= ϵ

and after

  SetFirstItem(*Lb*, 8),

Lb points to

  $[8] \rightarrow [10] \rightarrow [0] \rightarrow \epsilon$.

Then after

  SetTail(*La*, *Lb*),

Domain: as in Definition 9.1.

Syntax and semantics:

function *EmptyList:* ↑*listofT*;
   post *RESULT points to* ϵ;

function *IsEmpty*(*L*: ↑*listofT*): *boolean*;
   post *RESULT* ≡ *L points to* ϵ;

function *ConsList*(*x*: *T*; *L*: ↑*listofT*): ↑*listofT*;
   post *RESULT*↑$_{first}$ = *x* ∧ *RESULT*↑$_{tail}$ = *L* ∧
         RESULT is distinct from all other pointers to listofT s;
      {The name *ConsList* means "construct list"}

function *FirstItem*(*L*: ↑*listofT*): *T*;
   pre ∼*IsEmpty*(*L*);
   post *RESULT* = *L*↑$_{first}$;

function *Tail*(*L*: ↑*listofT*): ↑*listofT*;
   pre ∼*IsEmpty*(*L*);
   post *RESULT* = β *L*↑$_{tail}$;
      {Note that β means "name of"}

procedure *SetFirstItem*(*L*: ↑*listofT*; *x*: *T*);
   pre ∼*IsEmpty*(*L*);
   post *L*↑$_{first}$ = *x* ∧ *L*↑$_{tail}$ = *L*↑'$_{tail}$;

procedure *SetTail*(*L*, *Q*: ↑*listofT*);
   pre ∼*IsEmpty*(*L*) ∧ *the variable L*↑ *is not contained in Q*↑;
   post *L*↑$_{first}$ = *L*↑'$_{first}$ ∧ *Q*↑ = *Q*↑' ∧ β*L*↑$_{tail}$ = *Q*;

   {That is, the first item of *L*↑ is unchanged, *Q*↑ is
   unchanged, and *L*↑'s tail is now the list pointed to by *Q*}

Fig. 9.6 One possible set of specifications for linear lists

La points to

   [3] → [8] → [10] → [0] → ϵ

and the list pointed to by *Lb* is unchanged. Now finally, after

   SetTail(*Lb*, *Tail*(*Tail*(*Lb*))),

the list that Lb points to is

$[8] \rightarrow [0] \rightarrow \epsilon$

and the list that La points to has become

$[3] \rightarrow [8] \rightarrow [0] \rightarrow \epsilon.$

NOTE: Because Lb points to a list that is *part of* the list pointed to by La, the operation on Lb has affected the list pointed to by La. This is one example showing the difference between passing around references to lists and passing around the lists themselves. As another example, suppose we have a list pointed to by L,

$L: [3] \rightarrow [5] \rightarrow \epsilon,$

and we want to insert a 4 after the 3. What this means is that we want to *replace* the tail of $L\uparrow$ with the list

$[4] \rightarrow [5] \rightarrow \epsilon.$

The following operation does just that:

$SetTail(L, ConsList(4, Tail(L))),$

for

$Tail(L)$ points to $[5] \rightarrow \epsilon$

$ConsList(4, Tail(L))$ points to $[4] \rightarrow [5] \rightarrow \epsilon$

$SetTail(L, ConsList(4, Tail(L)))$ makes L point to

$[3] \rightarrow [4] \rightarrow [5] \rightarrow \epsilon.$

Now if L had originally pointed to the tail of some bigger list, say

$[2] \rightarrow [3] \rightarrow [5] \rightarrow \epsilon,$

then this bigger list is now

$[2] \rightarrow [3] \rightarrow [4] \rightarrow [5] \rightarrow \epsilon.$

All of which brings us to a crucial point. Suppose that L points to some list

$[0] \rightarrow \cdots.$

Consider the sequence of assignments

$La := ConsList(1, L);$
$Lb := ConsList(2, L).$

From the specifications, this implies that L points to the tail of *two different lists*.

La: [1]
　　　　　　[0] → · · ·.
Lb: [2]

That is, the lists pointed to by La and Lb *overlap*, and any changes to the tail of one may affect the other.

This is an example of *sharing*.   Its consequences can be serious. Changing a list may affect any other list of which it is a part.   We could avoid the problem altogether by removing the operations *SetFirstItem* and *SetTail*, so that no list could be changed once it was created.   To effect the change

$$SetTail(L, \ ConsList(4, \ Tail(L))),$$

above, we would then write

$$L := \ ConsList(FirstItem(L), \ ConsList(4, \ Tail(L))),$$

and this would affect no other list in the program.   Some experts argue that this sort of list manipulation—avoiding the equivalents of *SetFirstItem* and *SetTail*—is to be greatly preferred.   There are, however, reasons of efficiency for using these operations (see Exercise 9.16).   With a little care and discipline, moreover, it is possible to make safe and sparing use of them (see Exercise 9.17).

Finally, let us repeat that the set of operations given in Figure 9.6 is merely representative.   Much depends on the particular uses to which lists are to be put in a given program, as well as constraints imposed by the programming language being used.   As an example (which we will, in fact, have occasion to use,)   Fig. 9.7 shows the specifications for a *Replace* operation, which is meant as a substitute for *SetFirstItem* and *SetTail*.

procedure *Replace*(L, Q: *listofT*);
　　　pre $Q\uparrow$ does not contain $L\uparrow$;
　　　post $L\uparrow = Q\uparrow' \ \wedge \ Q\uparrow = Q\uparrow'$;

Fig. 9.7 *Replace*, an alternative to *SetTail*

9.1.4.2 Linear Lists with Predecessors

The function *Tail* will effectively give the next item on any list. Sometimes, it makes sense to talk about the *preceding* item on a list.   That is, given a list pointer, L, we might want to perform an operation *Pred*, that returns a pointer to the list of which L is the tail. More explicitly, we might try specifying

> **function** *Pred*(L: ↑*listofT*): ↑*listofT*;
>   *post* $L = \beta\, RESULT\!\uparrow_{tail}$;

(That is, L was the name of the result's tail). Unfortunately, the specification given just doesn't work; the predecessor function simply is not well defined. For example, after

> $La := ConsList(3, L); \quad Lb := ConsList(4, L),$

what is $Pred(L) - La$ or Lb? For that matter, what is $Pred(La)$ or $Pred(Lb)$? What is $Pred(EmptyList)$?

The only way to keep things well defined is to change the specifications of lists throughout. We will therefore introduce a new type, *plistofT* (predecessor list of T's). A *plistofT* looks like a *listofT*, with one small addition.

Definition 9.2. A *plistofT* either

- is empty, or
- consists of a data item of type T, a *plistofT* containing the succeeding elements of the list, and a reference to the list of which it is the tail or the empty list if there is no such list. □

We denote these parts of a list Q by

$$Q_{first},\ Q_{tail},\ \text{and}\ Q_{pred}.$$

Fig. 9.8 gives the syntax and semantics of a representative set of operations. Note that the "flavor" of these operations is different—there are, for example, fewer functions, and more work gets done by procedures.

We will denote predecessor lists with double arrows, indicating that one can move along them in two directions. For example, if *La* points to the list

> $\cdots \leftrightarrow [7] \leftrightarrow [10] \rightarrow \epsilon$

and L points to the predecessor of *La*,

> $\epsilon \leftarrow [5] \leftrightarrow [7] \leftrightarrow [10] \rightarrow \epsilon$

then after

> *AddPred*(La, 6),

function *EmptyPlist*: ↑*plistofT*;
   post *RESULT*↑ = ε;

function *Singleton*(*x*: *T*): ↑*plistofT*;
   post *RESULT*↑$_{pred}$ = ε and
      RESULT points to a new (unshared) list containing the one
      item, *x*;

function *IsEmpty*(*L*: ↑*plistofT*): *boolean*;
   post *RESULT* ≡ *L* points to the empty list;

function *FirstItem*(*L*: ↑*plistofT*): *T*
   pre ~*IsEmpty*(*L*)
   post *RESULT* = *L*↑$_{first}$

function *Tail*(*L*: ↑*plistofT*): ↑*plistofT*;
   pre ~*IsEmpty*(*L*);
   post *RESULT* = β*L*↑$_{tail}$;

function *Pred*(*L*: ↑*plistofT*): ↑*plistofT*;
   pre ~*IsEmpty*(*L*);
   post *RESULT* = β*L*↑$_{pred}$;

procedure *AddPred*(*L*: ↑*plistofT*; *x*: *T*);
   pre ~*IsEmpty*(*L*);
   post *x* is inserted immediately in front of *L*;

procedure *AddTail*(*L*: ↑*plistofT*; *x*: *T*);
   pre ~*IsEmpty*(*L*);
   post *x* has been inserted at the head of the tail of *L*↑,
      and *L*↑ is otherwise unchanged;

procedure *RemoveFirst*(*L*: ↑*plistofT*);
   pre ~*IsEmpty*(*L*);
   post the first element of *L'* has been removed.
      The pointer *L* is no longer valid;

Fig. 9.8 Syntax and semantics for predecessor lists

L points to

$$ε \leftarrow [5] \leftrightarrow [6] \leftrightarrow [7] \leftrightarrow [10] \rightarrow ε$$

and *La* points to the third item in *L* (the [7]). Now, after

   AddTail(*La*, 8),

we have *L* pointing to

   $\epsilon \leftarrow [5] \leftrightarrow [6] \leftrightarrow [7] \leftrightarrow [8] \leftrightarrow [10] \rightarrow \epsilon$

and *La* to

   $\cdots \leftrightarrow [7] \leftrightarrow [8] \leftrightarrow [10] \rightarrow \epsilon.$

9.2 Abstract Structured Types: Nonlinear Structures

The preceding material was concerned with *linear* structures—those in which there is a natural linear ordering among the elements of the structure. In sections 9.2.1 and 9.2.2 we will examine two very important nonlinear structures; in both of these structures there is *some* relation between the elements, but it is not a linear one. In section 9.2.3 we discuss a structure in which there is *no* relation between elements.

You are already familiar with at least two nonlinear structures: multidimensional arrays and graphs. In a two dimensional array, for example, the important relation is the adjacency between each element and each of its four nearest neighbors. In a graph each arc represents a relation between a pair of nodes; we will examine this abstraction in more detail below.

9.2.1 LABELED GRAPHS

You are already familiar with the mathematical notion of a directed graph; it is a pair consisting of a set (of *nodes*) and a relation or set of pairs (of nodes) called *arcs*. In addition, we have

> **Definition 9.3.** A *labeled directed graph*, *G*, consists of a set of *nodes*, designated *Nodes*(*G*), a set of *labels*, designated *Labels*(*G*), and, for each label, a corresponding directed graph on *Nodes*(*G*). For each label, *L*, we refer to the set of arcs (ordered pairs of nodes) in the corresponding relation as $Arcs_L(G)$. □

If *Label*(*G*) contains only one label, say *l*, there is only one set $Arcs_l(G)$, and we have, essentially, an ordinary (unlabeled) directed graph.

The graphs you are familiar with are static. For example, we used them to describe FSM's and flow charts—neither of which changes. Now, however, we want to consider graphs that do change dynamically. That is, both the number of nodes in a graph and arcs that connect them may change as a computation progresses.

The advantage of using a general graph structure in a program is simply that it permits us to express arbitrary relations between the data elements stored at the nodes of the graph, and to modify, extend, or shrink these relations during program execution.

Consider, for example, the personnel records for a company, and assume that information about each employee is stored at the node of a graph. Then, for example, the relation "X is Y's supervisor" or "X is married to Y" could easily be stored in the graph by creating arcs labeled "report to" or "is married to" As new employees were hired and others leave, the nodes of the graph would correspondingly increase or decrease. As people changed jobs within the company, got married or divorced, the relations between existing nodes would also change.

The employee record example is a good one for our purposes because it illustrates that several relations may be encoded in the same graph if the arc labels are used to distinguish between them. Figure 9.9 illustrates a portion of the employee-record graph and shows how both the "spouse" and "reports to" relations may be represented in one graph.

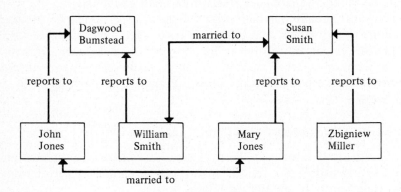

Fig. 9.9 An employee-record graph

One can imagine a rather large set of candidates for the operations on a graph; we will content ourselves with something like a minimal set:

NewNode:	*graph → graph × node*
DelNode:	*graph × node → graph*
SetSucc:	*graph × node × node × label → graph*
DelSucc:	*graph × node × node × label → graph*
Succ:	*graph × node × label → listofnode*

The operation *NewNode* creates a new node in the graph which is initially not related to any of the existing ones. The operation *DelNode* deletes a node and, of course, removes any relations (arcs) involving it. The operation *SetSucc* allows one to create an arc between two nodes and label it. The operation *DelSucc* deletes an arc with a given label between two nodes. Finally, the operation *Succ* allows one to find all nodes related to a given node along arcs with a given label.

> **function** *NewNode* (**var** *G*: *graph*): *node*;
> **post** $Nodes(G) = Nodes(G') - \{RESULT\}$
> and $RESULT$ not in $Nodes(G')$
> and for all q, $Arcs_q(G) = Arcs_q(G')$;

> **procedure** *DelNode*(**var** *G*: *graph*; *n*: *node*);
> **pre** n is in $Nodes(G)$;
> **post** $Nodes(G) = Nodes(G') - \{n\}$
> and for all q,
> $Arcs_q(G) = \{<x, y> \text{ in } Arcs_q(G') \mid x \neq n \text{ and } y \neq n\}$;

> **procedure** *SetSucc*(**var** *G*: *graph*; *n1*, *n2*: *node*; *L*: *label*);
> **pre** $n1$ is in $Nodes(G)$ and $n2$ is in $Nodes(G)$;
> **post** $Arcs_L(G) = Arcs_L(G') \cup \{<n1, n2>\}$
> and $Nodes(G) = Nodes(G')$
> and for all $q \neq L$, $[Arcs_q(G) = Arcs_q(G')]$;

> **procedure** *DelSucc*(**var** *G*: *graph*; *n1*, *n2*: *node*; *L*: *label*);
> **pre** $n1$ is in $Nodes(G)$ and $n2$ is in $Nodes(G)$;
> **post** $Arcs_L(G) = Arcs_L(G') - \{<n1, n2>\}$
> and $Nodes(G) = Nodes(G')$
> and $Nodes(G) = Nodes(G')$
> and for all $q \neq L$, $[Arcs_q(G) = Arcs_q(G')]$;

> **function** *Succ*(*G*: *graph*; *n*: *node*; *L*: *label*): *listofnode*;
> **pre** n is in $Nodes(G)$;
> **post** The set of nodes in the list $RESULT =$
> $\{x \text{ in } Nodes(G) \mid <n, x> \text{ is in } Arcs_L(G)\}$;

Fig. 9.10 Specifications for graphs

NOTE: We have shown in Figure 9.10 but one of many possible sets of operations on graphs. Indeed, the form of graph and, hence, the most appropriate operations are invariably application specific. Therefore, the definitions above should be considered illustrative rather than definitive.

9.2.2 TREES

Definition 9.4. A *tree* is a set of nodes, T, such that either

- T is empty, or
- T contains a distinguished node, r, called the *root of T*, and the other nodes of T are divided into zero or more *disjoint* sets, $T_1, \cdots,$ T_n, each of which is itself a tree. The sets T_i are called *subtrees* of T or of r.

A node having no nonempty subtree is a *leaf* or *terminal* node. All other nodes are *internal* or *nonterminal* nodes. The number of nonempty subtrees of any tree is called the *degree* of its root node. A collection of zero or more disjoint trees is called a *forest*.

Trees have an obvious graphical structure in which each node is connected by a directed *arc* to each of the roots of its subtrees. These arcs are also called *branches*. □

As usual, there are many possible variations on this definition. For example, many authors do not allow any of the subtrees to be empty.

Figure 9.11 illustrates a typical tree (with the nodes labeled so that we can talk about them). Notice that the root, A, is shown at the top and the branches of the tree are hanging down. Writing the root at the top allows us to talk about "subtrees" and "nodes deeper in the tree" without disrupting our intuitions.

The remaining nodes are partitioned into the sets $\{B\}$, $\{C, E, F, G\}$, and $\{D\}$. Each of these sets is a subtree of A; the roots of the subtrees are B, C, and D. The degree of A is 3; the degree of C is 2; the degrees of B and D are 0.

The standard terminology for relationships among nodes is taken from family trees. If a branch goes directly down from X to Y and Z, then X is the *father* (or *parent*) of Y and Z. Y and Z are *sons* of X. In addition, Y is a *brother* (or *sibling*) of Z and Z is a brother of Y. In the example above, G is a son of F, C is the father of E, A has no father, D has no sons, and C is a brother of D.

If there is a path from one node to another, the one higher in the tree is an *ancestor* of the other; the one lower in the tree is a *descendant* of the higher one. In the example, C is an ancestor of G, E, and F. A is an ancestor of everything else. G is a descendant of A, C, and F. There is no node that is a descendant of everything else. (See Exercise 9.27.)

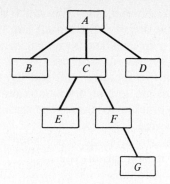

Fig. 9.11 A sample tree

Trees in which all nodes have degrees less than or equal to two form a very important special case.

Definition 9.5. A tree in which each node has exactly two subtrees is a *binary tree*. The subtrees of each node are called the *left subtree* and the *right subtree*. Their roots (if they are not empty) are called the *left son* and *right son* of the node. □

Let's consider an example. The binary tree in Figure 9.12 represents the arithmetic expression

$$(a * b + c) * (d * e + f)$$

Note that we have used operator precedence to sort out *'s from +'s and that the operators are at nonterminal nodes while operands are at terminal nodes. The tree is shown in Fig. 9.12.

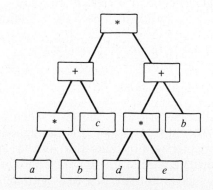

Fig. 9.12 A binary tree

Even if a node has only one son, the son must be either a left or a right son. In the example above, *d* is the left son of a * and *e* is the right son of the same *. The * in question is the left son of the right son of the root. You can see that we describe the position of any node in the tree as a series of applications of the operators left and right.

There are many possible sets of operations on trees. Usually, we use pointers to trees as arguments to these operations to make it easier to manipulate parts of the tree (see Exercise 9.31 for further discussion of this point). Tree nodes are of some type, *node*, which may or may not contain additional data, depending on the application. Figure 9.13 gives a representative set of operations. We have left off formal semantic specifications—leaving only a suggestive comment—and leave the complete specifications as an exercise.

function *EmptyTree*: ↑ *Tree*;
   {Returns a tree with no nodes}

function *Root*(*T*: ↑ *Tree*): ↑*node*;
   {Returns a pointer to the root of *T*↑}

function *Subtree*(*T*: ↑ *Tree*; *n*: *integer*): ↑ *Tree*;
   {Returns a pointer to the *n*th subtree of *T*↑}

function *NewRoot*(*r*: ↑*node*): ↑ *Tree*;
   {Returns a tree with *r*↑ the only node}

function *IsEmpty*(*T*: ↑ *Tree*): *boolean*;
   {true iff *T* points to an empty tree}

procedure *SetSubtree*(*T*, *S*: ↑ *Tree*; *n*: *integer*);
   {Make the tree pointed to by *S* the *n*th subtree of *T*}

Fig. 9.13 Specifications for trees

9.2.3 FINITE HOMOGENOUS SETS

Another familiar mathematical type is the set. When we begin to think of sets in programs, we must consider the limitations of programming languages. First, computers are of finite size, and the basic data structures we might use to implement sets are finite. Therefore we may want to impose a limit on the maximum size of a set. Second, languages generally require that the types of values be known. Thus if we want to *do* anything with an element after taking it out of a set we need to know its type. Since the elements of a set are unlabelled and unordered, the easiest way to keep track of the elements' types is to require that they be all the same—that is, to make the set homogeneous. It isn't important *which* type we choose, so we will simply denote it *Elt* and say no more about it. Finally, we must define a set of functions and procedures. We will choose operations that are commonly useful, and we will choose enough of them that we can write procedures, using defined functions, to do any of the operations we choose not to define as primitives.

Maxsize:	$\rightarrow$ integer
Clear:	SetofElts $\rightarrow$ SetofElts
Insert:	SetofElts $\times$ Elt $\rightarrow$ SetofElts
Remove:	SetofElts $\times$ Elt $\rightarrow$ SetofElts
Union:	SetofElts $\times$ SetofElts $\rightarrow$ SetofElts
Intersect:	SetofElts $\times$ SetofElts $\rightarrow$ SetofElts
IsMember:	SetofElts $\times$ Elt $\rightarrow$ boolean
IsSubset:	SctofElts $\times$ SetofElts $\rightarrow$ boolean
IsEmpty:	SetofElts $\rightarrow$ boolean

The functions and procedures of Fig. 9.14 provide the complete specifications. The constant function *Maxsize* gives the maximum size for any set of this type. Procedure *Clear* sets the value of its parameter to the empty set; it may thus be used for initialization. Procedures *Insert* and *Remove* add or delete individual elements; they alter the value of the input set. Predicates *IsMember*, *IsSubset*, and *IsEmpty* are sufficient to perform all the relational tests.

function *Maxsize*: *Elt*;
    post *RESULT* > 0;

procedure *Clear*(**var** *S*: *SetofElts*);
    post $S = \{\}$;

procedure *Insert*(**var** *S*: *SetofElts*; *x*: *Elt*);
    pre *cardinality*$(S - \{x\}) < maxsize$;
    post $S = S' \cup \{x\}$;

procedure *Remove*(**var** *S*: *SetofElts*; *x*: *Elt*);
    post $S = S' - \{x\}$;

Fig. 9.14 Specifications for sets (Part 1 of 2)

procedure *Union* (**var** *S*: *SetofElts*; *T*: *SetofElts*);
   pre *cardinality* $(S \cup T) <$ *maxsize*;
   post $S = S' \cup T$;

procedure *Intersect*(**var** *S*: *SetofElts*; *T*: *SetofElts*);
   post $S = S' \cap T$;

function *IsMember*(*S*: *SetofElts*; *x*: *Elt*): *boolean*;
   post $RESULT \equiv (x \in S)$;

function *IsSubset*(*S*, *T*: *SetofElts*): *boolean*;
   post $RESULT \equiv (S \subset T)$;

function *IsEmpty*(*S*: *SetofElts*): *boolean*;
   post $RESULT \equiv (S = \{\})$;

Fig. 9.14 Specifications for sets (Part 2 of 2)

9.3 Further Reading

Chapter 2 of Knuth, v. 1 (1973) spans less than 200 pages, yet is certainly the primary reference consulted for algorithms concerning data structures. However, it does not always separate the organization of data from its representation as we have tried to do. Thus we suggest that two readings of that chapter be made—the first, at this point, for another perspective on abstract organizational ideas, and the second, at the conclusion of Chapter 10, at which time you pay greater attention to the details of pointer manipulation.

A more recent book on the subject of data structures is Horowitz and Sahni (1976), which is perhaps less mathematically oriented. Together these two references cover most of the fundamental structuring algorithms.

Exercises

9.1 Write a procedure to print the elements of a linear structure in the order described below, using the indicated auxiliary storage structure.

   a) Print elements of a queue in forward order, using no auxiliary storage.
   b) Print elements of a queue in reverse order, using a stack for auxiliary storage.

c) Print elements of a stack in reverse order, using a stack for auxiliary storage.

d) Print elements of a deque alternately from the front and the back, using no auxilliary storage.

9.2 Suppose deque D has the initial value

$$D = <4, 7, 5>$$

What value does it have after each of the following series of operations? Start over with $D = <4, 7, 5>$ for each part.

a) *InsertFront*(D, 3); *InsertBack*(D, 6); $x := $ *RemoveFront*(D); $x := $ *RemoveFront*(D); $x := $ *RemoveFront*(D);

b) $x := $ *RemoveBack*(D); *Clear*(D); *InsertFront*(D, 3);

c) $x := $ *RemoveFront*(D); $x := $ *RemoveBack*(D); $x := $ *RemoveFront*(D);

d) *InsertFront*(D, 3); $x := $ *RemoveBack*(D);

e) $x := $ *RemoveFront*(D); $x := $ *RemoveFront*(D); *InsertBack*(D, 3); $x := $ *RemoveFront*(D); $x := $ *RemoveFront*(D);

9.3 Write formal specifications for the type *queue*. Use sequences for the model.

9.4 Suppose queue Q has the initial value

$$Q = <4, 7, 5>$$

What value does it have after each of the following series of operations? Start over with $Q = <4, 7, 5>$ for each part.

a) *Enq*(Q, 3); *Enq*(Q, 6); $x := $ *Deq*(Q); $x := $ *Deq*(Q); $x := $ *Deq*(Q);

b) $x := $ *Deq*(Q); *Clear*(Q); *Enq*(Q, 3);

c) $x := $ *Deq*(Q); $x := $ *Deq*(Q); $x := $ *Deq*(Q);

d) *Enq*(Q, 3); $x := $ *Deq*(Q);

e) $x := $ *Deq*(Q); $x := $ *Deq*(Q); *Enq*(Q, 3); $x := $ *Deq*(Q); $x := $ *Deq*(Q);

9.5 Suppose you are given a FIFO queue of elements of type *color*. Consider the problem of reordering the queue so that like-colored elements end up adjacent to one another. Any solutions are to use only the operations on deques, queues, stacks, and lists described in section 9.1. In each case you are allowed any number of scalar variables, and at most one auxiliary structure (deque, queue, stack, or list). In deciding upon an auxiliary structure you are to choose the least complex structure by the ordering (stack < queue < deque < list).

    a) Write a procedure to reorder the queue when type *color* = (red, blue).

    b) Write a procedure to reorder the queue when type *color* = (red, blue, green).

    c) Can you write this procedure when type *color* = (red, blue, green, yellow)? If so, do so.

    d) Suppose the input is a linear list as in section 9.1.4 instead of a FIFO queue. Redo (a). Can you do it without an auxiliary structure at all?

    e) Suppose there is a special color, black, that is used to mark the end of the input structure (queue or list). Redo parts (a) through (d).

9.6 Suppose stack S has the initial value

$$S = <4, 7, 5>$$

What value does it have after each of the following series of operations? Start over with $S = <4, 7, 5>$ for each part.

    a) *Push(S, 3); Push(S, 6); x := Pop(S); x := Pop(S);*
       x := Pop(S);

    b) *x := Pop(S); Clear(S); x := Top(S);*

    c) *x := Pop(S); x := Top(S); x := Pop(S);*

    d) *Push(S, 3); x := Pop(S);*

    e) *x := Pop(S); x := Pop(S); Push(S, 3); x := Pop(S);*
       x := Pop(S);

9.7 Arithmetic expressions are normally written in *infix notation*, that is, binary operators are interposed between their operands. For example, $1 + 2, 2 * (3 + 5), (2 + 3)/5 + 1$ are all expressed in infix.

    Postfix notation is another form in which one can write expressions. In postfix each operator occurs immediately to the right of its operands. For example, the above expressions are written in postfix as $1\ 2\ +, 2\ 3\ 5$ $+\ *$, and $2\ 3\ +\ 5\ /\ 1\ +$. An advantage of postfix notation is that parentheses are never necessary.

    Postfix expressions can be evaluated using a stack. The expression is scanned left-to-right. Operands are pushed on the stack. When an operator is encountered, the top two stack elements are popped, and the value of the operator applied to those elements is then pushed.

Write a procedure that uses a stack to evaluate postfix expressions. Assume you are given as input a FIFO queue. Each element of the queue is either an integer or an operator. The operators are just $(+, -, *, /)$. Use the stack and queue operations of this chapter only.

9.8 Repeat exercise 9.7 assuming that the input is a list instead of a queue. Use no auxiliary storage, but you may destroy the input list.

9.9 In the specification of linear lists given in Figure 9.6, the precondition of *SetTail* requires that the variable $L\uparrow$ not be contained in Q. Why is this restriction necessary? Give an example.

9.10 Suppose that T is *integer* and P and Q of type $\uparrow$ *listofT* point to the lists

P: $[2] \rightarrow [4] \rightarrow [6] \rightarrow \epsilon$

Q: $[3] \rightarrow \epsilon$

Refer to Fig. 9.6 for definitions of operations on type *listofT*. What are the values of P and Q after each of the following series of operations? (Start over with P and Q as above for each part.)

a) $P := EmptyList;$
b) $SetFirstItem(P, 5); Q := Tail(P);$
c) $P := ConsList(P, 5); Q := Tail(P);$
d) $SetFirstItem(P, 5); SetTail(Q, P);$
e) $P := ConsList(P, 5); Q := ConsList(FirstItem(P), Q);$

9.11 Use the operations on linear lists (type *listofT*) defined in Fig. 9.6 to write a function that concatenates two lists.

a) Write the specifications for *Concatenate*.
b) Write a program that uses the specified operations.
c) Show how your program works by tracing it on several examples. Be sure it works for empty and singleton lists and when the same list is named in both parameter positions.

9.12 Suppose that T is *integer* and P and Q of type $\uparrow$ *plistofT* point to the predecessor lists

P: $[2] \leftrightarrow [4] \leftrightarrow [6] \rightarrow \epsilon$

Q: $[3] \rightarrow \epsilon$

Refer to Fig. 9.8 for definitions of operations on type *plistofT*. What are the values of P and Q after each of the following series of operations?

(Start over with *P* and *Q* as above for each part.)

    a) $P := Singleton(5)$;

    b) $Q := Tail(P)$; $P := Tail(Q)$;

    c) $Q := Tail(P)$; $P := Pred(Q)$;

    d) $AddPred(Tail(P), 7)$;

    e) $AddTail(Q, 5)$;

    f) $AddTail(Tail(Q), 7)$;

    g) $RemoveFirst(Tail(P))$;

9.13 Use the operations on predecessor lists (type *plistofT*) defined in Fig. 9.8 to write a function that concatenates two lists.

    a) Write the specifications for *concatenate*.

    b) Write a program that uses the specified operations.

    c) Show how your program works by tracing it on several examples. Be sure it works for empty and singleton lists and when the same list is named in both parameter positions.

9.14 Consider the following data structure, which we call a *WindowedStream*. A *WindowedStream* is a sequence of values with a "window" that allows the user to access a single element in the middle of the stream. Thus the situation

$$S: \boxed{23 \mid 45 \mid 67 \mid 89 \mid\mid 42 \mid\mid 64 \mid 75 \mid 19 \mid 21}$$

may be described as:

    $S = \quad front \sim <window> \sim back,$

    where $front = <23, 45, 67, 89>,$

                $window = 42, \; back = <64, 75, 19, 21>$

Write formal specifications (precondition and postconditions) in terms of sequences for the following operations:

    function *Current*(*S*: *WindowedStream*): *ElementType*;

        post *RESULT* = value currently in the window of *S*;

    procedure *MoveLeft*(**var** *S*: *WindowedStream*);

        post moves the window of *S* left one element;

That is, tell what each operation does in terms of the initial (and for *MoveLeft*, the final) values of the *front*, *window*, and *back* parts of *S*, and any preconditions. You may call the pieces *S.front*, *S.back*, and *S.window*. How could you use queues to implement such a data type as *WindowedStream*? Answer in one or two prose sentences.

9.15 The underdeveloped but upwardly mobile country of Galosh had need of a secret code in which to express its diplomatic dispatches, so the head cryptographer proposed that first all sequences of nonvowels (including spaces and punctuation) be reversed. Then the entire message should be written backwards. The Prime Minister then sent the following message to an aide using this wonderful new code:

> rn.urtbae hes mevi ginoreppe. pesee chaxtret a thekam

The message was received by the aide who, having no paper and pencil, was forced to write a program to decode the message. (Upwardly mobile countries must obviously need lots of computers.) Unfortunately, this book is not yet available in Galosh, so the aide got all bogged down in irrelevant representational details before he could figure out the basic algorithm. You might say he just lost his head.

If you were the aide, what program would you have written? Assume that the input characters come from the standard PASCAL file *Input*, and that the end of file delimits the input.

9.16 Suppose that we have a list of integers pointed to by *L* and that we want to remove all 0s from the list. That is, when we finish, we want *L* to point to a list in the same order as the original, but with no 0s.

    a) Write a program to do this, using any of the operations in Figure 9.6.

    b) Write another to do it without using either *SetFirstItem* or *SetTail*.

    c) What can you say about the comparative speeds of your two programs? How do they depend on the number and placement of 0's in the original list?

9.17 Under what circumstances does it make a difference whether one uses the operations *SetFirstItem* and *SetTail* to effect changes in a list or whether one does not use them (as in exercise 9.16), and relies on the other list operations and assignments?

9.18 A list of lists is simply a list in which each element itself contains (or points to) a list. Use a list of lists to store names of people such that the major list contains one element for each distinct first letter of a name encountered, and the sublist for that element contains all the names that start with that letter. All lists are unordered. Write an abstract procedure that inserts a name in the structure, and one that determines whether or not a given name is present.

9.19 Repeat exercise 9.18 but this time keep all lists in alphabetical order.

9.20 Restate the specifications of the four graph operations in precise English.

9.21 The undirected graph in Fig. 9.15 depicts a segment of the highway system of Elk County, PA. Arcs are labeled to indicate the type of road: *T* for paved through-highway, *P* for other paved road, and *U* for unpaved roads. The nodes labeled J_i are unnamed junctions.

Labeled graphs can be used to represent networks of roads. Roads typically carry traffic in both directions. Each of the arcs in Fig. 9.15 therefore represents two arcs, one in each direction.

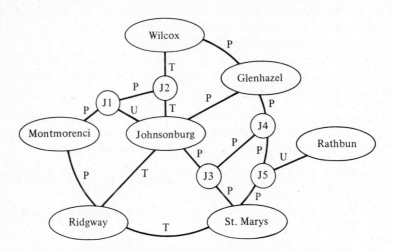

Fig. 9.15 Part of a road map of Elk County, PA

A much more efficient layout is that shown in Fig. 9.16. This layout is not actually feasible because of the terrain, but we ignore that difficulty for the sake of the problem. Write a sequence of graph operations to convert the first graph to the second.

9.22 Suppose a company starts in the state indicated by Fig. 9.9. Call this initial graph *E*.

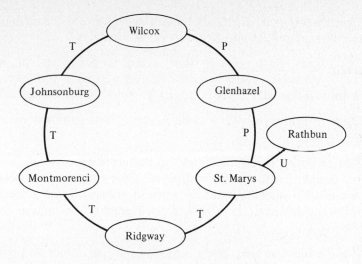

Fig. 9.16 Proposed reform of the Elk County road system

a) What are the values of
   i) *Nodes*(*E*)
   ii) *Labels*(*E*)
   iii) *Arcs*$_k$(*E*) for all *k* in *labels*(*E*)

b) The relation "reports to" is not symmetric but the relation "is married to" is symmetric. As we have defined graphs, all arcs are directed. What does this imply about employee graphs?

c) Suppose the following personnel actions occur. Express each of the actions in terms of one or more graph operations on E and show the cumulative effect on the graph:
   i) Elmer Wilson is hired to work for Susan Smith.
   ii) William Smith is fired.
   iii) Mary Jones is reassigned to Dagwood Bumstead.
   iv) John Jones and Mary Jones are divorced.
   v) Elmer Wilson marries Mary Jones.

9.23 In Figure 9.11 what are the degrees of nodes *E*, *F*, and *G*?

9.24 Determine which of the relations {father, son, brother} hold for the pairs <*B, D*>, <*A, C*>, <*E, F*>, <*E, G*>, <*C, A*>, <*A, G*>, and <*F, C*> of Figure 9.11. List the sons of *E* and the brothers of *G*.

9.25 Describe the positions of the nodes labeled *a, c* and *e* in Figure 9.12 in terms the operators *left* and *right* applied to the root.

9.26 Determine which of the four standard properties of relations (reflexive, symmetric, antisymmetric, transitive) hold for each of the relations "X is father of Y," "X is son of Y," and "X is brother of Y."

9.27 Draw a legal tree with a node which is a descendant of all other nodes in the tree.

9.28 The definition given for "trees" in Section 9.2.2 is stated in terms of a partition defined on the nodes of the tree. An equally good specification could be made by modeling trees in terms of graphs. Write a specification for "trees" that describes them as restricted graphs. Be precise about the restrictions.

9.29 Write a function that, given a graph as an parameter and using only the graph operations, determines whether that graph is a legal tree. (Note, you may need to add a field to each node.)

9.30 Replace the comments in the *Tree* specification of Section 9.2.2 with complete pre- and postconditions.

9.31 In Section 9.2.2, why did we have our operations manipulate ↑*Trees* rather than just *Tree*? Suppose we had removed all ↑'s from the specifications in Fig. 9.13. Now consider the following program

> *T2* := *Subtree*(*T1*, 1);
> *SetSubtree*(*T2, T3*, 1);

Has *T1*↑ changed? If we were using the original specifications—so that *T1, T2,* and *T3* were pointers—would *T1* have changed? What are the comparative advantages and disadvantages of the two versions of the specifications?

9.32 Specialize the specification of Section 9.2.2 to binary trees.

9.33 Suppose trees *H* and *J* have the initial values shown in Fig. 9.17. What values do they have after each of the following series of operations? Start over with *H* and *J* as in the figure for each part.

> a) *SetSubTree*(*H, J*, 1);
> b) *SetSubTree*(*J, SubTree*(*H*, 2), 5);
> c) *SetSubTree*(*SubTree*(*H*, 1), *Root*(*J*), 1);
> d) *SetSubTree*(*SubTree*(*SubTree*(*H*, 2), 3), *NewRoot*(*Root*(*J*)), 1);

9.34 A binary search tree of integers has the property that the value

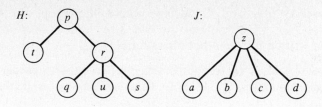

Figure 9.17

stored at a given node is greater than all values contained in its left subtree, and smaller than all values contained in its right subtree. For example, a binary search tree containing 1, 2, 4, 5, 7, 9, 11, 13 could be:

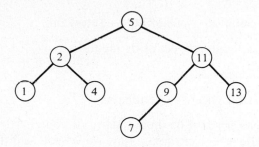

Figure 9.18

Write a procedure that takes a binary search tree and an integer and determines whether or not that integer is contained somewhere in the tree.

9.35 Restate the specifications of the set operations in precise English.

9.36 Suppose sets *A* and *B* have initial values

$$A = \{1, 2, 3, 4\}$$
$$B = \{1, 3, 5, 7\}$$

What values do they have after each of the following series of operations? (Start over with *A* and *B* as above for each part.)

    a) *Insert(A, 5); Insert(B, 5);*

    b) *Remove(A, 5); Remove(B, 5);*

    c) *Union(A, B); Union(B, A);*

    d) *Intersect(A, B); Intersect(B, A);*

9.37 Operations *Insert* and *Union* have preconditions that require the size of the output set to be within the set size limit. Why is no such precondition written for the other operations?

9.38 Write functions to test superset, proper subset, proper superset, equality, and inequality. Use only function *Issubset* and the boolean operators.

9.39 If we provided a function

Select: SetofElts → Elt

that returned as value some element of its input (but made no promises about which one), it would be possible to write procedures *Union* and *Intersection* using only *Clear, Insert, Remove, Ismember,* and *Select.* Do so.

9.40 Write a function

Cartesian: SetofElts × SetofElts → SetofPairs

to compute the Cartesian product of its inputs.

9.41 Write a function

powerset: SetofIntegers → SetofSets

to compute the power set of its input. (Hint: If you already had the power set of a subset with $n - 1$ elements, how would you find the power set of n elements ?)

9.42 Define—that is, formally specify—abstract data type *pairUV*, a pair consisting of a value of type U and a value of type V. Using this type and the type *set*, define type *function.* The only operation on a function should be *apply*(F: *function*, x: U): V—which "evaluates" the function F with argument x and returns a value of type V.

9.43 Declare types (including appropriate types for any components) that describe various attributes of

a) A television set (features that differ from brand to brand.)

b) A recorded tape (for example, what is on it?)

c) One student's semester course schedule.

d) A computer program (for example, who wrote it?)

e) A checkbook entry.

f) A piece of furniture.

Chapter 10

The Representation of Data

The specifications of the types presented in the preceding chapters defined *what* each operation does—but not *how* it does it. Now we will consider the question of how each operation is implemented. In the process of implementing the operations, the programmer must make a crucial decision: How should objects of the type be *represented*? That is, which of the types provided by the language should be used to hold the information content of the new type, and how should the operations be coded?

Let's illustrate the problem by showing one of the many possible representations for a stack. As you recall, a stack is a restricted deque—one in which both insertions and removals are made from the same end. The insertion and deletion operations are called *Push* and *Pop* respectively. Section 9.1.3 gave one complete set of specifications.

The representation we'll use consists of two data structures from the base language—a vector, *v*, and an integer, *sp*. The vector *v* holds items currently stored in the stack, and *sp* is the index of the most recently inserted item. Assuming that the items to be stacked are of type *T*, and that we know in advance that the depth of the stack will not exceed *maxstack* items, stacks may be represented by the structure

type *stack*=   **record** *sp*: *integer*;   *v*: **array** [1..*maxstack*] **of** *T* **end**;

The stack operations can then be coded as:

procedure *Clear*(**var** *S*: *stack*);
   begin   *S.sp* := 0 **end**;

procedure *Push*(**var** *S*: *stack*; *x*: *T*);
    begin   *S.sp* := *S.sp* + 1; *S.v*[*S.sp*] := *x* **end**;

function *Pop*(**var** *S*: *stack*): *T*;
    begin   *pop* := *S.v*[*S.sp*]; *S.sp* := *S.sp* − 1 **end**;

function *Top*(*S*: *stack*): *T*;
    begin   *top* := *S.v*[*S.sp*] **end**;

function *Empty*(*S*: *stack*): *boolean*;
    begin   *empty* := (*S.sp* = 0) **end**;

function *Full*(*S*: *stack*): *boolean*;
    begin   *full* := (*S.sp* = *maxstack*) **end**;

By providing these definitions, we explain how the abstract properties of stacks (as specified in section 9.1.3) can be *implemented* in the language. It is important to notice that we could completely specify the important characteristics of stacks long before deciding on an implementation. It is just as important to notice that we could have chosen many other equally good representations.

In this chapter we first explore some general schemes for representing one type in terms of others. We then use these schemes to represent some of the nonelementary structured types discussed earlier. In particular we will consider at least two or three representations for each of the following: deques, queues, stacks, sets, and arrays. We must stress that no one of these representations is "better" than another; the best choice depends upon the way the type will be used.

10.1 Representational Techniques for Data

There are four techniques that we shall use, either singularly or in combination, in the representations to follow:

- Encoding
- Packing
- Address arithmetic
- Linking (using references)

We shall discuss each of these techniques briefly before discussing some possible representations of the types discussed earlier.

10.1.1 ENCODING

Encoding schemes involve the direct mapping of values from one type onto values of another, previously defined type. For example, suppose we wish to represent the elements of an ordered enumerated type *day*:

Ordered Literals: *Sunday, Monday, Tuesday, Wednesday, Thursday, Friday, Saturday*

We may represent the values of this type by the mapping:

$$rep(Sunday) = 1, \ rep(Monday) = 2, \ \cdots, \ rep(Saturday) = 7,$$

where *rep* denotes "representation of" just as in Section 4.1. In this case the operations *succ*, *pred*, and $<$ are simply the integer operations "add 1," "subtract 1," and $<$ respectively (except, of course, we must be careful with *succ*(*Saturday*) and *pred*(*Sunday*), neither of which is supposed to be defined).

The most common example of encoding is the representation of integers inside a computer; indeed this example is so common that one is often inclined to forget that it *is* a representation. Nevertheless, until used in an integer context, the bits in a computer word are uninterpreted; only when used as an integer (for example, by an ADD instruction) is interpretation placed on the word. The representation of *positive* integers is given by:

$$rep(n) = \textbf{if } n > 2^{wordsize-1} \textbf{ then } ERROR \textbf{ else } b,$$

$$\textit{where } b_i = (n \textbf{ div } 2^i) \textbf{ mod } 2 \ \text{ for each } i.$$

Here, b_i is the ith bit of the word (numbering the bits from right to left and denoting the rightmost bit as b_0).

10.1.2 PACKING

The unit of information that can be read from or written into a computer memory—the word—is usually of a fixed size determined by the design of the computer. It often happens that the size of the unit of information actually needed to represent a value of a given type is smaller than this computer-specific word size. For example, a *boolean* value can be represented with a single bit; a character can be represented by 8 bits or less, depending on the size of the alphabet.

In order to save space it is common to *pack* several values into a single word. Thus, for example, it is common to store four 8-bit characters in a single 32-bit word. There is usually an additional cost (in for instance, time) to extract or insert a single unit of information when it is packed, but under suitable circumstances this additional cost is more than justified by the space saved. We will talk more later about how to make such time/space trade-offs rationally.

Suppose, for example, that you need to store records with two integer fields $v1$ and $v2$, and that you know the ranges of the fields are restricted to

$v1$: $0 \le v1 \le 255$

$v2$: $-100 \le v2 \le 100$

These two fields can be packed into a single word, v, using the rule

$$v = (v2 + 100) * 256 + v1$$

provided your machine's word size is large enough to store values in the range.

$$0 \le v \le 51455$$

Extracting the values of $v1$ and $v2$ is done using the rules

$rep(v1)$ $= v \,\mathbf{mod}\, 256$

$rep(v2)$ $= (v \,\mathbf{div}\, 256) - 100$

and storing new values of $v1$ and $v2$ is done using the rules

$rep(v1 := k) = v := ((v \,\mathbf{div}\, 256) - 100 + 100) * 256 + k$

$rep(v2 := k) = v := (k + 100) * 256 + v \,\mathbf{rem}\, 256$

The rules for storing new values are constructed from the basic packing rule by substituting the appropriate extraction rule for the value not being modified.

10.1.3 ADDRESS ARITHMETIC

The fact that integers are used as the names of locations in a computer memory makes it possible to perform a series of arithmetic operations and use the result as an address (that is, as a memory location name). Many examples of such *address computations* can be found, and many are extremely important. We illustrate this point with two of the simplest examples.

First, consider the representation of type $\uparrow T$ (reference to T). The implementation of a programming language usually represents a reference to an object as the address (which is representable as an *integer*) of the first word comprising that object. We can represent the value *nil* by some arbitrary integer, such as -1, as long as it cannot be mistaken for a valid address. If objects of type T are records, we can represent references to their components as constant displacements relative to the address of the beginning of the record. So, for example, if x has been declared

var x: **record** a, b, c: *integer* **end**

and each integer occupies one word, then

$\beta x.a = \beta x + 0$ [the address of $x.a$ is the address of x plus 0.]

$\beta x.b = \beta x + 1$

$\beta x.c = \beta x + 2$

Second, consider the represention of a one-dimensional array of integers in memory. We'll again assume that each integer requires one word. The declaration

var A: **array** [0..99] **of** *integer*;

can be represented by a sequence of 100 contiguous words. We can describe the memory of our computer as a single vector of words in SMAL:

var *MEM*: **array** [0..*memlimit*] **of** *word*

The array A has been allocated as a block of 100 words starting at location βA in *MEM*. To find the address (the relative location within *MEM*) of element i of A, whose address we shall denote $\beta A[i]$, we use the formula

$\beta A[i] := \beta A + i$

That is, to implement the statement

$A[i] := A[i] + 1$

We could generate the following instruction sequence in SMAL (the variable t is a temporary of type *word*):

var t: *word*;

.

.

.

$t := \beta A$;
$t := t + i$;
$MEM[t] := MEM[t] + 1$

Note that we have computed the sum of the *value* of i and the *address* of A in t. We then used this sum as the address of $A[i]$. This is a simple example, but it should illustrate the general notion of address arithmetic.

10.1.4 LINKING

The technique called *linking*, or "building linked data structures," is one of the most flexible representational schemes. It is extensively used for many of the more complex data structure representations—especially ones that grow (or shrink) during the course of a computation. In simple terms, a linked structure is one in which each data element contains one or more references to other data elements. These references explicitly encode the relations between the data elements. In the following we'll consider three common linking techniques: (1) single linking with "pointer swing," (2) single linking with "node overwrite," and (3) double linking with "pointer swing." Each of these has advantages for representing certain structures—and all three are used in practice.

10.1.4.1 Single Linking with Pointer Manipulation
Consider the linear lists defined in the previous chapter. One way to implement *listofT*s is as a record containing a data item (the *FirstItem*) and a reference to the tail.

> **type** *listofT* = **record** *first*: *T*;   *tail*: ↑*listofT* **end**;

This gives us no way to represent the empty list, however. Since we are dealing only with pointers, though, we may establish the convention that *nil* points at the empty list. We call the resulting data structure a singly linked list because each list element contains exactly one reference, or link—namely to its immediate successor. Figure 10.1 illustrates a typical singly linked list.

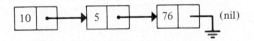

Fig. 10.1 A typical singly linked list

Implementation of the operations of Section 9.1.4.1 is now fairly straightforward, as shown in Figure 10.2.

10.1.4.2 Single Linking With Overwrite
The foregoing linear list implementation is sometimes called pointer swing, because nodes are inserted or deleted by moving only pointers. One slight problem, however, is that it is not always possible to delete a node without a pointer to its predecessor. In particular, consider the problem of deleting the *last* element without a pointer to its predecessor. There just isn't any way of setting the predecessor's *tail* field to *nil*.

```
function EmptyList: ↑listofT;
    begin  EmptyList := nil end;

function IsEmpty(L: ↑listofT): boolean;
    begin  IsEmpty := (L = nil) end;

function ConsList(x: T; L: ↑listofT): ↑listofT;
    var n: ↑listofT;
    begin
    new(n);  ConsList := n;
    n↑.first := x;  n↑.tail := L;
    end;

function FirstItem(L: ↑listofT): T;
    begin  FirstItem := L↑.first end;

function Tail(L: ↑listofT): ↑listofT;
    begin  Tail := L↑.tail end;

procedure SetFirstItem(L: ↑listofT; x: T);
    begin  L↑.first := x end;

procedure SetTail(L, Q: ↑listofT);
    begin  L↑.tail := Q end;
```

Fig. 10.2 Implementation of linear list operations with single linking

One remedy is to use a method called the node-overwrite technique, in which entire nodes are copied. The list representation remains the same, except that we change the representation of the empty list. Under this representation, every list ends with a special sentinel node, which does not contain an element of the represented sequence. It is recognizable as a sentinel node by a special value for its *next* field (such as *nil*) or by some other flag. The empty list then consists of just a sentinel node. This representation simplifies the implementation of some operators by guaranteeing that there will always be at least one element in the list.

This representation is most useful for implementing the alternative linear list specifications given in Fig. 9.7, in which *SetTail* is replaced with *Replace*. Figure 10.3 shows implementations of the basic list operations under the node-overwrite representation. Figure 10.4 illustrates some of the node-overwrite list operations.

function *EmptyList*: ↑*listofT*;
   var *n*: ↑*listofT*;
   begin
   new(*n*); *EmptyList* := *n*;
   n↑.*tail* := *nil*
   end;

function *IsEmpty*(*L*: ↑*listofT*): *boolean*;
   begin *IsEmpty* := (*L*↑.*tail* = *nil*) **end**;

procedure *Replace*(*L*, *Q*: ↑*listofT*);
   begin *L*↑ := *Q*↑ **end**;

{*SetTail absent*}

Fig. 10.3 Node-overwrite implementation of list operations (only the changed operations are listed)

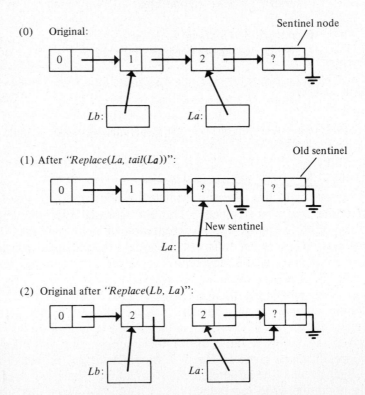

(0) Original:

(1) After *"Replace(La, tail(La))"*:

(2) Original after *"Replace(Lb, La)"*:

Fig. 10.4 Basic node-overwrite list operations

The node-overwrite technique has the disadvantages that it requires (1) an extra node in every list and (2) copying of the *first* fields of nodes during list manipulation. This latter may be serious, if *T*'s are big. Node-overwriting has the advantage of making the *Replace* operation convenient. Again, *Replace* may be difficult with pointer-swing, since in general, one needs a reference to the predecessor node in order to delete or replace the next one, for the predecessor's *tail* field must be updated. As always, the representation chosen depends on the application.

10.1.4.3 Double Linking

Lists can also be *doubly linked*, so that each element points to both its successor and its predecessor. This allows us to implement the *plistofT*s of section 9.1.4.2. A possible representation of *plistofT* is

> **type** *plistofT* =
>   **record** *first*: *T*; *pred, tail* : ↑*plistofT* **end**;

The implementation of the *plistofT* operations is given in Figure 10.5. It may look a bit formidable at first glance; pointers fly most alarmingly. You should try executing the operations "by hand" until you feel comfortable with them. See Fig. 10.6 for illustrations.

Despite the added complexity (and extra space for a second pointer in every node), the doubly linked list is the representational technique of choice in many circumstances. The primary advantage of double linking is that, given a reference to any node, all other nodes in the list can be easily found. The simplest case of this is illustrated by the ease with which *AddPred* and *AddTail* are implemented. There are more interesting cases, however, in which one wants to examine all the nodes on a list; with double linking, this can be done using any node as a starting point.

10.1.4.4 Circular Lists

Another common linked list structure is the *circularly linked list* or *circular list*, which comes in both singly and doubly linked forms. The essential property of circular lists is that the "end" elements point to the "other end" instead of containing *nil*. Consider, for example, a singly linked list. In the representations discussed before, the last element of such a list contained the "pointer to no element," *nil*. In a singly linked *circular* list, the last element will contain a pointer back to the first element of the list. In a doubly linked circular list, the *pred* field of the first element will point to the last, and the *succ* field of the last will point to the first.

```
function EmptyPlist: ↑plistofT;
   begin EmptyPlist := nil end;
function Singleton(x: T): ↑plistofT;
   var n: ↑plistofT;
   begin
   new(n); Singleton := n;
   n↑.first := x; n↑.pred := nil; n↑.tail := nil
   end;

function IsEmpty(L: ↑plistofT): Boolean;
   begin IsEmpty := (L = nil) end;

function FirstItem(L: ↑plistofT): T;
   begin FirstItem := L↑.first end;
function Tail(L: ↑plistofT): ↑plistofT;
   begin Tail := L↑.tail end;
function Pred(L: ↑plistofT): ↑plistofT;
   begin Pred := L↑.pred end;

procedure AddPred(L: ↑plistofT; x: T);
   var n: ↑plistofT;
   begin
   new(n); n↑.first := x; n↑.pred := L↑.pred; n↑.tail := L;
   if n↑.pred ≠ nil then n↑.pred↑.tail := n;
   L↑.pred := n;
   end;

procedure AddTail(L: ↑plistofT; x: T);
   var n: ↑listofT;
   begin
   new(n); n↑.first := x; n↑.pred := L; n↑.tail := L↑.tail;
   if n↑.tail ≠ nil then n↑.tail↑.pred := n;
   L↑.tail := n
   end;

procedure RemoveFirst(L: ↑plistofT);
   begin
   if L↑.pred ≠ nil then L↑.pred↑.tail := L↑.tail;
   if L↑.tail ≠ nil then L↑.tail↑.pred := L↑.pred;
   end;
```

Fig. 10.5 Implementation of *plistofT* with double linking

Original:

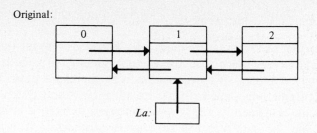

Original after *"AddPred (La, −1)"*:

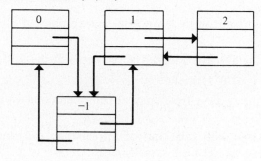

Original after *"RemoveFirst (La)"*:

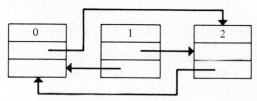

Fig. 10.6   Basic operations on doubly linked lists

The principal effect of circular linking is to simplify some of the operations. It turns out to be a bit harder to tell when you are at the end of such a list—but for many operations, this information is not required. Also, some of the types that might be represented with a circular list need no concept of an "end." The implementation of circular lists is left as an exercise—see Exercise 10.10.

10.1.5 CONCLUDING REMARKS ON REPRESENTATIONAL TECHNIQUES

In the preceding sections we have discussed four common techniques used in the representation of data: encoding, packing, address arithmetic, and linking. The remainder of this chapter is devoted to the use of these techniques to represent some of the more common data types used in real computations.

Perhaps one of the most frequent misconceptions about the representation of data is that there is a "best" representation for each data type. That is simply not true. For each data type there are generally several possible representations, and each has its advantages and disadvantages. In order to decide intelligently which representation to pick, it is necessary to look at the way the data type will be *used.* It is necessary to ask questions such as

- Is there a fixed upper bound on the number of elements in the data structure?
- What is the average number of elements in the structure?
- How big is each element?
- What is the relative frequency of the operations on the type?

And so on. Only with this kind of information can one really determine the best representation for a specific case.

10.2 Representation of Stacks

At the beginning of this chapter, we presented a representation of stacks using vectors. Let us now consider an alternative—a linked representation.

A stack will consist of a list of elements of type T:

type *stack* $=$ L: $\uparrow listofT$

We implement the stack operations as in Fig. 10.7.

As an example, suppose that S points to

[3] $\rightarrow$ [7] $\rightarrow$ [10] $\rightarrow$ ϵ.

Then *Pop(S)* returns 3, and afterward, S points to

[7] $\rightarrow$ [10] $\rightarrow$ ϵ.

Then, after *Push(S*, 2), S points to

[2] $\rightarrow$ [7] $\rightarrow$ [10] $\rightarrow$ ϵ.

This list implementation has several advantages over the vector representation—most notably that there is no fixed upper limit on the

> **procedure** *Clear*(**var** *S*: *stack*);
>   **begin** *S*:= *EmptyList* **end**;
>
> **procedure** *Push*(**var** *S*: *stack*; *x*: *T*);
>   **begin** *S*:= *ConsList*(*x*, *S*)   **end**;
>
> **function** *Pop*(**var** *S*: *stack*): *T*;
>   **begin** *pop*:= *FirstItem*(*S*); *S*:= *Tail*(*S*) **end**;
>
> **function** *Empty*(*S*: *stack*): *Boolean*;
>   **begin** *empty*:= *IsEmpty*(*S*) **end**;
>
> **function** *Full*(*S*: *stack*): *Boolean*;
>   **begin** *full*:= *false* **end**;

Fig. 10.7 Implementation of stacks using lists

number of elements. Moreover, it uses less space if, on the average, the vector is less than half full (if we assume that *T*'s and addresses are the same size). On the other hand, each operation may be a bit more expensive.

10.3 Representation of Deques and Queues

Stacks are, you will recall, a restricted form of deque. Because of these restrictions, it is possible to obtain the simple vector and list implementations described above. Without these restrictions, life gets a bit more complicated. The next two sections present vector and list implementations of deques and queues.

10.3.1 VECTOR REPRESENTATION: CIRCULAR BUFFERING

The vector representation of stacks took advantage of the fact that the "bottom" of the stack is never manipulated. Alas, queues and deques do not have this property; both insertions and deletions are made at both ends. If we attempt to extend the implementation we used for stacks in the obvious manner, introducing a new pointer to point to the "other end" of the queue or deque, we run into a serious problem: we will run out of space in the vector even though the total number of elements in the queue or deque is less than the number of vector components.

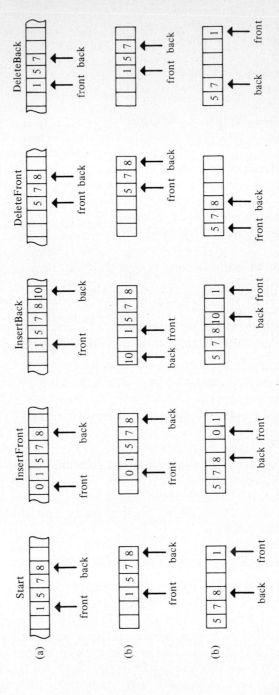

Fig. 10.8 (a) Deques using infinite vectors, and (b) two examples of deques using circular buffers. Each column shows the result of applying the indicated operation to the "Start" configuration.

The problem arises in the following way: any number of insertions are possible on one end of a queue or deque, with no intervening deletions from that end. As long as there are deletions from the other end, the total amount of data in the queue or deque at any one time will remain bounded. In the generalization of our stack implementation, however, the pointer advances whenever there is an insertion. This will cause the pointer to go off the end of the data vector—even with fewer elements in the queue than there are elements in the vector. The first part of Fig. 10.8 illustrates the effects of the four basic deque operations, each performed to a deque in the state labeled "start."

One way to remedy this is to keep the elements of the deque or queue "packed" at the front of the array. To do so, each time there is a deletion we move all of the remaining elements "down one." The new front element will then be at the front of the array again. This procedure will work, but it is more expensive than necessary. In fact, it is often true that if your solution to a problem requires you to copy data from one place to another, there is probably another, better algorithm that doesn't.

To solve these problems, we use a technique called *circular buffering*. Two integer pointers, one to the front and one to the end of the queue or deque, "chase" each other around the vector. One pointer is decremented for insertion and incremented for deletions. The other is incremented for insertions and decremented for deletions. The vector is treated as though it is circular: when a pointer increments or decrements past either end of the vector, it *wraps around* to the other end, as illustrated in the second part of Fig. 10.8. As before, each operation is applied to the deque in state "start."

Since a queue is a restricted deque and can be implemented identically, we give only the implementation of *deque* here. The data structure is

> **type** *deque* =
>   **record**
>   *fp, bp*: *integer*;
>   *V*: **array** [0..*maxdeq*] **of** *T*
>   **end**;

Here, *maxdeq* is an integer constant greater than 1. The operations are implemented as shown in Fig. 10.9. Figure 10.8 illustrates the effects of the various operations.

This circular buffer representation is a bit more subtle than it may, at first, appear. Note, for example, that the array has *maxdeq* + 1 elements (numbered 0 to *maxdeq*)—this ought to be a tipoff that something suspicious is happening. The problem is in correctly representing the empty and full cases, and the extra word in the vector lets us do it. To see

```
procedure Clear (var D: deque);
    begin D.fp := 1; D.bp := 0 end;

procedure InsertFront(var D: deque; x: T);
    begin
    D.fp := D.fp − 1; if D.fp < 0 then D.fp := maxdeq;
    D.V[D.fp] := x
    end;

procedure InsertBack(var D: deque; x: T);
    begin
    D.bp := D.bp + 1; if D.bp > maxdeq then D.bp := 0;
    D.V[D.bp] := x
    end;

function RemoveFront(var D: deque): T
    begin
    RemoveFront := D.V[D.fp];
    D.fp := D.fp + 1; if D.fp > maxdeq then D.fp := 0
    end;

function RemoveBack(var D: deque): T
    begin
    RemoveBack := D.V[D.bp];
    D.bp := D.bp − 1; if D.bp < 0 then D.bp := maxdeq
    end;

function Empty(D: deque): boolean
    begin  empty := (D.fp = (D.bp + 1) mod maxdeq ) end;

function Full(D: deque): boolean;
    {Exercise 10.24 asks you to write the omitted function body}
```

Fig. 10.9 Implementation of deques using arrays

why the problem exists, note first that the two pointers, *fp* and *bp*, point at the front and back elements of the deque. So the deque with just one element is represented as shown in Fig. 10.10.

Now, suppose that the operation *DeleteFront* were applied to this configuration; the resulting representation of the empty deque is shown in Fig. 10.11. Examine this figure carefully. It is the *same* configuration that would result if we allowed a full deque to use all the elements of the

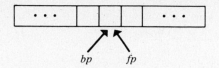

Fig. 10.10 A deque with one element

vector.   The representation we have given, however, avoids this situation—it never, allows all the vector elements to be in use at once. Instead, the full deque is represented as shown in Fig. 10.12.

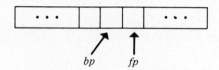

Fig. 10.11 An empty deque

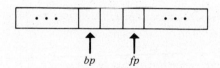

Fig. 10.12 A full deque

10.3.2 LINKED IMPLEMENTATION

Let's now consider a representation of deques containing objects of type T, using a *plistofT*. A deque will have two ends serving as sentinels, so that the empty deque will be as in Fig. 10.13. The data structure is

> **type** *deque* = **record** *front, back*: ↑*plistofT* **end**,

where the *front* and *back* fields point to the two sentinels. The operations are now easy to implement, as shown in Fig. 10.14.

    The pros and cons of the vector and list implementations for deques and queues are very similar to those for stacks. The vector implementation is generally faster, although the tests for the special cases at the ends of a vector make the case for vectors weaker. The list implementation is generally simpler to understand, and may save space if the average

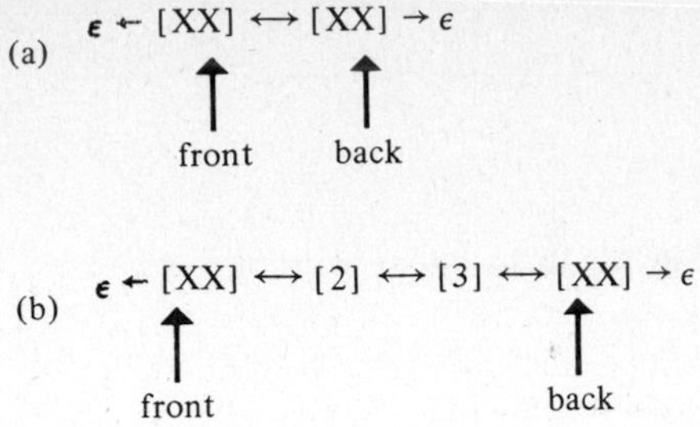

(a) $\epsilon \leftarrow [XX] \longleftrightarrow [XX] \rightarrow \epsilon$

front   back

(b) $\epsilon \leftarrow [XX] \longleftrightarrow [2] \longleftrightarrow [3] \longleftrightarrow [XX] \rightarrow \epsilon$

front   back

Fig. 10.13 (a) The empty deque, and (b) a deque containing $<2, 3>$

```
procedure Clear(var D: deque);
   begin
   D.front := Singleton(XX);
   AddTail(D.front, XX);   D.back := Tail(D.front)
   end;
            {where XX is any arbitrary T value}

procedure InsertFront(var D: deque; x: T);
   begin  AddTail(D.front, x)  end;

procedure InsertBack(var D: deque; x: T);
   begin  AddPred(D.back, x)  end;

function RemoveFront(var D: deque): T;
   begin
   RemoveFront := FirstItem(Tail(D.front));
   RemoveFirst(Tail(D.front))
   end;

function RemoveBack(var D: deque): T;
   begin
   RemoveBack := FirstItem(Pred(D.back));
   RemoveFirst(Pred(D.back))
   end;

function Empty(D: deque): Boolean;
   begin  Empty := (D.back = Tail(D.front))  end;
```

Fig. 10.14 Implementation of deques using lists

number of elements per deque is significantly smaller than the maximum number. Again, as always, the best choice depends on the structure's pattern of use.

10.4 Representation of Trees and Graphs

The technique of linking (Section 10.1.4) gives an obvious representation for trees and graphs: with each node, we associate pointers to its successor nodes. By this time, you have had enough experience with type implementations to fill in the implementations of operations on graphs and trees. Therefore, we will describe the data structures used and leave implementations as exercises.

10.4.1 LINKED UNLABELED GRAPH REPRESENTATIONS

To see what must go into a graph's structure, let's consider the operations that we must be able to perform. To simplify matters, we'll consider only unlabeled graphs, so that arguments of type *label* are omitted from all the graph operations that were shown in Fig. 9.10.

First, although we have said that graphs contain nodes, we have never said anything about constitutes a node. In general, we can expect that there is some data incidental to the graph structure associated with each node. For the purposes of this section, it is sufficient to treat that data as a single value of some type, T, and to assume that there are operations on nodes for fetching and storing the data.

Second, we note that the *Succ* operation must return a list of *nodes*. At first, this might seem to suggest that a node could have a structure such as

> **type** *node* = **record** *data*: T; *succ*: ↑*listofNode* **end**,   (10.1)

where *listofNode* is a list type as described in Section 9.1.4. The problem with this representation is that it fails to capture an essential property of graphs, the fact that a node can have multiple predecessors. That is, if a node appears on the *succ* lists of two different nodes, then each of those nodes has a *copy* of that node's data and successor list. If a node were changed—if its data or its set of successors were modified—it would be necessary to update *all* copies of the node, so that no matter by what path a program encountered a node, it would "see" the same thing.

Therefore, we'll keep just one copy of each node, and turn successor lists into lists of *pointers* to nodes. This gives us the representation

> **type** *node* = ↑*NodeData*;   (10.2)
> **type** *NodeData* = **record** *data*: T; *succ*: ↑*listofNode* **end**.

This data structure is illustrated in Fig. 10.15.

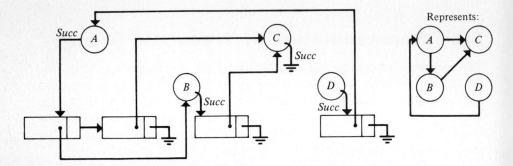

Fig. 10.15 Unlabeled graph representation

The declaration (10.2) for the type *node* allows us to implement *NewNode*, *SetSucc*, *DelSucc*, and *Succ*, but not *DelNode*. The reason is that to implement *DelNode*, we must delete all links *to* the node being deleted. Under the representation above, this would require getting to the *predecessors* of the node to be deleted. A more complete representation is therefore the following.

> **type** *node* = ↑*NodeData*,   (10.3)
> **type** *NodeData* = **record** *data: T*; *succ,pred:* ↑*listofNode* **end**.

There are other alternatives as well. Of course, if *DelNode* is not needed, the data structure (10.2) is adequate.

You may have noticed that we have not had to put anything into the type *graph* itself; so far, its representation has an empty domain and no operations. This may seem a little suspicious. Indeed, if it were possible to take a node and to have it a member of two different graphs simultaneously, we would be in trouble, and would have to put some data into the type *graph*. Fortunately, we defined graphs in such a way that a node can belong to only one graph and the representations (10.2) or (10.3) are adequate.

10.4.2 LINKED TREE REPRESENTATION

Of course, we can use the preceding representation for trees as well, since they are also graphs. However, since a tree node has exactly one predecessor, we can use the representation (10.1) from Section 10.4.1, to give rise the structure illustrated in Fig. 10.16.

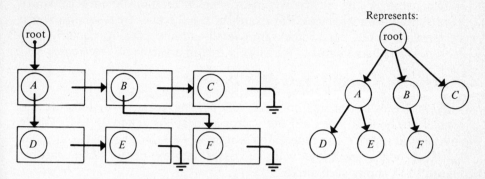

Fig. 10.16 Linked tree representaion: arbitrary degree

If there is a fixed maximum number of descendents of a node, we can put pointers to all sons in each node. For example, we may represent a binary tree with nodes defined

   type *binaryTreeNode* =
      record *data*: *T*; *LeftSon, RightSon*: ↑*binaryTreeNode* **end**

or

   type *naryTreeNode* =
      record *data*: *T*; *sons*: **array**[1..*n*] **of** ↑*naryTreeNode* **end**

Thus, for binary trees we get structures such as those shown in Fig. 10.17.

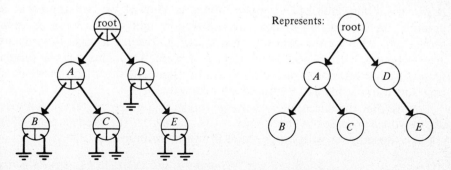

Fig. 10.17 Linked binary tree representation

10.4.3 VECTOR IMPLEMENTATION OF BINARY TREES

The linked representation of trees and graphs is so obvious that many people overlook the alternative array implementations—some of which have substantial advantages. For example, a full binary tree—one in which all terminal nodes are at the same level—admits of a particularly nice contiguous implementation. We store a tree in a vector of *TreeNodes*,

    type *tree* = **array** [1..*maxtree*] **of** *TreeNode*

where a *TreeNode* contains no links, just auxiliary data. We store the two sons of the node at index *i* in the vector at locations $2i$ and $2i + 1$. Given any node at index *i*, its predecessor is then at *i* **div** 2. The root is at index 1. Leaf nodes are those with successor indices greater than *maxtree*. Figure 10.18 shows such a tree.

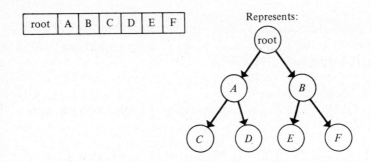

Fig. 10.18 A binary tree stored in a vector

10.4.4 ARRAY REPRESENTATION OF UNLABELED GRAPHS

If the maximum number of graph nodes is known, we can represent a graph by its "connection matrix." A connection matrix is simply a boolean array, say C, having the property that $C[i, j]$ is *true* just in case there is an arc in the graph from node *i* to node *j*. Figure 10.19 illustrates a simple graph and its connection matrix (we use the standard shorthand in which 1 indicates a *true* value and 0 indicates a *false* value.)

Note that the successors of the ith node of a graph, g_i, are just those nodes with *true* values in the ith row of C. Similarly, the predecessors of g_i are just those nodes with a *true* value in the ith column of C.

$$succ(g_i) = \{g_j \mid C[i, j] = true\}$$
$$pred(g_i) = \{g_j \mid C[j, i] = true\}$$

The coding of the operations on graphs using this representation is left as an exercise. Note that some operations are both faster and conceptually easier using this representation.

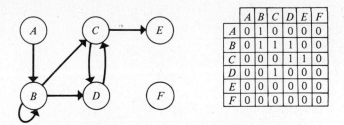

	A	B	C	D	E	F
A	0	1	0	0	0	0
B	0	1	1	1	0	0
C	0	0	0	1	1	0
D	0	0	1	0	0	0
E	0	0	0	0	0	0
F	0	0	0	0	0	0

Fig. 10.19 A graph and its connection matrix

10.5 Representation of Sets

All of the data structures presented previously have focused on how the relations between elements, the *structure*, will be encoded in the representation. Sets present a slightly different problem: there is no relation between elements, and the only relation between an individual element and the entire set is *membership*. On the other hand, any element in the set may be deleted at any time, so we must find a way to keep track of exactly which elements are currently in the set.

We present two representations of sets here and a third in Chapter 20. All three are reasonable representations; different circumstances will favor different ones.

10.5.1 BIT-VECTOR REPRESENTATION OF SETS

We frequently want to define sets whose members are drawn from a relatively small group of values, all of which are known in advance. The elements might, for example, be values from an enumerated type, called the *base type*. This case is common enough to be provided as a primitive data type in some languages (for example, PASCAL). We can represent such a set by imposing a linear order on the members and representing the set as a boolean vector, S. $S[i]$ indicates whether the ith element of the enumeration is, or is not, in the set.

Thus the values are encoded by the positions in the boolean vector to which they correspond. The boolean vector may then be packed into machine words, using one bit for each vector element. If we choose this representation, we must limit the size of a set, and hence the number of values in the enumerated type, to the number of bits in a word.

Assume now that we are constructing sets of integers in the range $[0..K - 1]$ where K is the number of bits in a word. The operations can be coded as in Figure 10.20.

```
type SetofElts = word;

function Maxsize;
    begin Maxsize := K end;

procedure Clear(var S: SetofElts);
    begin S := 0; end;

procedure Insert (var S: SetofElts; x: integer);
    begin S := S ∨ 2^x end;

procedure Remove (var S: SetofElts; x: integer);
    begin  S := S ∧ ~2^x end;

procedure Union(var S: SetofElts; T: SetofElts);
    begin S := S ∨ T end;

procedure Intersect (var S: SetofElts; T: SetofElts);
    begin S := S ∧ T end;

function IsMember(S: SetofElts; x: integer): boolean
    begin  IsMember := ((S ∧ 2^x) ≠ 0)  end;

function IsSubset(S, T: SetofElts): boolean;
    begin  IsSubset := (S = (S ∧ T))  end;

function IsEmpty(S: SetofElts): boolean;
    begin  IsEmpty := (S = 0)  end;
```

Fig. 10.20 Implementation of sets using bit vectors

These programs rely on several of the basic operations upon values of type *word*:

- $\wedge$, the *bit-wise and* operator. $S1 \wedge S2$ is a word whose i^{th} bit is a 1 iff the ith bits of both $S1$ and $S2$ are 1.

- V, the *bit-wise or* operator. $S1 \vee S2$ is a word, whose i^{th} bit is 1 iff either the ith bit of $S1$ or $S2$ (or both) is 1.
- 2^x, whose value is the word whose x^{th} bit is 1 (counting from 0), and whose other bits are 0.

To explain this representation of sets, we will discuss one operation, *IsMember*, in detail. Suppose that $K=8$ and the set S contains 0, 2, and 5. Then the binary value of S is

$$S = 00100101$$

We have chosen to assign the bits to integer values starting at the right. The reason is that we can obtain the representation of the set $\{j\}$ by constructing a binary number with a bit in the jth position. For example,

$$rep(\{4\}) = 2^4 = 00010000$$

Thus in function *IsMember*, 2^x has a 1 digit only in the xth position. Therefore, $S \wedge 2^x$ will be 0 unless the xth bit of S is a 1.

10.5.2 VECTOR-POINTER REPRESENTATION OF SETS

Suppose now that you want to operate on sets of values, but the number of possible values is very large even though the actual set sizes are reasonably small.

This would be the case, for example, if we wanted to represent a class of students as a set of Social Security numbers. The representation of Section 10.5.1 would require 10^9 bits. However, if no class has over 300 students, we can store the complete list of social security numbers in 300 9-digit chunks. Machine words often have about 30 bits or more, so a social security number can be comfortably stored in one word. Even with the wasted bits, this takes only about 9000 bits. This is about a factor of 10^5 smaller—certainly enough to care about.

We can also make a reasonably good estimate of the size of each set, and the maximum is close enough to the actual working size that the unused space is not a major consideration. Under these circumstances it may be useful to use a vector to store the set. We will use an additional integer, *CurSize*, to keep track of how many elements are actually in use, and we will require that these be stored, *without duplication*, in positions 1, 2, $\cdots$, *CurSize*. Since we never look at any elements whose indices are greater than the current *CurSize*, duplications at the other end of the vector don't matter. As before, consider the set $\{0, 2, 5\}$ and a maximum set size of 8. This will be represented as any of the following (and many others).

$$Vector = \langle 2, 5, 0, 9, 11, 23, 46 \rangle \qquad CurSize = 3$$
$$Vector = \langle 0, 2, 5, 8, 16, 9, 7, 8 \rangle \qquad CurSize = 3$$
$$Vector = \langle 0, 5, 2, 29, 47, 86, 4, 3 \rangle \qquad CurSize = 3$$
$$Vector = \langle 5, 0, 2, 0, 5, 2, 5, 2 \rangle \qquad CurSize = 3$$

In each of these and in Exercise 10.39, the first $CurSize$ (in this case, the first 3) elements is some permutation of $\langle 0, 5, 2 \rangle$, and the other values are irrelevant.

Note that this design permits many different representations for the same set. This is an entirely reasonable thing to do, but it means we must take care when we implement the operations. For example, if we want an equality test, we are not able to simply test to see if all the components of one set are equal to the corresponding components of the other.

The operations on sets, using the vector-pointer representation, may be coded as in Figure 10.21. The programs for *Insert* and *Union* illustrate an important point. Both increment $CurSize$, but neither tests to be sure $CurSize$ does not exceed *Maxsize*. This is safe in this case, but only because both operations have preconditions (see section 9.2.3) that prohibit their use if the result would be too large. As a result, programs that call *Insert* and *Union* must make sure that the preconditions of those routines are true. If the preconditions did not guarantee that the results will be small enough, it would be necessary for the implementation to test and ensure the safety of expanding the set.

10.5.3 LINKED REPRESENTATION OF SETS

By now it should be fairly obvious that any structure that can be represented with an array can also be represented with a list—and that there may be advantages to doing so under some circumstances. This is certainly the case for sets; a set may be represented as the linear list of its members. We leave this implementation to you as Exercise 10.46.

10.6 The Representation of Multidimensional Arrays

There are two general categories of representation for multidimensional arrays: contiguous, or vector, and linked implementations. The former are used for moderately sized and "dense" arrays. Here, "dense" means "having mostly nonzero elements." Linked representations are appropriate for large, sparse arrays—those having many zero elements. This breakdown reflects a trade-off between space considerations (linked

```
type SetofElts =
    record
        CurSize: integer;
        Vec: array [1..MaxSize] of integer
    end;

function Maxsize;
    begin maxsize := k; end;

procedure Clear(var S: SetofElts);
    begin S.CurSize := 0; end;

procedure Insert(var S: SetofElts; x: integer);
    var i: integer;
    begin
    for i := 1 to S.Cursize do
        if S.Vec[i] = x then goto Hit;
        S.CurSize := S.CurSize + 1;  S.Vec[S.CurSize] := x;
Hit:    end;

procedure Remove(var S: SetofElts; x: integer);
    var i: Integer;
    begin
    for i := 1 to S.CurSize do
        if S.Vec[i] = x then
            begin
            S.Vec[i] := S.Vec[S.CurSize];
            S.CurSize := S.CurSize - 1;
            goto Hit
        end;
Hit:    end;

procedure Union(var S: SetofElts; T: SetofElts);
    var i: integer;
    begin
    for i := 1 to T.CurSize do Insert(S, T.Vec[i])
    end;

procedure Intersect(var S: SetofElts; T: SetofElts);
    {Exercise 10.41 asks for the body of this procedure.}
```

Fig. 10.21 Implementation of sets using arrays (Part 1 of 2)

```
function IsMember(S: SetofElts; x: integer): boolean;
  var i: integer;
    begin
    Ismember := false;
    for i := 1 to S.CurSize do
        if S.Vec[S.CurSize] = x then
            begin  Ismember := true;  goto Done end;
Done:
    end;


function IsSubset(S, T: SetofElts): boolean;
    {Exercise 10.42 asks for the body of this function.}


function IsEmpty(S: SetofElts): boolean;
    begin  IsEmpty := (S.CurSize = 0)  end;
```

Fig. 10.21 Implementation of sets using arrays (Part 2 of 2)

representations take up less space) and speed considerations (linked representations may be more complicated and slower to access).

10.6.1 MORE ON VECTOR REPRESENTATION: DESCRIPTORS

In Section 10.1.3, we discussed the implementation of simple vectors using address arithmetic. There we considered only the case where the vector has a lower bound of 0 and each element occupied one word. Let's first generalize this simple one-dimensional case to vectors with arbitrary lower bounds and whose elements may occupy several contiguous locations.

Suppose that vector A is declared

var A: **array** $[lb..ub]$ **of** T

and suppose that each element of type T takes up k words of storage. We may then compute the address of $A[i]$ according to the formula.

$$\beta A[i] = \beta A + (i - lb) \cdot k$$

where, as in Section 10.1.3, βA is the address of the first element of A (that is, of $A[lb]$). A trivial rearrangement of terms gives us

$$\beta A[i] = (\beta A - lb \cdot k) + i \cdot k$$

The quantity $(\beta A - lb \cdot k)$ is a constant called the *virtual origin* of A. Intuitively, it is the address that $A[0]$ would have if it existed (see Fig. 10.22 in which we assume $k = 1$).

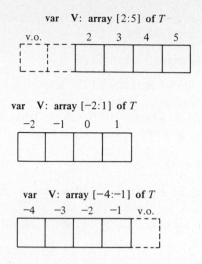

Fig. 10.22 The virtual origin of an array

For various reasons—principally because we will have to pass vectors to and from procedures—it is convenient to encapsulate the data needed to do address calculations on vectors in a record called a descriptor. For example,

> **type** *VectorDescriptor*=
>   **record**
>   *VirtualOrigin*: *integer*;   *mult*1: *integer*;
>   *LowerBound*1, *UpperBound*1: *integer*
>   **end**;

Here, *mult*1 contains k (the reason for the new name will become apparent later) and *LowerBound*1 and *UpperBound*1 contain *lb* and *ub*. Note that these last two are not necessary for the computation and could be eliminated. They can be useful, however, if we wish to do bounds checking, checking to see that array indices are legal.

If the *VectorDescriptor Adesc* contains the appropriate data for the vector A, above, we may write the accessing algorithm as

$$\beta A[i] = Adesc.VirtualOrigin + i \cdot Adesc.mult1$$

10.6.2 ADDRESS COMPUTATIONS FOR MULTIDIMENSIONAL ARRAYS

Suppose that B is an array declared

> **var** B: **array** [*lb*1..*ub*1, *lb*2..*ub*2] **of** T

We can store this array in a vector, using the algorithms of Section 10.6.1 by treating B as a *vector of vectors*:

array $[lb1..ub1]$ **of array** $[lb2..ub2]$ **of** T

The type T of section 10.6.1 is replaced by "**array** $[lb2..ub2]$ **of** T." If one T takes up k words of storage, a vector of $ub2 - lb2 + 1$ of them will take up $(ub2 - lb2 + 1) \cdot k$ words. Therefore, the address of the first element of the ith row of B, $B[i, lb2]$, is given by

$$\beta B[i, lb2] = \beta B + (i - lb1) \cdot [(ub2 - lb2 + 1) \cdot k]$$

where B is the address of $B[lb1, lb2]$. Now that we have the base address of row i of B, which is an "**array** $[lb2..ub2]$ **of** T," we can calculate the address of its element number j using the same algorithm.

$$
\begin{aligned}
\beta B[i, j] &= (\beta B + (i - lb1) \cdot (ub2 - lb2 + 1) \cdot k) + (j - lb2) \cdot k \\
&= [\beta B - k \cdot (lb1 \cdot (ub2 - lb2 + 1) + lb2)] \\
&\quad + i \cdot [(ub2 - lb2 + 1) \cdot k] \\
&\quad + j \cdot [k]
\end{aligned}
$$

The bracketed formulas in the last equation are all constants. As before, the first is called the virtual origin of B. Again, we may encapsulate these constants as a descriptor:

type *ArrayDescriptor* =
    record
    VirtualOrigin: *integer*; *mult1*, *mult2*: *integer*;
    LowerBound1, *UpperBound1*, *LowerBound2*, *UpperBound2*: *integer*
    end;

An *ArrayDescriptor Bdesc*, describing the array B, would contain the following information:

$Bdesc.VirtualOrigin = \beta B - k \cdot (lb1 \cdot (ub2 - lb2 + 1) + lb2),$

$Bdesc.mult1 = (ub2 - lb2 + 1) \cdot k,$

$Bdesc.mult2 = k,$

$Bdesc.LowerBound1 = lb1, \quad Bdesc.LowerBound2 = lb2,$

$Bdesc.UpperBound1 = ub1, \quad Bdesc.UpperBound2 = ub2$

and the address computation becomes

$$\beta B[i, j] = Bdesc.VirtualOrigin + i \cdot Bdesc.mult1 + j \cdot Bdesc.mult2$$

We arrived at these results by treating B as a vector of rows. That is, we stored the array B in *row major order*. Naturally, we could just as easily have stored B in *column major order* as a vector of columns.

var B: **array** [$lb2..ub2$] **of array** [$lb1..ub1$] **of** T

so that $B[i, j]$ is stored in element $B[j][i]$.

10.6.3 A SLIGHT VARIATION: TRIANGULAR ARRAYS

The arrays considered so far have all been *rectangular*, with each row taking the same amount of space. We sometimes have use for *n*-by-*n* arrays in which we use only, say, the *upper triangle*—that is, only elements [i, j], such that $j \geq i$. If Q is such an array, we could, of course, declare it as

var Q: **array** [$1..n, 1..n$] **of** T

but this would waste nearly half of its elements. Instead, we *pack* the portions of the rows we use into a single vector so that, for example, $Q[2, 2]$ is stored immediately after $Q[1, n]$, giving us the representation illustrated in Figure 10.23. That is, we let

type *triangarray* = **array** [$1..n2$] **of** T;

Where $n2$ is large enough to accommodate all the elements in the upper triangle.

$$n2 = (n^2 + n)/2$$

The address calculation for Q of type *triangarray* is now

$$\beta Q[i, j] = (\beta Q - n - 1) + i \cdot (2 \cdot n + 1 - i)/2 + j$$

Note that this algorithm requires more computation than the algorithm for a rectangular array, but it saves space. We often find such time/space trade-offs in representation decisions (recall, for example, the discussions of vector and list representations of deques, queues, sets, and so forth).

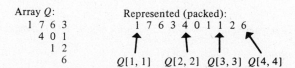

Fig. 10.23 Packing an upper triangular array

10.6.4 REPRESENTATION OF SLICES

One of the great advantages to using descriptors to contain addressing data is that it is easy to perform *slicing* of arrays (see section 8.3.1). Suppose, for example that we have a function, $SUM(v)$, that will return the sum of the elements of its vector argument, v. Let B be an array declared

   var B: **array** [$lb1..ub1$, $lb2..ub2$] **of** *integer*

It would be convenient to be able to use the SUM function to compute the sum of the elements of a single row of B. Some languages permit us to specify a *slice* of an array—an array of smaller dimensionality whose elements are taken from the original array in a specified fashion. For example, we might write

   $SUM(B[i, *])$

where the notation $B[i, *]$ means the ith row of B, to obtain the desired summation.

The representational problem is, "How we can arrange matters so that SUM can be used both for this cross section and for ordinary vectors?" If arguments are passed to SUM as descriptors, nothing could be easier. Suppose that $Bdesc$ is an *ArrayDescriptor* of B as described in Section 10.6.2. Then we can compute a new *VectorDescriptor*, $Browdesc$, to be passed to SUM as follows.

   $Browdesc.VirtualOrigin := Bdesc.VirtualOrigin + i \cdot Bdesc.mult1;$

   $Browdesc.mult1 := Bdesc.mult2;$

   $Browdesc.LowerBound1 := Bdesc.LowerBound2;$

   $Browdesc.UpperBound1 := Bdesc.UpperBound2$

The descriptor $Browdesc$ describes a vector dimensioned

   array [$lb2..ub2$] **of** *integer*

Similarly, if we want the sum of the column j of B,

   $SUM(B[*, j])$

we pass the *VectorDescriptor* $Bcoldesc$ to SUM, where $Bcoldesc$ is set up as follows.

   $Bcoldesc.VirtualOrigin := Bdesc.VirtualOrigin + j \cdot Bdesc.mult2;$

   $Bcoldesc.mult1 := Bdesc.mult1;$

   $Bcoldesc.LowerBound1 := Bdesc.LowerBound1;$

   $Bcoldesc.UpperBound1 := Bdesc.UpperBound1$

The descriptor *Bcoldesc* describes a vector dimensioned

    array [*lb*1..*ub*1] **of** *integer*

If you work out the address arithmetic, you will find that, indeed, the address of the jth element of the row described by *Browdesc* is $\beta B[i,j]$, as is the address of the jth element of the vector described by *Bcoldesc.*

10.6.5 ILIFFE VECTORS

Occasionally, we have a need to represent arrays whose rows are not all of the same length. One way to do so is a combined linked and contiguous representation known as an *Iliffe* vector (named after their inventor, John Iliffe). An Iliffe vector is a vector of descriptors. If each descriptor is a *VectorDescriptor*, describing a row of an array, the whole structure represents an irregular two-dimensional array. A typical Iliffe vector (for an array *B*) is shown in Fig. 10.24. The descriptor *Bdesc* of *B* describes a vector of descriptors stored contiguously, each of which describes and, in particular, points to a vector of integers, also stored contiguously. Accessing element $B[i, j]$ consists of using *Bdesc* to access the descriptor of $B[i, *]$, and then using that to access $B[i, j]$. Obviously, we can extend this idea by making the elements of $B[i, *]$ be descriptors themselves, giving rise to higher-dimensional arrays.

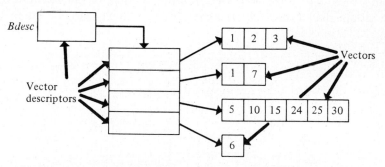

Fig. 10.24 Typical Iliffe vector for a two-dimensional array

10.6.6 LINKED REPRESENTATION OF SPARSE ARRAYS

When most of the elements of a large multidimensional array are known to contain some specific value (usually 0), it is wasteful to reserve space for all the elements containing that value. Such arrays, known as "sparse matrices," are common in a number of very important mathematical problems.

Using a linked representation, we can get away with storing only nonzero elements. Consider the representation of an *m* by *n* array of *reals.* Each element of the array is stored in an *arrayelement.*

type *arrayelement* =
    record
        val: *real*; *row*, *col*: *integer*;
        nextright, nextleft, nextup, nextdown: ↑*arrayelement*
    end;

The fields *row* and *col* contain the row and column indicies. The "next" fields point to the next nonzero element to the left, right, etc., of the element.

An entire *m* by *n* array is represented as a vector of pointers to the rows and a vector of pointers to the columns.

type *SparseArray* =
    record
        rows: **array** [1..*m*] **of** ↑*arrayelement*;
        columns: **array** [1..*n*] **of** ↑*arrayelement*
    end;

To access the [*i, j*] element of a *SparseArray*, *S*, we pick up the ith row or the jth column header and search across or down the row or column for the element with *row* = *i* and *col* = *j*. If there is no such element, the value of *S*[*i, j*] is 0. Otherwise *S*[*i, j*] is the *val* field of the element found. Figure 10.25 illustrates the representation, which can be seen as a specialized form of labeled graph.

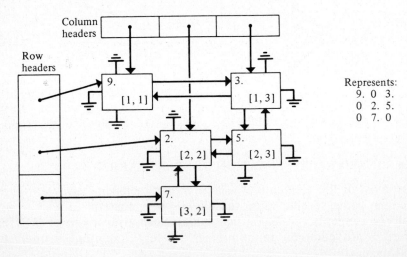

Fig. 10.25 A sparse array

10.7 Further Reading

The primary additional references for both Chapters 9 and 10 are Knuth, v. 1 (1973) and Horowitz and Sahni (1976).

Exercises

10.1 FORTRAN provides only integers, reals, and arrays and does not support enumerated types. If you were using FORTRAN, how would you achieve the effects of the types mentioned in section 8.1.1? In general, what advantages does the facility to define enumerated types provide?

10.2 Three different internal binary representations are commonly used for integers on modern computers. They are

- *Sign-magnitude*: One bit is designated as the *sign* bit and the remaining bits represent the magnitude of the number, with individual bits corresponding to the coefficients of powers of 2. (The representation of the magnitude is independent of the sign.)
- *One's-complement*: Positive numbers are represented in the same way as magnitudes in sign-magnitude representation, but the negation of a positive number is formed by complementing every bit.
- *Two's-complement*: Positive numbers are represented as above, but the negation of a positive number is formed by complementing every bit and adding a 1 in the unit's position.

Compare these representations in the following ways:

a) Assume integers are to be represented in four bits (including sign). List all 16 bit patterns and indicate which values they represent in each of the three representations.

b) Treat the internal representation of an integer as a vector of bits, R, with $R[0]$ corresponding to the unit's position. Write a formal definition of each representation in terms of sums of powers of two.

c) Devise addition algorithms for all three representations.

10.3 The familiar decimal system is a convenient shorthand for denoting numbers. We start with the ten numerals (0 through 9) denoting the numbers 0, 1, 1+1, 1+1+1, etc. We get larger numbers using successive multiplication by ten. Thus, "195" is short for

$$1 \cdot 10^2 + 9 \cdot 10^1 + 5$$

Because the meaning of each digit of 195 depends on its position (9 denotes $9 \cdot 10$ because it appears second from the right), the decimal system is called a *positional number system*. Ten is called the *base* or *radix* of the decimal system. The obvious question is, of course, "Why ten?" Indeed, some civilizations in the past have used other bases. (The Babylonians, for example, used radix 60 for some calculations; the Aztecs used radix 5 and radix 20; and the Coahuiltecans used radix 3. The Encounter Bay Australians used radix 2, but they didn't count very high). When dealing with computers, in particular, it is often useful to deal in bases which are powers of two, notably two (the *binary* system), eight (the *octal* system), and sixteen (the *hexadecimal* system).

a) Write routines to perform input and output for arbitrary number bases up to 64, using ordinary integers as an internal representation.

b) Note that if there is an *i* such that $a = b^i$, then numbers can be converted from base *a* to base *b* by converting each digit of the base-*a* representation directly to *i* digits of the base-*b* representation. Conversely, numbers in base *b* can be converted directly to base *a* by converting groups of *i* digits directly to the corresponding base-*a* digit. Use this fact to write a conversion routine that converts numbers directly from base *a* to base *b* when $a = b^i$ (Do not use the machine integers as an internal representation.)

10.4 Let *X* and *Y* be variables. Ordinarily, to exchange their contents, one uses a third variable, as a temporary location, as in

$$T \leftarrow Y;$$

$$Y \leftarrow X;$$

$$X \leftarrow T;$$

The original value of *Y* is saved in *T* while *Y* acquires the value of *X*. This exercise concerns an alternative method that uses *no other storage location* than the two variables to be exchanged. Instead, it relies on an *encoding* technique to eliminate the temporary variable. We will develop the technique in a series of exercises:

a) Write the truth table for $\otimes$, exclusive or:

$$p \otimes q \text{ iff } (p \vee q) \wedge \sim (p \wedge q)$$

Add to your truth table a column for $(p \otimes q) \otimes q$.

b) The idea for the algorithm consists of 3 exclusive or's, the results of which may only be stored in the original 2 variables. Let those variables be X and Y. Fig. 10.26 shows the idea for the algorithm, with boxes you must fill in. The boxes on the left represent the remaining details of the exchange algorithm. Fill them in, deciding which exclusive "or" of X and Y should be assigned to X and which to Y. The boxes on the right are to hold logical expressions (for example, $p \otimes q$) characterizing the contents of X and Y after each of the 3 steps. Use them (as a symbolic form of test data) to demonstrate that the algorithm you completed does, in general, work.

c) In order to use this space-saving algorithm, what additional representational assumptions about X and Y have to be made?

d) Devise a similar scheme that uses integer operators $+$ and $-$, in place of exclusive "or."

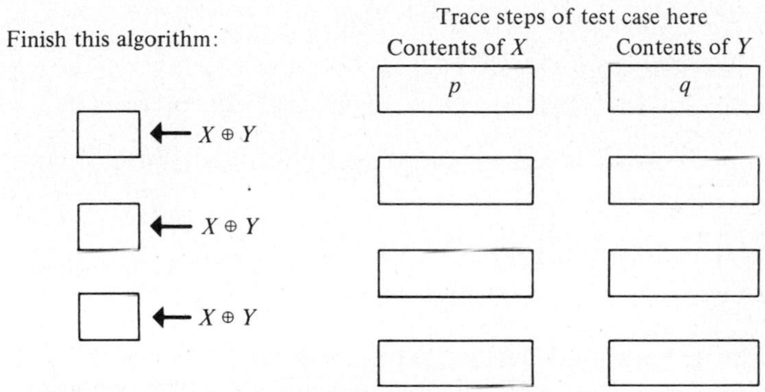

Finish this algorithm:

Trace steps of test case here

Contents of X Contents of Y

$\boxed{} \leftarrow X \oplus Y$ $\boxed{p}$ $\boxed{q}$

$\boxed{} \leftarrow X \oplus Y$ $\boxed{}$ $\boxed{}$

$\boxed{} \leftarrow X \oplus Y$ $\boxed{}$ $\boxed{}$

$\boxed{}$ $\boxed{}$

Fig. 10.26 Fill in the boxes

10.5 Suppose you have a computer capable of representing integers in the range $-2^{31} < x < 2^{31}$, a large number of 6-bit values to store, and a compiler that offers no facility for packing values. How would you use integer operations to substantially improve your storage utilization above the one-value-per-word representation your compiler provides?

10.6 It is occasionally necessary to compute with exact integer values when the values are substantially much larger than can be stored in one word. You can support such computations by using two or more normal integers to represent each large value. We can call the resulting type *biginteger*.

a) Write specifications for the type *biginteger*. Abstractly, *bigintegers* are arbitrarily large integers. What do you do about the fact that PASCAL does not allow us to define operators such as + or *?

b) Implement these specifications. Write routines to perform input and output on *bigintegers*.

10.7 Assume that linear lists are implemented with pointer-swing techniques. Suppose that T is *integer* and P and Q of type ↑*listofT* point to the lists

P: [2] → [4] → [6] → ϵ

Q: [3] → ϵ

Show the representations of P and Q before and after each of the following series of operations. Start over with P and Q as above for each part.

a) $P := EmptyList$;

b) $SetFirstItem(P, 5)$; $Q := Tail(P)$;

c) $P := ConsList(P, 5)$; $Q := Tail(P)$;

d) $SetFirstItem(P, 5)$; $SetTail(Q, P)$;

e) $P := ConsList(P, 5)$; $Q := ConsList(FirstItem(P), Q)$;

10.8 Suppose that T is *integer* and P and Q of type ↑*plistofT* point to the lists

P: [2] ↔ [4] ↔ [6] → ϵ

Q: [3] → ϵ

(Both P and Q have predecessors, which may be ϵ.) Show the values of P and Q before and after each of the following series of operations. Start over with P and Q as above for each part.

a) P: $Singleton(5)$;

b) $Q := Tail(P)$; $P := Tail(Q)$;

c) $Q := Tail(P)$; $P := Pred(Q)$;

d) $AddPred(Tail(P), 7)$;

e) $AddTail(Q, 5)$;

f) $AddTail(Tail(Q), 7)$;

g) $RemoveFirst(Tail(P))$;

10.9 Suppose that x contains a pointer to a *listofT* in some list L. Also suppose that we perform only inserts and deletes on the elements of L (never changing the data fields directly), and that we never delete the node whose data is originally contained in *x*. What can you say about the data at node *x* if we are using a pointer swing implementation? Can we say the same for node-overwrite implementation? Why? Give a sample of a program that would work using pointer swing, but not node overwrite.

10.10 Consider a doubly linked circular list in which we guarantee that there is always at least one node. How can we simplify the operations of insertion and deletion on such a list? Write new implementations of the *plistofT* operations that use a circular list representation.

10.11 Suppose that we have defined the type *listofInt* as in the text.

> **type** *ListofInt* = **record**
> > *First*: *integer*;   *Tail*: ↑*ListofInt*
> **end**;

We wish to implement the procedure *Copy*, which is specified

> **procedure** *Copy*(*OldList*: ↑*listofInt*;   **var** *NewList*: ↑*listofInt*);
> > **post** The sequence of values in *NewList* is the same as that in
> > *OldList*, and the original *OldList* is unchanged and
> > does not overlap *NewList*.

That is, *Copy* is supposed to create a new list which contains exactly the same values (the same integers in the *First* fields) as the original. For example, if *x* points to the list

> [4] → [7] → [0] → ε

Then after the call *Copy*(*x*, *y*) (where *y* is also declared to be a ↑*listofInt*,) we get

> *x* points to [4] → [7] → [0] → ε

and

> *y* points to [4] → [7] → [0] → ε

and the lists do not overlap.

> a) Write the body of *Copy* so that it fulfills these specifications. Note that it makes perfect sense to copy an empty list.
> b) Someone gives you a version of *Copy* that doesn't work. It does not produce any run-time error messages—just wrong answers. You trace the call *Copy*(*x*, *y*) by hand, where *x* and *y* are as above. You find that this *Copy* procedure executes the following operations (*P* and *Q* are local variables of type ↑*listofInt*).

new(P); $Q := x$; $y := P$;

$P\uparrow := Q\uparrow$;

new(P); $Q := Q\uparrow.tail$;

$P\uparrow := Q\uparrow$;

. . .

Draw a diagram of the situation that results from executing these seven instructions. What is wrong?

10.12 If L points to a list, then the command

Replace(L, *Tail*(L))

will remove the first item from L and also remove that item from any list containing L. Why can't we use the singly linked, pointer swing operation to do this? For example, what's wrong with this?

SetFirstItem(L, *FirstItem*(*Tail*(L))); *SetTail*(L, *Tail*(*Tail*(L)))

10.13 In the representation of stacks at the beginning of this chapter, sp is the index of the current top of the stack. Sometimes it's better for sp to point at the next free element. Rewrite the implementation the latter way.

10.14 Assume that stacks are represented with vectors and pointers as at the beginning of this chapter. Suppose stack S has the initial value

$S = \;<4, 7, 5>$

Show the representation of S before and after each of the following series of operations. Start over with $S = \;<4, 7, 5>$ for each part.

    a) *Push*(S, 3); *Push*(S, 6); $x := $ *Pop*(S); $x := $ *Pop*(S);
       $x := $ *Pop*(S);
    b) $x := $ *Pop*(S); *Clear*(S); $x := $ *Top*(S);
    c) $x := $ *Pop*(S); $x := $ *Top*(S); $x := $ *Pop*(S);
    d) *Push*(S, 3); $x := $ *Pop*(S);
    e) $x := $ *Pop*(S); $x := $ *Pop*(S); *Push*(S, 3); $x := $ *Pop*(S);
       $x := $ *Pop*(S);

10.15 Redo Exercise 10.14 with the linked-list implementation of Section 10.2. Assume a pointer-swing implementation of lists.

10.16 Use the abstract specification of a stack given at the beginning of this chapter, but only implement *Push*, *Pop*, and *Top*. Assume that you want a stack of boolean values, and that the stack will never get very deep (say 10–15 elements).

a) Using only the operations add 1, multiply by 2, integer divide by 2, and a test for oddness, show by implementing it that this stack can be represented by a single integer.

b) Explain the correspondence between the implementation and the specification.

c) How large, or deep, may the stack be permitted to grow?

10.17 Redo exercise 10.16, but assume the values to be stacked are integers in the range 0–63.

10.18 Two implementations of type *stack* were described in the text: one used an array, the other a linked list. Yet another is suggested in Exercise 10.16. What questions would you need to ask in order to determine which is the better implementation for a given situation? What sort of answers would lead you to pick one implementation over another?

10.19 Can you think of a way to put two stacks in the same vector so that they share space efficiently (that is, so that if one grows small, the other can use the resulting extra space)? Describe your method. Can you do the same with deques? Why?

10.20 Assume that deques are represented in vectors as in Section 10.3.1. Suppose deque D has the initial value

$$D = <4, 7, 5>$$

Show the representation of D before and after each of the following series of operations. Start over with D = $<4, 7, 5>$ for each operation.

a) *InsertFront*(D, 3); *InsertBack*(D, 6); $x :=$ *RemoveFront*(D); $x :=$ *RemoveFront*(D); $x :=$ *RemoveFront*(D);

b) $x :=$ *RemoveBack*(D); *Clear*(D); *InsertFront*(D, 3);

c) $x :=$ *RemoveFront*(D); $x :=$ *RemoveBack*(D); $x :=$ *RemoveFront*(D);

d) *InsertFront*(D, 3); $x :=$ *RemoveBack*(D);

e) $x :=$ *RemoveFront*(D); $x :=$ *RemoveFront*(D); *InsertBack*(D, 3); $x :=$ *RemoveFront*(D); $x :=$ *RemoveFront*(D);

10.21 Redo Exercise 10.20 with the linked-list representation of Section 10.3.2. Assume a node-overwrite implementation of lists.

10.22 Our vector implementation of deques in Section 10.3.1 wasted one element of the vector. A vector of size *maxdeq* contained only *maxdeq* − 1 elements). Why? If you can do better, do so.

10.23 You could implement deques and queues as shown in the first part of Figure 10.8 if you had vectors whose bounds could change one step at a time as long as the total number of elements remained less than some bound (maxdeq). Using the ideas behind circular buffering, tell how you would implement such a "movable vector." Be sure to provide the accessing algorithm.

10.24 We did not implement the *Full* operation for deques using the circular buffer representation. Do so now.

10.25 A doubly-linked list can be represented with a single link field if that field is used to store the *sum* of the left and right links. In order to recover one of these values, it is necessary to store the other. First, discuss the ways this process violates the type rules of normal languages. Second, ignore the type problem and write the bodies of *AddPred* and *AddTail*.

10.26 Suppose you had to represent a directed graph structure with n nodes. The number n is fixed. There will be about $2n$ edges out of a possible n^2. Consider how to save space in representing the graph. Assume that the word size is 16 bits and that a word can hold one address.

    a) If n is "small" (but saving space is still important because, for example, there are hundreds of such graphs), what representation would you use?

    b) If n is "large," what representation would you use?

    c) Make precise *your* particular use of "small" and "large": Provide a mathematical condition that determines whether the representation you have chosen for "small" and "large" values of the constant n should be used.

10.27 Fill in the implementations of the basic operations on graphs given in Section 9.2.1 (but assume unlabeled graphs). Use the representation of Section 10.4.1.

10.28 Devise a new representation for unlabeled graphs in which the type *node* is simply a pointer to the incidental data (of type T) and all of the graph structure—the arcs—is in the representation of the type *graph*. Implement the operations on type *graph* using your representation.

10.29 Extend the linked representation of graphs to handle labeled graphs.

10.30 Fill in the implementations of the basic operations on trees given in Section 9.2.2. Use the representation of Section 10.4.2.

10.31 Consider an inplementation of graphs that will be used to store information about cities. Each node in the graph will represent a city; arcs, which will be labeled, will represent certain relations between these cities.

a) Consider first a graph in which each node contains descriptive information about a city—its name, population, latitude, longitude, etc. The arcs between nodes have one of two labels—flight time and driving time between the respective cities. Of course, pairs of cities for which there is no direct airline connection will not have an arc of the first type.   Similarly, pairs of cities without direct road connections will have no arcs of the second type. Devise an implementation for this case.

b) Consider another graph with the same node information, but with a much larger set of possible arcs. In addition to the two arc types in (a) there might be passenger rail travel time, freight rail travel time, freight barge travel time, "plays in baseball," "is upstream of," and so on. In particular, assume that the number of arc labels is not *a priori* known—new labels can be invented at whim. Devise an implementation for this case.

c) For *both* (a) and (b) implementations, write (and run) a program that determines the minimal travel time between an arbitrary pair of cities. (Of course, your program will have to know the possible modes of transportation; in the first case, only automobile and airplane travel are possible; but in the second case you must also consider the other possibilities.)

10.32 Fill in the implementations of the basic operations on trees given in Section 9.2.2. Use the representation of Section 10.4.3.

10.33 Generalize the vector representation of trees given in Section 10.4.3 to *n*-ary trees (for fixed *n*). Implement the basic operations for this representation.

10.34 Fill in the implementations of the basic operations on graphs given in Section 9.2.1 (but assume unlabeled graphs). Use the representation of Section 10.4.4.

10.35 Extend the array representation of graphs to handle labeled graphs. Assume that the set of labels is small and known in advance.   Each element of the connection matrix may have to store more than one label. Why? What representation of sets of labels is indicated?

10.36 Assume that trees in Exercise 9.33 are represented as the linked structures described in Section 10.4.2; use the implementation you designed for Exercise 10.30. Suppose trees *H* and *J* have the initial values shown in Fig. 10.27. Show the representations of *H* and *J* before and after each of the following series of operations. Start over with *H* and *J* for each part.

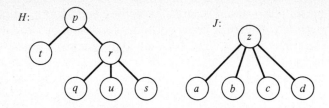

Fig. 10.27 Initial values for *H* and *J*

a) *SetSubTree*(*H*, *J*, 1);

b) *SetSubTree*(*J*, *SubTree*(*H*, 2), 5);

c) *SetSubTree*(*SubTree*(*H*, 1), *Root*(*J*), 1);

d) *SetSubTree*(*SubTree*(*SubTree*(*H*, 2), 3), *NewRoot*(*Root*(*J*)), 1);

10.37 To a genealogist, a "family tree" is a chart showing a group of people and their relationships. There are two kinds of links: marriage links and offspring links.

a) Design a data structure that can be used to represent a simple family tree. By "simple," we mean that each person marries only once, and that no person in the tree marries incestuously. (Hint: represent each person with a record; each record should have a *spouse* field, a *parent* field, etc.)

b) Modern families are not so simple. Family trees are often family DAG's. People divorce and remarry. Revise your data structure to be able to handle multiple marriages. Suppose that a person marries his first cousin. Make sure your data structure will be able to handle this.

c) Write a PASCAL program that will build this data structure. Its input should be a series of lines identifying events: "John marries Susan"; "Bill is born to John and Susan." Feel free to devise your own syntax to make the input processing as easy as possible.

d) Write a relationship-finding program. This program will tell you the relationship, in English, between any two people in the graph. For example, if you input the names "John" and "Mary," it might print out "John is Mary's father" or "John is

Mary's 3rd cousin 2 times removed." (Hint: to find the relationship between nodes A and B, find the nearest common ancestor, say C. Let X be the number of generations between A and C, and Y be the number of generations between B and C. If X is equal to Y, then A and B are $(X-1)^{th}$ cousins. If X is not equal to Y, then without loss of generality assume X less than Y; A and B are then $(X-1)^{th}$ cousins $Y-X$ times removed. Some relationships have special names: a zeroeth cousin is a brother or sister, a first cousin once removed is an aunt, uncle, nephew, or niece.)

10.38 Is it necessarily true that a reasonable *representation* of sets will never contain duplicate values among the elements that may be inspected by the operators? If so, say why. If not, describe a counter-example.

10.39 Write four more legal values for the set $\{0, 2, 5\}$ in the representation of Section 10.5.2. Write as many representations of the same set as you can using the representation of Section 10.5.1.

10.40 In Section 10.5.2, when procedure *Remove* finds the element it's trying to delete, it copies the $CurSize^{th}$ element into that position. Is this correct if the element to be removed is the one in the $CurSize^{th}$ position? Why?

10.41 The implementation of the *Intersect* operation for Section 10.5.2 is straightforward. Use *Insert* and *Ismember* to write this body.

10.42 The implementation of *Issubset* for Section 10.5.2 is also straightforward. Write it.

10.43 Rewrite the bodies of *Insert* and *Union* (Section 10.5.2) to eliminate the need for the precondition, adding tests to be sure the set does not overflow. Do not prohibit anything that's now legal. Is it necessary to rewrite these operations for the bit-vector representation of Section 10.5.1? Why?

10.44 Assume that sets in Exercise 9.36 are represented as in Section 10.5.1. Suppose sets A and B have initial values

$$A = \{1, 2, 3, 4\}$$
$$B = \{1, 3, 5, 7\}$$

Show the representations of A and B before and after each of the following series of operations. Start over with A and B as above for each part.

    a) *Insert*$(A, 5)$; *Insert*$(B, 5)$;

b) *Remove*(*A*, 5); *Remove*(*B*, 5);

c) *Union*(*A*, *B*); *Union*(*B*, *A*);

d) *Intersect*(*A*, *B*); *Intersect*(*B*, *A*);

10.45 Redo Exercise 10.44 with the vector-pointer representation of Section 10.5.2.

10.46 Write a linked implementation of sets.

10.47 For each of the following cases, assume that the first word of storage allocated is decimal location 4000. Give descriptors and numbers of words of storage required for each of the following arrays. Assume that *R*'s require one word each and *S*'s require two words each.

a) **var** *A*: **array** [1..3] **of** *R*

b) **var** *B*: **array** [0..2] **of** *R*

c) **var** *C*: **array** [−5..5] **of** *R*

d) **var** *D*: **array** [1..3] **of** *S*

e) **var** *E*: **array** [0..2] **of array** [−5..5] **of** *S*

f) **var** *F*: **array** [0..2] **of array** [−5..5] **of** *R*

10.48 For each of the following cases, assume that the first word of storage allocated is decimal location 4000. Give the number of words, the virtual origin, and the location of M[2, 3] in each declared array. Assume that *R*'s require one word each and that *S*'s require two words each.

a) **var** *M*: **array** [1..3, 1..3] **of** *R*

b) **var** *M*: **array** [0..3, 0..4] **of** *R*

c) **var** *M*: **array** [−1..2, 2..4] **of** *R*

d) **var** *M*: **array** [1..3, 1..3] **of** *S*

e) **var** *M*: **array** [0..3, 0..4] **of** *S*

f) **var** *M*: **array** [−1..2, 2..4] **of** *S*

10.49 Generalize address computations for array accessing to three dimensions.

10.50 Suppose *V* is declared

var *V*: **array** [1..5, 1..5] **of** *integer*

and $V[2, 3] = 10$, $V[3, 4] = 12$, $V[4, 5] = 14$, $V[5, 1] = 16$, and $V[i, j] = 0$ for all other i, j. Draw descriptor, Iliffe-vector, and sparse representations of *V*.

10.51 One trivial operation on vectors is "bounds revision." If *A* is dimensioned

> **var** A: **array** $[lb1..ub1]$**of** T

then

> $A[*@n]$

represents A, but with a lower bound of n instead of $lb1$ and an upper bound of $n + ub1 - lb1$ instead of $ub1$. If *Adesc* is the descriptor of A, how do you compute *Amovedesc*, the descriptor of $A[*@n]$?

10.52 Section 8.3.1 described the general operation of slicing, taking a section of a vector. We have already discussed the special case of cross-sectioning. Suppose that *Adesc* is a *vectordescriptor* for a vector A. How do you find the descriptor *Asubdesc* for the *subvector* $A[j..k]$ of A?

10.53 Give a contiguous representation and an accessing algorithm for *n*-dimensional rectilinear arrays, where n is any fixed integer.

10.54 Give a contiguous packed representation and an accessing algorithm for lower-triangular square arrays (that is, where we access only elements $[i, j]$ for which $i \geq j$).

10.55 What are the components of a three-dimensional array's descriptor? How would you build the descriptor of a *two-dimensional* cross section of an array, such as $A[*,*,i]$?

10.56 Suppose you are writing a program to operate on a number of $N \cdot N$ matrices with three special properties:

- N is fixed.
- The matrices are symmetric ($M[j, k] = M[k, j]$ for $1 \leq j, k \leq N$).
- You know in advance that q fixed elements will always be zero, for $0 \leq q \leq N^2$.

Give four possible representations, including at least one from each of the styles suggested by the phrases

- sequential allocation
- Iliffe vectors
- sparse

10.57 If *xdesc* is the descriptor of an array slice (subvector or cross section,) how must the cross-sectioning procedure be changed to take slices of the array?

10.58 Fill in the body of the procedure *getele*

>   **function** *getele*(*S*: *sparsearray*; *i, j*: *integer*): $\uparrow$*arrayelement*;
>   **begin** $\cdots$ **end**;

so that *getele* returns the [*i, j*] element of *S* or *nil* if this element is not present (that is, is 0).

10.59 Fill in the body of the procedure *putele*

>   **procedure** *putele*(*var S*: *sparsearray*; *i, j*: *integer*; *x*: *real*);
>   **begin** $\cdots$ **end**;

so that it places the value *x* in position [*i, j*] of *S*.   (Remember that element [*i, j*] may not be present!   Also remember that if *x* is 0, the element [*i, j*] is to be removed.)

10.60 Certain mathematical problems involve arrays in which the only nonzero elements are either on the main diagonal or immediately adjacent to the main diagonal.   These are called tridiagonal arrays; they are characterized by the predicate

$$(abs(i - j) > 1) \supset (T[i, j] = 0)$$

An example of a 5×5 tridiagonal array is

```
1 3 0 0 0
5 1 2 0 0
0 4 2 5 0
0 0 9 3 2
0 0 0 3 1
```

Clearly, it is a waste of space to store all the zeros in the positions away from the three central diagonals.   You could instead use a $3 \times n$ array, storing each diagonal in one row.   For example, the 5×5 array above becomes

```
3 2 5 2 0
1 1 2 3 1
0 5 4 9 3
```

Write a type declaration for this representation of tridiagonal arrays (of size *n*, where *n* is a given constant), and a function, *GetVal*, which takes a tridiagonal array, *A*, and two indices, *i* and *j*, and returns *A*[*i, j*]. Note that if [*i, j*] represents a point that is not on one of the diagonals, *GetVal* should return 0.

10.61 The goal of this exercise is to develop a system for maintaining a "data base" of information about locks, keys, and the people who have the keys. For purposes of this problem, we define

- A *key* is a five-tuple, $<k_1, k_2, k_3, k_4, k_5>$, of integers. Each integer, k_i, lies in the range $1 \leq k_i \leq 10$.

- A *lock* is a five-tuple, $<l_1, l_2, l_3, l_4, l_5>$ of nonempty sets of integers (which also lie in the range 1–10).

- A key, k, *opens* a lock, l, iff k_i is in l_i, $1 \leq i \leq 5$

- A key, X, can be *modified* into a key Y iff $X_i \leq Y_i$ $(1 \leq i \leq 5)$.

In addition, every key and every lock is identified by a unique identification number. Figure 10.28 illustrates the workings of a simplified ordinary lock. You should study this figure and be sure that you understand the correspondence between these definitions and the physical implementation of a lock.

Read all the following before beginning.

a) Specify data types for locks and keys that, at least would allow you to construct new locks and keys, to test if a given key opens a given lock, to test if two locks or two keys are the same, and to test whether a given key can be modified (that is, if the serrations can be filed down) to open a given lock. Note that because pins can have multiple breaks, a given key may open several different locks. Assume that all locks have the same number, of pins *npins* (In Fig. 10.28, *npins* = 3.)

b) Specify data types that allow you to store information about sets of keys and and locks. It must at least be possible to insert keys and locks together with unique identification numbers, to find the key or lock having a given identification number, to find all keys or locks, and to find all keys fitting a given lock or all locks opened by a given key. Write a function in a useful form that solves all these key and lock problems.

c) Specify data types that associate users (each with a unique identification number) with the keys they own and that associate locks with their room numbers. Again, it must at least be possible to retrieve all user numbers who have a given key, all keys owned by a given user, the lock of a given room and the room with a given lock, as well as all user numbers and all rooms.

d) Design a program that accepts commands telling it to

- Create keys and locks;
- Assign keys to users and locks to rooms;

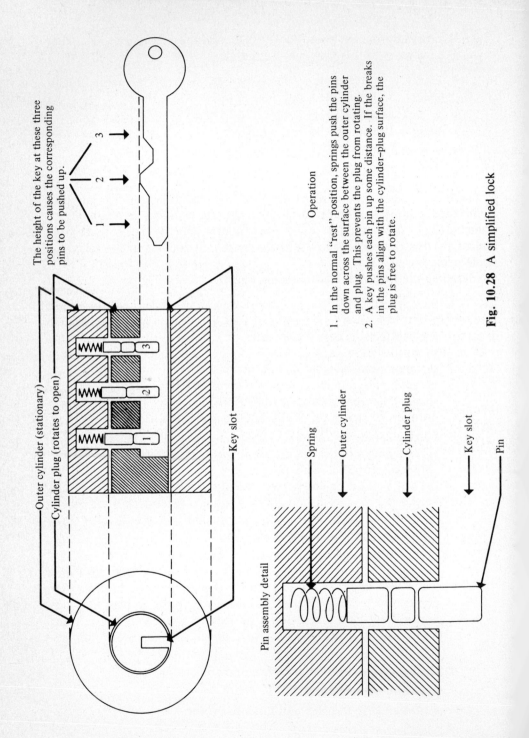

The height of the key at these three positions causes the corresponding pins to be pushed up.

Outer cylinder (stationary)

Cylinder plug (rotates to open)

Key slot

Pin assembly detail

Spring

Outer cylinder

Cylinder plug

Key slot

Pin

Operation

1. In the normal "rest" position, springs push the pins down across the surface between the outer cylinder and plug. This prevents the plug from rotating.

2. A key pushes each pin up some distance. If the breaks in the pins align with the cylinder–plug surface, the plug is free to rotate.

Fig. 10.28 A simplified lock

- Discover who can get into what rooms;
- Change key and lock assignments;

and any other operations that seem reasonable. Make this program as general and useful as you can. (This is a difficult specification problem; think about it carefully.) Implement this program using the data types defined in previous parts of this exercise. What new operations, beyond those suggested, do you find it useful to add?

e) Implement the data types and make the whole program run.

10.62 Most compilers provide an option by which it is possible to obtain a listing of the machine language version of your source program. Use this facility to experiment with various basic data types in your language (arrays, records, etc), and determine the representation the compiler uses for these types.

10.63 In Exercise 9.42 in the last chapter, you were asked to formally specify type *function*. Now, devise an implementation of type *function*. Program this implementation; in particular, also implement an input routine that will read the definition of a function, check it for legality, and store it. Provide convincing tests to show that the implementation works.

10.64 Write an implementation of the routines designated as Parnas modules in Exercise 7.23. Note that you will have to introduce global variables for use in those routines only. To make things simple, substitute "*integer*" for "*string*" and some appropriate constant for *Undefined*.

Chapter 11

Correctness of Data Representations

When we write a procedure, we have certain intentions for what it is to do; we may express these intentions as preconditions and postconditions. What the computer executes, however, is the body or implementation of the procedure. There is no automatic guarantee that this implementation matches our intentions; this is something we must test or prove. Likewise, when we define a new data type, we intend it to have certain properties; we express these intentions using the methods of Section 7.4. There remains the task of showing that the implementation we write for this type has the desired properties. That task is the subject of this chapter.

Intuitively, verifying the correctness of a data type implementation involves two steps. First, we must show that the representation of this type in terms of data structures of other types is adequate; second, we must show that the implementation of each operation does what the specifications say it does. The first step is accomplished by defining a function that explains how to interpret a value of the representation data structure (for example, what its value is in the abstract model). If every desired abstract value that can be constructed with the available operations is represented by some value of the representation, then we say the representation is adequate. The second step is simply an application of the methods we've used before to show that the operations are correctly implemented.

The issues of correct representation and correct implementations of the operations are connected by the notion of an invariant assertion for the

315

data type. An invariant assertion for a data type is a predicate that describes the set of legal values of the abstract model or of the representation data structure. An invariant that limits the values of an abstract model is called an *abstract invariant*; for example, when we model stacks as sequences of limited length, the abstract invariant says that the length of the sequence must be between 0 and the maximum stack size. An invariant that limits the values of the data in the representation is called a *concrete invariant*; for example, when a stack is implemented as a vector and an integer, the concrete invariant says that the integer must lie between 0 and the length of the vector, and it allows any values (of the proper type) to be stored in the vector.

These invariants help to organize the problem of showing that a representation is adequate. In addition, the invariants help to organize the verification of individual operations because they describe assumptions about the values that may be assumed at the beginning of every operation and that must, in turn, be restored at the end of every operation.

We examine the issue of verifying the correctness of the implementation of a data type by first considering an example, then stating the general strategy for such proofs. The closing section of this chapter returns to some unfinished business—the formal semantics of reference types. Although we won't use them in this chapter, we need these semantics in order to verify other data type implementations.

11.1 An Example: Proof of Type Stack

To illustrate what it means to prove that the implementation of a data type is correct, let's consider an almost classic example—the implementation of the type *stack*.

In Section 9.1.3, we gave a specification for finite depth stacks of objects of type *T*; we used the abstract modeling specification technique. At the beginning of Chapter 10, we gave one possible implementation for stacks with these specifications—namely an implementation in terms of a vector and an integer.

Specifications such as those in Section 9.1.3 are sometimes said to describe an abstract data type; the domain specification describes abstract objects and the operation descriptions describe abstract operations on the type. The word *abstract* is used to emphasize that this domain and these operations are not what actually gets implemented. To further emphasize the same point, and to clarify discussions, the word *concrete* is used to describe the data structure and procedures that actually implement the type. Thus, for example, the concrete representation for stacks that we will verify is

     type *stack* = **record** *sp*: *integer*;   *v*: **array** [1..*maxstack*] **of** *T* **end**.

and the concrete operations are as shown at the beginning of Chapter 10.

11.1.1 WHAT WE MUST PROVE

We want to encourage the user of the data type to think in terms of the astract domain and operations— *not* in terms of the concrete representation. In simple cases such as the current one, it may be relatively easy to think directly in terms of the implementation of a type, but this will not always be the case (see the representation of sets in Chapter 20, for example). The major reason for verifying the correctness of a type implementation is to permit the user to use its specifications confidently without having to resort to examining the implementation. Hence, to prove that an implementation is correct, we must prove that each property promised by the specification is fulfilled by the implementation. Figures 11.1 through 11.3 should help to illustrate what must be done.

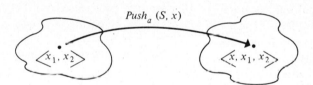

Fig. 11.1 Abstract *Push* operation

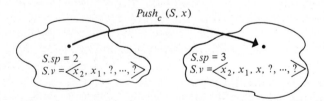

Fig. 11.2 Concrete *Push* operation

Figure 11.1 illustrates the set of possible "abstract" values of a stack by the two "clouds" on either side of the figure. The abstract values are formally specified by the abstract invariant. Each element of these sets is a sequence, and each is the abstract representation of a stack. Remember that, because the stacks are of bounded size, each such sequence is also of

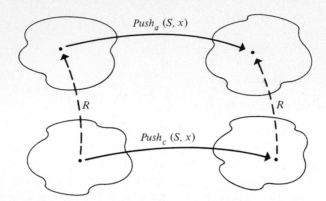

Fig. 11.3 Relation between abstract and concrete operations

bounded length. The operation *Push*(*S*, *x*) maps the particular value of the stack *S* into another—namely the one with the value of *x* appended to the sequence.

Figure 11.2 is similar to Fig. 11.1 except that the "clouds" represent the set of possible "concrete" values actually used to represent the stacks; these concrete values are records with an array and an integer component. The concrete values are formally specified by the concrete invariant. The concrete implementation of *Push* maps a particular record value into another particular record.

Figure 11.3 shows how the previous two figures must be related—that is, it shows what we must prove.

First, as illustrated by the figure, we must state how the concrete objects from Fig. 11.2 relate to the abstract ones of Fig. 11.1. We do this with a *representation mapping*, *R*, that *maps* particular concrete objects into particular abstract objects.* Obviously, we will have to show this mapping represents each abstract object by some concrete one. (In many cases we will find that several concrete objects represent the *same* abstract object; this is, in fact, the case with the representation of stacks that we are using here.)

Second, we must show that each operation works correctly. This is the same as saying that we must show that the object resulting from a concrete operation is the representation of the abstract object that would have resulted from the corresponding abstract operation. We can state this a bit more formally. Suppose that S_a is an abstract stack and that S_c is its concrete representation—that is, $R(S_c) = S_a$. Further, suppose that $Push_a$

---

* The name "representation mapping" would be better given to a function that maps abstract objects into concrete objects. However the mapping of interest is from concrete to abstract, and it is counterintuitively called a representation mapping by all authorities.

is the abstract push operation and that $Push_c$ is the corresponding implementation. Then, by definition,

$$\sim full(S_a) \ \{Push_a(S_a, x)\} \ S_a = <x> \sim S_a'$$

so that the concrete operation must satisfy

$$\sim full(R(S_c)) \ \{Push_c(S_c, x)\} \ R(S_c) = <x> \sim R(S_c')$$

That is, assuming that the precondition is satisfied, the concrete operation produces the representation of "the right thing"—the thing that the corresponding abstract operation would have produced. In technical terms, what we are doing is proving that the diagram in Fig. 11.3 *commutes*.

11.1.2 ADEQUACY OF THE REPRESENTATION

The type of proof illustrated in Fig. 11.3 is what will be required for *any* data type; however, for the moment let's continue with the finite-depth stacks. As we noted, the first step is to prove that the concrete data structure is adequate—and for that we must first establish the correspondence between the abstract and concrete domains. In other words, we must show that the abstract domain is equivalent to the concrete domain, in the sense that every element in the concrete domain matches one element in the abstract domain and every element in the abstract domain has at least one matching element in the concrete domain. There are three aspects to establishing this correspondence:

- An abstract invariant, A.
- A concrete invariant, C.
- A representation mapping, R.

As you will recall, not all sequences are abstract representations of stacks. In particular, since we are only interested in finite depth stacks, the abstract domain consists of only those sequences that satisfy

$$A: \ 0 \leq length(S_a) \leq maxstack$$

A predicate such as this is called an *abstract invariant* because it is true of *all* and *only* those sequences that are valid stacks. Together with the statement that sequences are used as the abstract model of stacks, the abstract invariant actually defines the abstract domain—the "clouds" in Fig. 11.1.

The *concrete invariant*, C, is a predicate that is true of all initialized concrete objects. In our case,

$$C: 0 \leq S.sp \leq maxstack$$

Just as the abstract invariant defines the abstract domain, the concrete invariant defines the concrete domain—the "clouds" in Fig. 11.2.

The representation mapping maps concrete objects into corresponding abstract ones. In this case, what we need is a mapping that takes a pair $<sp, v>$ into a corresponding sequence—where sp is an *integer* and v a vector of T. It is not difficult to see that

$$R(<sp, v>) = <v[sp], v[sp-1], \cdots, v[1]>$$

will do the job admirably. ($R(<0, v>) = <>$ but if $k < 0$, $R(<k, v>)$ is undefined.) Under the mapping R, each abstract object (satisfying A) has a corresponding concrete object (satisfying C) that maps to it. The representation is therefore adequate.

11.1.3 CORRECTNESS OF THE OPERATIONS

The second step of the proof process requires us to show that each of the operations behaves as specified. We will illustrate these proofs with that for the *Push* operation and leave the others as an exercise. We will divide the correctness proof into two parts:

- Proof that the concrete invariant is indeed invariant, and
- Proof that the pre- and postcondition specifications are satisfied.

11.1.3.1 Invariance of C

The predicate C is clearly intended to be true of any *stack* variable once it has been initialized (in the case of *stack*s, *Clear* is the initializing operation). Before proceeding with the main proof, we must verify that this intention is satisfied by the implementation.

It is easy to see that the initializing operation establishes C; that is

$$\{Clear(S)\}\ 0 \leq S.sp \leq maxstack$$

since the body of *Clear* simply sets $S.sp$ to zero. The *Push* operation *maintains* the invariant. That is, just as

$$A \wedge \sim full(S)\ \{Push(S, x)\}\ A, \tag{11.1}$$

in the abstract domain, we also have

$$C \wedge \sim full(R(<S.sp, S.v>))\ \{S.sp := S.sp + 1;\ S.v[S.sp] := x\}\ C$$

in the concrete domain. This is clear because we can see from the definition of R that

$$length(S) = length(R(<S.sp, S.v>)) = S.sp,$$

and, therefore, (11.1) is equivalent to

$$C \wedge S.sp \neq maxstack \; \{S.sp := S.sp + 1; \; S.v[S.sp] := x\} \; C, \quad (11.2)$$

which is clearly true. Similarly, we can show that *Pop* maintains *C*. Since *Clear*, *Push*, and *Pop* are the only operations on *stacks* which can change them (*Top*, *Empty*, and *Full* are functions without side effects), all this simply goes to show formally what we already knew informally: Once a *stack* variable is initialized, the predicate *C* remains true for it forever.

11.1.3.2 Precondition/Postcondition Satisfaction

We must show that the body of the concrete operation satisfies the specifications. In each case it is fair to assume that the invariant is true before the operation. (In the previous section we showed that the invariant is reestablished afterward.) For the *Push* operation, in the abstract domain,

$$\sim full(S) \wedge A \; \{Push(S, x)\} \; S = <x> \sim S',$$

or, from the definition of the prime notation,

$$S = S' \wedge \sim full(S) \; \wedge \; A \; \{Push(S, x)\} \; S = <x> \sim S'. \quad (11.3)$$

Using the specifications of *Full* we may translate condition (11.3) to

$$
\begin{aligned}
&S = S' \wedge (length(S) \neq maxstack) \; \wedge \; A \\
&\quad \{Push(S, x)\} \\
&S = <x> \sim S'.
\end{aligned}
\quad (11.4)
$$

The pre- and postconditions are written with the assumption that *S* and *S'* are sequences, whereas the program text treats *S* as a record containing an integer variable and a vector. The mapping *R* allows us to translate between the two views and treat *S* uniformly as a record. This also changes the abstract invariant *A* to the concrete invariant *C*. Thus, in the concrete domain, we must have

$$
\begin{aligned}
&S = S' \wedge (length(R(<S.sp, S.v>)) \neq maxstack) \wedge C \\
&\quad \{S.sp := S.sp + 1; \; S.v[S.sp] := x\} \\
&R(<S.sp, S.v>) = <x> \sim R(<S'.sp, S'.v>)
\end{aligned}
\quad (11.5)
$$

To prove (11.5) we must use the concrete invariant, *C*, to guarantee that $R(<S.sp, S.v>)$ is defined. Replacing *C* with the formula given above, we obtain

$$
\begin{aligned}
&S = S' \\
&\wedge (length(R(<S.sp, S.v>)) \neq maxstack) \wedge 0 \leq S.sp \leq maxstack \\
&\quad \{S.sp := S.sp + 1; \; S.v[S.sp] := x\} \\
&R(<S.sp, S.v>) = <x> \sim R(<S'.sp, S'.v>)
\end{aligned}
\quad (11.6)
$$

Referring to the definition of R, we can write (11.6) as

$$S = S' \wedge 0 \leqq S.sp \leqq maxstack \wedge S.sp \neq maxstack$$
$$\{S.sp := S.sp + 1; \ S.v[S.sp] := x\} \tag{11.7}$$
$$<S.v[S.sp], \cdots, S.v[1]> = <x> \sim <S'.v[S'.sp], \cdots, S'.v[1]>$$

Condition (11.7) should now appear reasonable. It simply states that if we assign a value, x, to the next location in an array, the reversed sequence of values in that array will be the same as before, but with the value x prefixed. Let's plug in our verification rules to see how the proof works mechanically. From (11.7) we get as the verification condition,

$$0 \leqq S.sp \leqq maxstack \wedge S.sp \neq maxstack \wedge S'.v = S.v \wedge S'.sp = S.sp$$
$$\supset <v^*[S.sp + 1], \cdots, v^*[1]> =$$
$$<x> \sim <S'.v[S'.sp], \cdots, S'.v[1]> \tag{11.8}$$
$$\text{where } v^* = \alpha(S.v, S.sp + 1, x)$$

(since $S = S'$, $S'.v = S.v$ and $S'.sp = S.sp$.). Verification condition (11.8) reduces to

$$0 \leqq S.sp < maxstack \supset <v^*[S.sp + 1], \cdots, v^*[1]>$$
$$= <x> \sim <S.v[S.sp], \cdots, S.v[1]> \tag{11.9}$$

The properties of α imply that

$$v^*[i] = S.v[i], \text{ except at } i = S.sp + 1, \text{ where } v^*[i] = x.$$

From this, we immediately deduce that condition (11.9) is equivalent to

$$0 \leqq S.sp < maxstack \supset$$
$$<x, S.v[S.sp], \cdots, S.v[1]>$$
$$= <x> \sim <S.v[S.sp], \cdots, S.v[1]>$$

which is obviously true.

11.2 The General Approach

The detailed proof of *Push* given in the previous section may seem like a great deal of trouble to invest in two assignment statements. Part of the complexity arose because we were simultaneously discussing the method of proof and carrying it out. However, much of the difficulty is actually inherent in any rigorous proof of a program. At the present state of the art—lacking automated aids to carry out the proof details—it is seldom economically justifiable to verify any but the most simple data type definitions.

Despite the tedious complexity of such proofs, there are two strong reasons why such proof techniques should be part of a programmer's repertoire of tools. First, in some cases the cost of verification is more than justified. Such proofs need to be done only once—then the type definition can be added to a library and used confidently by others. A library entry that is certifiably correct is *much* more valuable than one that is not. Second, the mathematical treatment gives us a framework for organizing data type construction and informally convincing ourselves that its implementation is correct. Within that framework, our proofs can be as rigorous or as informal as we like.

To recap, we performed the following steps in the verification of type *stack*:

- Discover a representation mapping, R, that maps concrete objects onto corresponding abstract ones.

- Discover abstract and concrete invariants, A and C, that describe properties that are true of all initialized objects of the type.

- Prove that each initialization operation establishes the concrete invariant. Prove that each noninitializing operation preserves the concrete invariant. That is, for all initializing operations, INIT, and all noninitializing operations, OP, show

  $$\{INIT\}\ C\ \text{and}\ \ C\{OP\}\ C$$

- Prove each operation correct with respect to its pre- and post-conditions. In doing so, we use R to translate the concrete representation values (for *stack*, vectors and integer pointers) into the abstract values used in the pre- and postconditions (for *stack*, sequences).

The last two steps can be combined; that is, if P and Q are the pre- and postconditions, then we can show

$$P\{INIT\}\ C \wedge Q\ \text{and}\ \ C \wedge P\{OP\}\ C \wedge Q$$

These steps will, in fact, establish the commutativity of the diagram in Fig. 11.3. That is, they prove that the implementation faithfully mimics the behavior described by the specifications.

The foregoing example was developed with the specification technique called abstract modeling. An equivalent verification technique exists for algebraic specifications:

- Discover any suitable concrete invariants.

- Prove that each operation preserves any established invariants and that each initialization operation establishes any invariants.

- Prove that each axiom holds for each operation as implemented, assuming any suitable invariants.

These steps look simpler because the algebraic specification technique imposes less structure on the proof process than the abstract modeling technique. In practice, neither technique is consistently simpler to use or to verify. Like so many other things discussed in this book, the choice of the best specification method can only be decided on an instance-specific basis.

11.3 Weakest Preconditions for Dereferenced Assignments

Many of the data type implementations discussed in Chapter 10 used references, pointers, in their implementation. In order to be able to verify these implementations we must return to an unfinished topic—the formal semantics of reference types, and specifically the semantics of assignment through such types. So, suppose that we declare

var y, z: $\uparrow integer$

Given this declaration, it is *not* valid to write

$$z\uparrow = 5 \ \{y\uparrow := 6\} \ y\uparrow = 6 \wedge z\uparrow = 5, \quad \text{(wrong)}$$

even if we assume that y and z contain valid, non-*nil* pointers. The counterexample is the case where y and z contain a reference to the same object, in which case $y\uparrow$ and $z\uparrow$ would both be set to 6. On the other hand, it *is* valid to write

$$y \neq z \wedge z\uparrow = 5 \ \{y\uparrow := 6\} \ y\uparrow = 6 \wedge z\uparrow = 5$$

where $y \neq z$ means that the *reference* values of y and z—the objects they point to—differ.

The notation $y\uparrow := 6$ is confusing because it suggests that y gets changed. In fact, the value of y—the *name* of the object it points to—does *not* change; rather, the *value* of that object changes. Therefore, we can conclude that

$$wp(y\uparrow := 6, Q) \equiv Q'$$

where Q' denotes "Q with all terms referring to the same object as y replaced by 6." We must now explore how to make this statement more rigorous.

When we write

$$y\uparrow := 6$$

we are changing, not y, but some other object without a variable name. Much of our problem, in fact, is that we don't have a fixed name by which we can always refer to this object. The situation is much like the problem we had with assignment to array elements in Section 7.2.2; and, in fact, the solution will be much the same too. In particular, we'll introduce one large structured object whose parts are "all the integer variables that any variable of type $\uparrow integer$ can reference." When we say $y\uparrow := 6$, one of the variables in the structured object gets changed. Let us denote this structured object

ColofInt: The collection of all integer variables that can be referenced.

Each of the variables in *ColofInt* has a name. Thus, we may think of the variables in *ColofInt* as being indexed by their names. This is, in fact, almost exactly the view presented back in Section 7.2.2.

This particular way of viewing referenced objects is highly reminiscent of arrays (see Section 8.3.1). We have an index set—the set of names— and a structured object whose parts are indexed by the set—*ColofInt*. It is as if *ColofInt* were declared

    var *ColofInt*: **array** [$\uparrow integer$] **of** *integer*;

and $y\uparrow$ were a shorthand notation for

    $ColofInt[y]$,

so that $y\uparrow := 6$ becomes

    $ColofInt[y] := 6$.

In fact, this is a completely correct and rigorous explanation of dereferenced assignment.

We have, then, explained the semantics of dereferenced assignment in terms of subscripted array assignment. From this point of view, the dereferencing operator, $\uparrow$, is a selector. Looking back at Section 11.3, we can now define

$$wp(y\uparrow := E, Q) \equiv wp(ColofInt[y] := E, Q(ColofInt))$$
$$\equiv Q(\alpha(ColofInt, y, E)).$$

To see this formula in action, let's consider computing

    $wp(y\uparrow := 6, y\uparrow = 6 \wedge z\uparrow = 5)$

First, we remove all shorthand to get

    $wp(ColofInt[y] := 6, ColofInt[y] = 6 \wedge ColofInt[z] = 5)$

Applying the rule for array assignment, we get

$$\alpha(ColofInt, y, 6)[y] = 6 \wedge \alpha(ColofInt, y, 6)[z] = 5$$

Using the properties of α given in Theorem 7.1, this reduces to

$$6 = 6 \wedge (\text{if } y = z \text{ then } 6 \text{ else } ColofInt[z]) = 5$$
$$\equiv (y = z \wedge 6 = 5) \vee (y \neq z \wedge ColofInt[z] = 5)$$
$$\equiv y \neq z \wedge ColofInt[z] = 5$$

Reintroducing the standard shorthand gives

$$wp(y\uparrow := 6, y\uparrow = 6 \wedge z\uparrow = 5) \equiv y \neq z \wedge z\uparrow = 5$$

which makes perfect sense: In order for one object $(y\uparrow)$ to have the value 6 and another object $(z\uparrow)$ to have the value 5 after the first object's value is set to 6, the two objects must be different $(y \neq z)$, and the second must already have the value 5 $(z\uparrow = 5)$.

11.4 Further Reading

We have presented a rather cursory discussion of the subject of verifying data representations. This is partly because the mathematics is frequently at much too high a level for a course such as this. However, another good reason not to delve too deeply into the subject at this point is that this aspect of data representation verification is still being actively researched, and it is probably premature to set too many current (and probably preliminary) ideas down in a textbook. For those very interested, a representative sample of articles consists of Hoare (1972), Wulf, et al (1976), and Guttag (1977).

Exercises

11.1 It is possible for a single abstract value to have more than one corresponding concrete representation. Show this is true by giving three representations of the stack $<8, 2, 5>$ in a vector-pointer representation with $MaxSize = 6$.

11.2 Some people think that a new abstract type should automatically "inherit" the assignment and equality operators from its underlying representation. Thus, for example, if stacks are implemented as an integer plus a vector—stack assignment would automatically mean to use the integer and vector assignment operations to copy these two components;

stack equality would automaticcally mean that the integers were equal and the stacks were equal. In many cases, these are "just what you want"— but you must be careful. To illustrate the point, give examples of abstract types and representations for which

- a) Unequal representations may imply equal abstract values.
- b) Assignment is not desirable *at all*.
- c) Equal representations may imply unequal, even invalid, abstract values.
- d) The assignment of representation values would invalidate an invariant (either abstract or concrete) of the type.

11.3 Show that the *Pop* operation as given in the beginning of Chapter 10 maintains the invariant *C*.

11.4 Prove that the implementations of *Pop*, *Top*, *Empty*, and *Full* given in the beginning of Chapter 10 satisfy their specifications.

11.5 Could the representation mapping, *R*, map abstract objects into concrete ones, rather than vice versa as presented, and still allow a valid proof? Explain.

11.6 Consider the implementation of deques in terms of vectors as given in Section 10.3.1.

- a) Write a representation mapping for this implementation.
- b) Write abstract and concrete invariants for this implementation.
- c) Show that *Clear* establishes the invariants and that the rest of the deque operations preserve them.
- d) Show that each of the deque operations satisfies its specifications.

11.7 Consider the implementation of sets in terms of vectors of values given in Section 10.5.2.

- a) Write a representation mapping for this implementation.
- b) Write abstract and concrete invariants for this implementation.
- c) Show that *Clear* establishes the invariants and that the rest of the set operations preserve them.
- d) Show that each of the set operations satisfies its specifications.

11.8 Consider the implementation of sets in terms of bit-vectors given in Section 10.5.1.

- a) Write a representation mapping for this implementation.

b) Write abstract and concrete invariants for this implementation.

c) Show that *Clear* establishes the invariants and that the rest of the set operations preserve them.

d) Show that each of the set operations satisfies its specifications.

11.9 Suppose p, q, and t are all of type $\uparrow$ *integer.* What is the following?

$$wp(t\uparrow := p\uparrow; \ p\uparrow := q\uparrow; \ q\uparrow := t\uparrow, \ p\uparrow = 2 \wedge q\uparrow = 3)$$

11.10 Consider the implementation of stacks in terms of lists given in Section 10.2. Apply the techniques of Section 11.3 to the following tasks:

a) Write a representation mapping for this implementation.

b) Write abstract and concrete invariants for this implementation.

c) Show that *Clear* establishes the invariants and that the rest of the deque operations preserve them.

d) Show that each of the deque operations satisfies its specifications.

11.11 Exercise 10.61, in the last chapter, asked you to design and implement several abstract data types connected with the representation of locks and keys. Verify your representations.

11.12 In Chapter 7, we defined the notion of a mathematical sequence algebraically. Add an axiom to that definition that restricts sequences to a maximum finite size, N. Devise an implementation of these restricted sequences using vectors. (For this, its fair to assume that you are dealing with sequences of integers.) Verify your implementation using the method outlined for algebraic specifications.

11.13 In this chapter we asserted that if one performs certain steps, then the implementation of an abstract data type is "correct." In fact, we oversimplified a bit; all we *actually* argued was that the implementation of each operation, *applied once*, was "correct." That doesn't guarantee that a *sequence* of applications of the operations will be correct. State and prove a theorem to this effect. [Hint: Use induction on the number of operations in the sequence.]

11.14 In Exercises 9.42 and 10.63 you were asked first to specify, then implement the abstract type *function*. Now verify your implementation.

Chapter 12

Space Requirements

In Chapter 6 we discussed ways to evaluate the *speed* of programs; we now turn to the problem of estimating *size*, or storage utilization, of the data used in a program.

It is characteristic of modern computers that programs and data are both stored in the same memory. Indeed, programs are often treated as if they were data (by the operating system, for example). Thus a complete description of the amount of storage required to run a program must allow for both the program and its data. The size of the executable form of a program is usually proportional to the size of the program's source text; size is usually provided by the compiler, and it usually does not change during the execution of the program.* The amount of storage required for data, however, depends strongly the program's declarations; programs of similar length may have wildly different data space requirements.

The space required for data is determined both by the size of individual data objects (*static* space requirements) and by the number of

---

* However, *overlay* techniques can be used to reduce the size of a running program. In an overlay scheme, portions of the program are moved to and from secondary storage; only those portions that are needed at a given instant are in main memory. Overlaying differs from "paging," a newer technique that achieves some of the same effect, in two ways: first, the programmer must explicitly organize and invoke the overlays, whereas paging is automatic; second, overlaying generally replaces completely the entire portion of memory devoted to program, whereas paging operates in smaller, independent units.

them that exist at one time (*dynamic* space requirements). We analyze space utilization with the help of a set of formulas for computing the individual requirement for each data type. For composite types, of course, this formula will depend on the requirements for the individual components. The analyses, however, will differ from the ones we did for analyzing computation time; unlike time, storage space can be reused. Thus, in space analyses we are generally concerned with the amount of space required at a particular point in the program—and usually the point at which the maximum amount is used.

12.1 Static Space Requirements

Static storage requirements, or the amount of space needed for a single instance of a data type, are easy to estimate: you need to know something about how each type is represented, and you need to know what variables are declared.

We express the space consumption, *sc*, of each data type as a function of the type itself, the types of its components, and the values of existing variables. We assume costs, s_t, for each primitive type, t; thus, the formulas are all relative to the basic storage costs of a particular language implementation.

To illustrate the kind of static analysis that we want to do, let's assume that

- An integer requires one word of storage.
- A vector of length N requires
  - N times the size of one of its elements, and
  - 3 words of bookkeeping "overhead"

Then the storage required to satisfy the declarations

> **var** *i, j*: *integer*;
> **var** *V*: **array** $[1..n]$ **of** *integer*;

can be determined as follows. Integers require one word, so *i* and *j* require one word each. Vectors require three words of overhead, plus storage for all the elements. *V* has *n* elements, all integers; integers require one word each; therefore *V* requires $n + 3$ words. Hence, the whole set of declarations requires a total of $n + 5$ words.

We can describe these costs more effectively with formulas than with prose. The rules in Fig. 12.1 give the storage cost *sc* for the basic declaration forms in our language.

Scalar	$sc(\textbf{var } I: st) = s_{st}$
Pointer	$sc(\textbf{var } I: \uparrow T) = s_{\uparrow}$
Record	$sc(\textbf{var } I: \textbf{record } D \textbf{ end}) = sc(D) + s_{record}$
Array	$sc(\textbf{var } I: \textbf{array}[e..f] \textbf{ of } T) =$
	$\quad (f - e + 1)\, sc(\textbf{var } i: T) + s_{array}$
Type	$sc(\textbf{type } DT = TDef) = 0 \quad \text{and}$
	$\quad s_{DT} = sc(\textbf{var } I: TDef)$
Defined Type	$sc(\textbf{var } I: DT) = s_{DT}$
Id List	$sc(\textbf{var } I_1, \cdots, I_n: T) =$
	$\quad sc(\textbf{var } I_1: T) + \cdots + sc(\textbf{var } I_n: T)$
Multiple Declaration	$sc(\textbf{var } I_1: T_1; \cdots; I_n: T_n) =$
	$\quad sc(\textbf{var } I_1: T_1) + \cdots + sc(\textbf{var } I_n: T_n)$
Constant	$sc(\textbf{const } I = e) = sc(\textbf{var } I: T_e)$

where

e, f denote expressions

I_i denotes an identifier

T_i denotes any type

st denotes any scalar type (integer, real, $\cdots$)

D denotes a declaration

DT denotes a user-defined type: $DT = TDef$

Fig. 12.1 Rules for data costs

These formulas state fairly obvious storage requirements. The requirement for each built-in scalar type or pointer depends directly on the implementation. For records and arrays, there may be a small fixed amount of overhead, but the costs are determined primarily by the storage required for the elements. A type declaration does not itself cost anything, but it serves as an abbreviation for a longer declaration; it may be regarded as determining a new constant, s_{DT}, that can be used when variables or constants of the new type are declared. Multiple declarations and identifier lists are simply abbreviations; the costs of the variables so declared are the same as if the variables had been declared separately. Finally, costs for declared constants are the same as for variables. (This may not be true in some implementations that embed constant values known at compile-time directly in the program.)

Of course there is a great deal of variation between implementations, but at least one version of PASCAL has the following values for s_i (in words of storage). These values are, at least, the right order-of-magnitude for most implementations.

$$s_{integer} = 1$$
$$s_{real} = 1$$
$$s_\uparrow = 1$$
$$s_{record} = 0$$
$$s_{array} = 3$$

These rules are illustrated in the program shown in Fig. 12.2.

```
begin
type
    complex = record        0 (and s_complex ≡ 2 s_real + s_record)
        RealPt,ImagPt: real
        end;
    elt = record            0 (and s_elt ≡ s_complex + s↑ + s_record)
        k: complex;
        r:↑elt
        end;
var
    i, j, n: integer;        3 s_integer
    x, y: complex;           2 s_complex
    rt, ru: ↑elt;            2 s↑
    V: array [1..n] of elt;  s_array + n * s_elt
    M: array [1..n] of       s_array +
        array of [1..m] of ↑elt;   n * (s_array + m * s↑)
    <program>
end
```

Fig. 12.2 An example of a program with data cost annotations

The costs lead to the following expression for the total cost:

$$nm\, s_\uparrow + n\, (s_{array} + s_{elt}) + 3\, s_{integer} + 2\, s_{complex} + 2\, s_\uparrow + 2\, s_{array}$$

This expression depends on the defined constants that give the sizes of elements of the defined types. We can eliminate these constants by substituting their definitions. We expand first s_{elt}, obtaining

$$nm\, s_\uparrow + n\, (s_{array} + s_{complex} + s_\uparrow + s_{record})$$
$$+\, 3\, s_{integer} + 2\, s_{complex} + 2\, s_\uparrow + 2\, s_{array}$$

We then eliminate $s_{complex}$, giving the final result

$$nm\, s_\uparrow + n\, (s_{array} + 2\, s_{real} + s_\uparrow + 2\, s_{record})$$
$$+\, 3\, s_{integer} + 4\, s_{real} + 2\, s_{record} + 2\, s_\uparrow + 2\, s_{array}$$

As in Chapter 6, we can treat this in as much detail as desired. We might, for example, substitute specific values for s_i—like those given above. Doing so leads to the value

$$n \cdot m + 6\, n + 15$$

Alternatively, even if we didn't know the relative sizes of the various types, we could still make a reasonable estimate of the space requirement using order-of-magnitude arithmetic. You will often care only about the order of magnitude of the result, or at most only about the first term and the order of the remainder. If this is true, you can save a great deal of work by considering only the "big" declarations: the ones that contribute to the major term. With static structures, these will be the composite structures (the array declarations with the largest number of dimensions and any large records). You must, of course, remember that programmer-defined type names often abbreviate large definitions. Thus the important part of the result found above can be obtained by checking only the lines

     V: **array** $[1..n]$ **of** *elt*;
     M: **array** $[1..n]$ **of array of** $[1..m]$ **of** $\uparrow$*elt*;

12.2 Dynamic Space Requirements

If all the data space required by a program were allocated at once—before the program began execution, for example—the static analysis illustrated in the previous section would be sufficient. Certain languages, such as FORTRAN, force this to be true. However, many languages provide for *dynamic* allocation and deallocation of storage: space is allocated only when it is actually needed and then deallocated as soon as it is no longer required.

In this section we will examine two such cases. The first arises in block-structured languages, where the space used for data declared in an inner block is no longer needed when the block ends. The second arises when the sizes of list structures change dynamically and new elements are explicitly requested.

12.2.1 SPACE REQUIREMENTS IN BLOCK-STRUCTURED LANGUAGES

In a language with restricted variable scopes, the storage required for variables declared in a scope is no longer required when control leaves that scope. We can take advantage of this fact to reuse the storage for such variables; when we do, the space requirement for the program is just the maximum amount of space in use at one time, not the sum of the space requirements for all declarations.

This situation occurs most commonly in block-structured languages, in which the scopes are nested in such a way that no two scopes ever overlap. For such languages, the space requirements can be analyzed by keeping a running tally of the space required, incrementing the tally when the declarations for a block are processed and decrementing it by a corresponding amount when the block is exited.

The skeletal program in Fig. 12.3 shows how this works. (Note that we are using ALGOL-like syntax here, in which declarations occur inside **begin–end** blocks. This departure from PASCAL is necessary to discuss nested blocks comfortably.) The code in each block has been replaced by a comment that shows the current space requirement. In the analysis, we use the constants given in the discussion of static analyses, Section 12.1.

The program in Fig. 12.3 contains four blocks, nested three deep. The space for $V1$ (8 words) is required throughout the program. The array $V2$ requires 13 words, and the total space required while the block that declared $V2$ is executing is thus 21 words (8 for $V1$ and 13 for $V2$). When the block that declared $R1$ starts, $V2$ is no longer needed, so the old $V2$ space can be reused for $R1$, i, j, n, and other variables. In this block, 14 words are required: 8 for $V1$, 4 for $R1$, and one apiece for i and j. Then, in the innermost block, $V3$ must also be allocated 10 words, for a total requirement of 24 words. The maximum instantaneous space requirement is thus 24 words, even though the total required by all the declarations is 37 words. The fact that the same storage locations may be used at one time for a vector of reals and at another time for a record, two integers, and part of another vector causes absolutely no difficulty.

12.2.2 DYNAMIC DATA STRUCTURE ALLOCATION

If you are building lists or trees or other dynamic structures, the situation becomes more complex. Cells in these structures are created in response

```
begin
var V1: array [1..5] of integer;
   begin
   var V2: array [1..10] of real;
   · · · {21 words required} · · ·
   end;

   begin
   var R1: record a, b: real; c, d: integer end;
   var i, j: integer;
   · · · {14 words required} · · ·
      begin
      var V3: k[array] [1..7] of integer;
      · · · {24 words required} · · ·
      end;
   · · · {14 words required} · · ·
   end;
· · · {8 words required} · · ·
end;
```

Fig. 12.3 Storage requirements for nested blocks

to explicit requests (via the *new* function). When these cells are no longer accessible or when you explicitly release them, they become available for reuse. Thus, although you can estimate the sizes of various individual nodes, as described in Section 12.1, the number of nodes actually in use will fluctuate as the program is executed.

If you can describe the circumstances under which data will be inserted into a structure or removed from that structure, you may be able to get good estimates of the amount of storage required. Depending on the kind of information you have available, the estimates might be upper limits on the size or predictions about the average size.

Since the storage requirements are not static, but grow and shrink as the program executes, you should expect that the analysis of the storage requirements would also be dynamic rather than static. This is indeed the case; if you wish to obtain a precise specification of the space requirements for dynamic data structures, you must be prepared to invest the same kind and amount of effort as you would to verify assertions about the values the progam computes.

If you are using an ideal storage management system, analysis of the creation and elimination of program variables will give very good estimates of storage utilization. In many storage management systems, however, storage that has been recovered from cells that have been used and released is not reused with 100% efficiency. Storage analysis for such systems is very much dependent on each individual system's special characteristics.

12.3 Design Decisions and Representation Trade-Offs

If you make predictions about storage requirements while you are designing a program, they may help you make decisions about data structures. In this section we consider a simple example that involves only the costs of storage.

Suppose that you plan to use a stack of integers in your program, and suppose that your major concern about efficiency is about space rather than time. Chapter 10 showed two representations for stacks: one used a vector; the other, a linked list. Let's consider the storage requirements for these two representations.

For the vector implementation, you need to decide in advance how deep the stack will get; we call this depth *maxstack*. The type declaration for this representation of stacks of integers is

    type *Stack* = **record** *sp: integer; v:* **array** [1..*maxstack*] **of** *integer* **end**;

The space required by such a stack is *maxstack* + 4—or one word for each potential element plus 4 words of overhead.

For the list implementation, new cells are generated as needed, so you don't need an *a priori* bound on the size. However, the overhead is roughly one link for each element in the stack, rather than the 4 words required for the vector implementation. More precisely, we look at the declarations, which are

    type *Stack* = ↑*listofInt*;
    type *listofInt* = **record** *first: integer; tail:* ↑*listofInt* **end**;

This implementation requires one word for the head of the list plus two words per element in acutal use.

The space costs of these two implementations cannot be compared directly. The vector implementation has a constant cost that depends on the maximum stack depth, whereas the list implementation has a variable cost that depends on the number of elements in the stack. For simplicity, assume that you are concerned with the average space requirement during the execution of the program, and assume that you can predict that the average stack size will be *AvgStack*. Then the costs are

Vector implementation: *maxstack* + 4

List implementation: 2 · *AvgStack* + 1

It is clear that the storage requirement for the list implementation is smaller exactly when

$$AvgStack < (maxstack + 3)/2.$$

12.4 Further Reading

Storage requirements for various representations are often analyzed in the standard texts on data structures. Chapter 2 of Knuth, v. 1 (1973) pays particular attention to this. Representations that correspond to standard abstract types are given in Hoare (1972); this development pays particular attention to the savings that can be obtained through packing.

Exercises

12.1 In the representation of Section 10.5.2, what happens if the sets are kept sorted? In particular, how do the operations of finding, inserting, and deleting elements change?

12.2 Write a rule for the space required for multidimensional arrays. [Hint: **array** [*a..b,c..d*] **of** *T* means **array** [*a..b*] **of** array [*c..d*] **of** *T*.]

12.3 Write a rule for the space required for a *union type*, or *variant record*, as defined in Section 8.4.1.

12.4 Extend the rule for defined types to deal with parameters (for example, size parameters) to the type definitions.

12.5 Determine the space required to execute the programs of Exercises 8.23, 8.22, and 8.26.

12.6 Several types are declared in Exercises 8.6, 8.11, and 8.26. Determine the space required for one instance of each of these types.

12.7 The program of Exercise 8.12 allocates space dynamically. Determine the amount required.

12.8 In Chapter 10, several representations of each of several data types were discussed. For this problem you are to compare the space used for these various representations. In making these comparisons, you will need to account for factors such as "average size," "maximum size," "element size," and so on. In some of these types there will be representations that are preferable (smaller) under some circumstances and not preferable under others. Be sure to state the precise point at which the cross-over occurs. The type representations are:

a) Binary trees: Linked versus vector representations.

b) Labeled graphs: Linked versus connection-matrix representations.

c) Deques: Linked versus vector representations.

12.9 Exercise 10.56 asks for four representations of matrices under the assumption that three special properties hold. Under what assumptions about operation cost, storage cost, and values of q will each of your representations be preferred? State your reasoning carefully.

12.10 The observation that it may be desirable to have different implementations for different patterns of use applies to finite sets of integers. Consider the three representations of sets given in Chapter 9, but consider only the operations *Insert*, *Remove*, and *IsMember*. Consider the use patterns given by

Element range	Maximum/Average elements per set	Relative frequency of insert/remove/test
1–10	10 / 5	1 / 1 / 1
1–10^5	10^5 / 5	1 / 1 / 1
1–10^5	10^5 / 1000	10 / 1 / 1000

Choose and defend a representation for each case. Justify your decisions by analyzing the time and space costs of each one for the average case under the usage assumptions given.

12.11 Most compilers permit the user to print a symbolic version of the machine-language program that is generated from your high-level language source text. Use this facility of your compiler to determine the representation the compiler uses for the basic types in your programming language—and, hence, determine the storage costs, *sc.*

12.12 In Exercises 9.42 and 10.63 you were asked to first specify, then implement abstract type *function*. Now analyze the space requirements of your implementation. Did you make good representational choices?

Part Three

The Interaction of Control and Data

Parts One and Two of this textbook have considered control and data issues separately. In this part we explore some of the issues that arise from their interaction.

As usual, we start with a set of mathematical models. These models will relax some of the finiteness constraints that limited the power of FSM's. In fact, the way we relax the constraints is by adding an explicit data memory to the FSM model. One of these models, the Turing machine, is generally believed to be capable of carrying out *any* effective computation (unfortunately, there is no way to prove this is so). We will also see, however, that we can describe some computations that even the Turing machine is incapable of performing. Therefore we come to the somewhat surprising conclusion that there are computations that cannot be performed on a computer.

We then explore the ways in which these same finiteness constraints are relaxed in programming languages—namely by introducing dynamic data types and by permitting recursive subroutines. Both of these changes create opportunities for different styles of programming, and we explore a few of these styles, introducing new representational, verification, and cost-analysis problems where appropriate.

Chapter 13
Models of Computation and Grammars

In Chapter 1 we presented a simple model of computation, the FSM, and showed its equivalence to several other simple models, including (in Section 2.2) programs whose variables take on values from a finite set. Even before we had finished this presentation, however, we showed that FSM's were *inadequate* for describing all possible computations. Because of the equivalence arguments, these results applied to programs with finite memory as well. The limitation turned out to be the finite memory restriction in each model. In this chapter, we extend these simple models to encompass a much larger set of computations, the set of all *effective procedures*.

We intend the term effective procedure to include all computations that we can imagine being performed on any present or imagined computer. Many mathematicians have proposed essentially equivalent definitions to capture this intuitive notion; one such is

Definition 13.1. An *effective procedure*, or algorithm, is a "well-specified" and finite description of a computation that can be performed in finite time. □

Note that although the *description* of an algorithm must be finite, we do not put any bounds on how long any *particular* computation using that algorithm may run, or on how much memory it may use in computing its results. Also note that in the process of executing the algorithm, only some finite amount of time will have passed, and only some finite amount of memory will have been used. Nothing is ever *infinite* in the course of a

341

computation; it is only *unbounded*. That is, at any point it can get more time and memory if necessary.

Of course, any real computer has only a finite amount of memory. Indeed, we can even go so far as to compute a finite bound on the maximum amount of memory constructable out of the material composing the known universe. We might therefore be tempted to claim that, for all practical purposes, FSM's serve as models of effective procedures. Unfortunately, the finiteness of a FSM is not always a mathematically *useful* concept. The finiteness constraints can often get in the way of a concise, understandable description of an algorithm. We often, for example, write programs in which we assume that all our intermediate results will fit in their respective variables (that our integers will not overflow, for example). Even though the assumption may not always be strictly justified, by making it we greatly reduce the amount of detail we have to handle—and this is certainly desirable. It is just this suppression of detail that we are trying to accomplish by extending our FSM model.

We extend the FSM model in two steps, both of which add a potentially infinite memory to an FSM-like control mechanism. The first, Push-Down Automata (PDA's), have only very restricted access to this memory. Although PDA's turn out to be more powerful than FSM's, they are not as powerful as the second model—Turing machines (TM's). TM's are believed to be capable of performing any effective procedure.

Just as regular expressions turn out to be a language description tool closely related to FSM's, production grammars are language-description tools closely related to our extended computational models. In their full generality, production grammars are equivalent to TM's. A special kind of production grammar, Backus-Naur form (BNF), turns out to be very useful in describing the syntax of programming languages and is equivalent to nondeterministic PDA's.

Finally, we touch on the halting problem, which has the property that no effective procedure exists to solve it. By showing that there is no effective procedure for solving this particular problem, we demonstrate that there exist problems that one can never hope to solve with a computer. This result is the cornerstone of the theory of *computability*, an abstract but nonetheless fascinating branch of computer science and formal logic.

13.1 Pushdown Automata

In this section we explore one way of relaxing the fixed memory bound on FSM's. The computational models we introduce, while more powerful than the FSM, are still not entirely adequate for describing all effective procedures.

13.1.1 DPDA'S AND NDPDA'S

In Section 1.5.1 we saw that there are languages that cannot be recognized by an FSR. We considered the language called NMN that consisted of the set of strings $\{1^n0^m1^n \mid n, m > 0\}$, and we argued that an infinite number of states would be required for its recognizer. An even simpler example of a language that cannot be recognized by an FSR is the language of all strings from the set $\{a^nb^n \mid n \geq 0\}$, that is, $\{\epsilon, ab, aabb, aaabbb, \cdots\}$. Let's review the reasoning we used to conclude that no single FSR can recognize a language such as a^nb^n. When the FSR has read all the a's and is about to read the b's, it must somehow "remember" how many a's have been read. Since the only "memory" of an FSR is the internal state, there must at least be a state that corresponds to "k a's" for any positive value of k. Thus there would have to be an infinite number of states in this "finite state" recognizer, which is, of course, not permitted. (See Section 1.5.1 for a more formal proof.)

We may, therefore, imagine a more powerful group of machines that are capable of recording an infinite number of different situations. Suppose we were to record each of these situations in an internal state, analogous to that of the FSR model. We would now have an Infinite State Automaton like the one in Fig. 13.1.

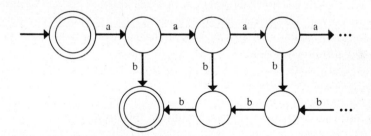

Fig. 13.1 Infinite-state automaton

In theory, such a model would suffice, but machines in the model would have infinitely large descriptions, when fully written down. That is not very satisfying. Our mathematical sense of elegance rebels at ellipses in solutions (and responds with utter horror to the thought of machines like this for more complex problems). In addition, this formulation does not seem to tell us anything fundamental about computation. Let's look further into our own intuitions about computing to find a clue to a better model—one that is clean, closed, and informative.

How would a human computing machine (you, for example) recognize a member of $\{a^n b^n \mid n \geq 0\}$, honoring the restriction that it must be read one character at a time, left to right? Here are some possibilities:

- You could count the number of a's, count backwards through the b's, and "accept" the string if the result is 0 after the last b.

- You could make a vertical mark for each a, cross one of the vertical marks with a horizontal mark for each b, and "accept" the string if all marks have been crossed when no b's remain.

- You could place a stone in a basket for each a, remove one for each b, and "accept" the string if the basket is empty after the last b.

(In each of these, the human computer would also have been checking that no letters other than a and b have occurred, and that once a b was seen, no a's occurred.)

Note that each of these possibilities is a finite description of a process. Each description involves only a handful of words with no ellipses or etceteras. On the other hand, each method requires that something be unbounded when the method is carried out—the unrestricted size of the number, the size of the strip of paper, or the supply of stones. Note again that at each point during the execution of one of these methods, the "something" (number, paper, or stones) will be finite; in fact, the solution of any particular problem will only use a finite number of configurations of that "something." However, there will always be problems that need more configurations than any given upper limit—that's why we need an *unbounded* number of them.

This distinction is crucial to extending our model of computation. We have separated the machine solution into two pieces, a finite "control" piece, expressed in a handful of words, and an unbounded "data" piece, a device that can contain an arbitrary number of configurations.

FSR's already contain a finite control. To get the effect of unbounded memory, we first try adding a stack to this finite control. Why a stack? A stack is conceptually simple, and it is appropriate for the problem. Only one element is ever accessible (the top), there is only one place to erase data (from the top), and there is only one place to add more data (to the top). In spite of these restrictions, many interesting problems can be solved with only this kind of memory. For example, a machine that recognizes valid programs in our programming language (except for checking types and name declarations) can be built using only a stack memory.

Here is an example of the kind of machine we will build. These machines are called push-down automata (PDA's), a reference to their push-down stack. They come in two kinds, deterministic (DPDA's) and nondeterministic (NDPDA's).   Figure 13.2 shows a state-transition diagram for a DPDA that recognizes the language $\{a^n b^n \mid n \geq 0\}$.

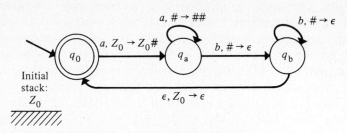

Fig. 13.2 Recognizer for $\{a^n b^n\}$

The PDA in Fig. 13.2 looks very much like an FSR.   First, it has a finite set of states, one of which is designated the starting state (by the "arrow coming from nowhere"), and some of which are designated final states (by the double circles). In this case, the starting and final states are the same, but of course they needn't be.   There are also transitions between various states (the labeled arcs), and these, too, are like those of an FSR. The only real difference is that the labels on the arcs are more complex. This complexity permits us to examine and modify the contents of the stack that we have added to the model.

The labels on the state transitions are of the form

$$x, A \rightarrow B_1 B_2 \cdots B_n$$

The x character is the input string character, just as for an FSR. If x is specified as ϵ, the transition is called an ϵ-transition (or null transition) and no input character is read from the input string.   The A is the character that must appear at the top of the stack; it will be removed from the stack when the transition is executed.   Thus the input "symbol" is really the pair <next-input-character, top-of-stack>. The B_i are elements to be pushed onto the stack, B_1 first and B_n last, such that B_n is the new top of the stack.   The sequence $B_1 \cdots B_n$ therefore replaces A on the stack.   The sequence of B_i's can also be given as ϵ.   This permits a transition to pop the top of the stack without replacing it with anything. This is the only way that the stack can shrink.   The set of symbols that may be pushed on the stack is known as the *stack alphabet*, and it must be finite.

In a DPDA, only one transition out of any state may be labeled with the same pair of input and stack characters. No transition may be labeled "$\epsilon, Q \rightarrow \cdots$" if there is another transition out of the same state with stack symbol Q.

The interpretation of a DPDA is analogous to that of an FSR. By convention, the stack is initialized to contain only the distinguished element Z_0. The machine begins in the "start" state, and at each cycle it selects the appropriate transition, modifies the input pointer and stack as specified by the transition label, and goes into the new state selected by the transition. A DPDA *accepts* a string if it is in a final state after its input is exhausted and it has taken any final null transitions. A DPDA *rejects* a string if it does not end up in a final state, if its stack empties before the input is exhausted, if it goes into an infinite loop of null transitions, or if it "jams"—that is, if it gets to a state where there are no applicable transitions and the input is not exhausted.

For an NDPDA, the interpretation is almost the same. As in a nondeterministic FSR, there may be more than one feasible transition out of a state. The NDPDA is presumed to follow all of them "simultaneously." It accepts a string iff *some* sequence of choices the machine could make leaves it in a final state when it exhausts its input.

A *configuration* of a PDA consists of its internal state, the input pointer, and its stack. Figure 13.3 is a trace of the configurations of the machine in Fig. 13.2 for the input string "aaaabbbb."

State:	q_0	q_a	q_a	q_a	q_a	q_b	q_b	q_b	q_b	q_0
					#					
				#	#	#				
Stack:			#	#	#	#	#			
		#	#	#	#	#	#	#		
	Z_0	Z_0	Z_0	Z_0	Z_0	Z_0	Z_0	Z_0	Z_0	
Input read:		a	aa	aaa	aaaa	aaaab	aaaabb	aaaabbb	aaaabbbb	aaaabbbbb

Fig. 13.3 Configurations of machine in Fig. 13.2 on "aaaabbbb"

Let's consider another example, a recognizer for properly nested sequences of brackets from the set $\{(,), [,]\}$. Figure 13.4 shows a machine that recognizes strings such as $[(())]([]([]))$, but not $[(()$ or $[)$. Note that this machine will jam if it encounters any invalid sequence of brackets. If it reads the wrong type of closing bracket, the stack will contain the wrong character at the top. If a closing bracket has no corresponding opening bracket, the stack will have Z_0 at the top. Only when the string has been completely scanned with all brackets matched can the machine stop in state q_0.

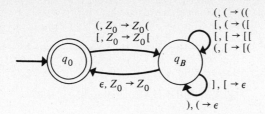

Fig. 13.4 PDA for properly nested brackets

Another example: Recognize a pattern of 0s and 1s, a marker, and the same pattern in reverse, (for example, 10110c01101). Technically, this is the set

$\{wcw^R$ in $\{0, 1, c\}^*$ | w is in $\{0, 1\}^*$ and w^R is w reversed$\}$.

A DPDA for recognizing this set is shown in Fig. 13.5.

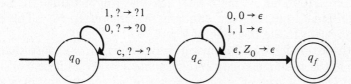

Fig, 13.5 PDA for recognizing wcw^R

(For brevity we have adopted the convention that "?" means "any character." When it appears on the left side of a label, it matches any input character. When it appears on the right side of a label, it must also have appeared on the left and it means "whatever was matched on the left." Thus, "0,?→?0" can be considered an abbreviation for the three separate transitions "0, 1→10," "0, 0→00," and "0, $Z_0 \rightarrow Z_0$0.") Note that in q_0, the machine builds up a record of the "w" part of the string on the stack. When it reads "c," the machine switches from building up the stack to tearing it down by reading the reversed string.

The preceding examples have all been deterministic PDA's; at any point, at most one transition is possible. If we relax this restriction, as we did for FSR's, the result is a nondeterministic PDA, or NDPDA.

Example: Recognize $\{ww^R$ in $\{0, 1\}^*$ | w^R is w reversed$\}$. This is similar to the previous example. The difference is that the "c"

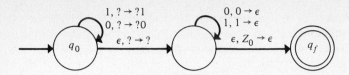

Fig. 13.6 NDPDA for recognizing ww^R

middle-of-string marker is not used. An NDPDA for recognizing this set is given in Fig. 13.6.

We have simply changed "$c, ? \rightarrow ?$" to "$\epsilon, ? \rightarrow ?$." This introduces nondeterminism because the machine is free to take this null transition at any time.

When we studied FSR's, we found that NDFSR's were often simpler and smaller than FSR's for the same problem, but that the two models were of equivalent power. That is, for any NDFSR there was always an FSR that would recognize the same set of strings, although the FSR might possibly have an exponentially larger number of states. An immediate question comes to mind: Is there a DPDA that will recognize the same set as any NDPDA? Interestingly enough, the answer is *No*. There is no DPDA equivalent to the machine of Fig. 13.6.

Let's try to understand intuitively why this is so. Suppose that in the last example the machine is currently reading a pair of adjacent 0's or 1's. These could both be part of w, or both part of w^R, or the end of w and the beginning of w^R. Since the stack must be increased to represent an addition to w and decreased to match a piece of w^R, the decision to increase or decrease the stack must be made as the string is read. If the DPDA were to wait until the entire string had been read, it would either have to have a potentially infinite stack or have the ability to read the bottom of the stack; neither, of course, is permitted. A DPDA must make the decision irrevocably; a NDPDA tries all possibilities simultaneously.

13.1.2 LIMITATIONS OF PDA'S

Although PDA's offer a vast increase in power over FSM's, they are not powerful enough to model all of what we consider effective procedures. For instance, we know intuitively that in order to "look at" something buried in the stack, a PDA must necessarily "throw away" everything on top of it. There are many more ways PDA's can be inadequate to express effective procedures, and we describe one simple case in the following theorem.

Theorem 13.1. The language L consisting of all strings containing any number of a's followed by the same number of b's and then the same number of c's cannot be recognized by a PDA. Formally,

$$L = \{a^n b^n c^n \mid n > 0\} = \{abc, aabbcc, aaabbbccc, \cdots\} \quad \square$$

(Note that the set $\{a^n b^n \mid n > 0\}$ *can* be recognized by a PDA. It's the sequence of c's that makes the difference.)

We won't prove this theorem, but it clearly has two parts. First we must show that there is *some* effective procedure for recognizing L; then we must show that no PDA can do it. The first part is easy. (Try extending the methods for recognizing $a^n b^n$ by using some stones and two baskets, or one basket and two colors of stones.) An intuitive explanation of the second part can be derived by examining the PDA solution for $\{a^n b^n\}$. In that solution we emptied the stack as b's were input, and thus we "threw away" the knowledge of the original number of a's. Any attempt to extend the solution must have that same property—and hence we won't have the information necessary to determine whether the right number of c's are present.

13.2 Turing Machines

The memory introduced in Section 13.1.1 is unbounded, but access to it is quite restricted; it is impossible to examine the middle of a stack without destroying what is above it. This limitation is the reason that PDA's are incapable of performing all effective procedures. In this section, however, we remove this access restriction to arrive at our final computational model, the *Turing machine* (TM). TM's are of considerable theoretical interest—both because of their simplicity and because of Church's Thesis. A famous logician named Alonzo Church has conjectured that

> Any computation for which there is an effective procedure can be realized by a Turing machine.

This famous proposition is by nature unprovable since we have no rigorous definition of "effective procedure." However, no counterexample has yet been found, and many alternative definitions of computability have been shown to be equivalent to TM's. In fact, the conjecture is so widely accepted that we often define "effective procedure" as "realizable by a TM."

The simplicity of TM's makes them useful in arguments of the form, "There is no effective procedure for doing X." If a TM cannot do X and if we believe Church's thesis, then we believe that X cannot be done at all, *effectively.*   In addition, it is sometimes possible to use TM's to prove statements of the form, "Machine Y cannot do computation X on all data in less than time T." For if we know how long it takes a TM to do X for each possible input and we know how much faster machine Y is than a TM, we can establish a bound on how fast Y can do X.   We have simplified the problem by breaking it into two smaller pieces.   Unfortunately, the problem of establishing these lower bounds on how complex a procedure X is (how much time it takes) is extremely difficult even with such simplifications.

13.2.1 DEFINITION OF A TURING MACHINE

A Turing machine, like a PDA, has a finite state control part, an input tape, and an unbounded memory.   Its memory consists of an additional work tape divided into squares that it can read and write using symbols from a fixed, finite tape alphabet.   Each square of the work tape contains a blank symbol when new.   You may think of the work tape as infinite, or you may think of there being a tape-manufacturing facility nearby that adds squares to the ends of the tape as needed.   Figure 13.7 illustrates the configuration.

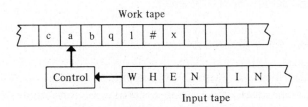

Fig. 13.7 A typical Turing machine

As in all the other simple models we've seen, a TM works by a series of transitions.   Each transition consists of

- Going to a new state,
- Possibly writing something on the square of the work tape currently under the read/write head, and
- Possibly moving the read/write head one square left or right.

Exactly which transition the TM takes depends on

- Its current state,

- The next input symbol, and
- The current work tape symbol.

A TM may be used to *recognize* a language by designating a set of final states, just as with FSR's and PDA's. If such a TM is in a final state when it exhausts its input and halts (taking all remaining null transitions), then it accepts its input. Otherwise, and in particular, if it goes into an infinite loop and never halts, it rejects its input.

A TM may also compute a result, yielding more than a simple yes/no answer. To do so, we merely define the final contents of its work tape as its output and build the TM to develop its answer on the work tape.

We describe TM's with state-transition diagrams much as we have done for FSR's and PDA's. For example, the TM shown in Fig. 13.8 accepts exactly the language $\{a^n b^n c^n \mid n > 0\}$ that our PDA's could not.

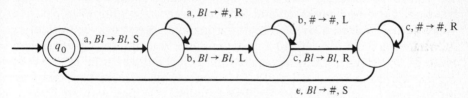

Fig. 13.8 A TM to accept $a^n b^n c^n$ for all n

Note the strong similarities between the notation above and that for FSR's and PDA's. The label "$x, y \rightarrow u, v$" on an arc means

Take this transition if the input symbol is x and the current work tape symbol is y. Replace the y on the work tape with u and move the read-write head one square in direction v, where v is L for left, R for right, or S for "stay here" (that is, "don't move").

The input symbol Bl stands for the *blank* work tape symbol (remember, the work tape starts out blank).

Let's examine how the machine of Fig. 13.8 works. It first puts # marks on its work tape for each of the a's in the input, except the first. When it finds one of the b's, it starts back along the work tape, checking that there is one # for each b except the first. It does not erase the #'s from the tape. When it finds a c, it starts right again along the work tape, checking that there is # for each c except the first. After reading the final c, the read/write head should be over a blank, so the machine can enter the accepting state and write a # to prevent further transitions. Try "running" this TM with various legal and illegal inputs, noting the various ways it can halt in a nonfinal state.

Our definition of TM's is by no means the only one. There are dozens of possible variations (multiple or two-dimensional work tapes, one-ended work tapes, etc.), none of which affects the set of possible computations. As was the case for FSR's, nondeterminism adds no power, although it may simplify machine descriptions and increase speed. Thus we need only consider simple, deterministic TM's.

We have introduced the TM model to complete our collection of computational models. They have no analogs in everyday programming, unlike FSM's and, as we will see in Section 14.3, DPDA's. Building TM's can be absorbing, since doing so amounts to writing clever programs for a very primitive machine, but one for which every computation is possible, and none is easy.

13.2.2 INTERPRETERS AND UNIVERSAL TURING MACHINES

We chose to describe TM's by state-transition diagrams; we could, however, have used any of several forms of description. For example, just as we earlier defined a simple programming language that is equivalent to FSM's, we could have defined a programming language that is equivalent to TM's. Programs in this language would be just a finite string of symbols. This would mean you could put a TM program on an input tape and feed this tape to a TM (or to another TM program). In fact, you are already familiar with a class of programs called *compilers* that do take other programs as their inputs. For TM programs, we can design a TM (or write a TM program) that simulates the actions of the TM described by a TM program. We'll call such a program a *TM interpreter.*

A TM interpreter takes as its input a TM program followed by the data (the input tape) for that program. It then executes the program on the data. If the program halts with an ACCEPT command, so does the interpreter. If it halts with a REJECT, so does the interpreter. If the program never halts, neither does the interpreter.

At first, such an interpreter program might seem impossible to write ("How can a compiler for a language be written in the same language?"). In fact, it is not impossible—just extremely tedious. We are not going to build a TM interpreter here, not even as an exercise. The details quickly become unbearably dull. It is only necessary to realize that its construction is possible. Historically, this construction was made using the TM's themselves rather than programs. A TM that can interpret any TM given to it as input is called a *universal Turing machine.*

13.2.3 COMPUTABILITY

Even with the addition of a potentially infinite memory, there still remain problems that cannot be solved even by Turing machines. According to

Church's thesis, these are precisely those problems for which there are no effective procedures. The purpose of this section is simply to indicate the existence of a particular unsolvable problem that is of great significance to both practical and theoretical computer science.

13.2.3.1 The Halting Problem

Consider the problem, "If program P is given the symbols Q as input, determine if P will ever finish or if it will go into an infinite loop." This is called the *Halting problem.* It has the fascinating property that no TM program exists that can solve the halting problem for all P and Q. We can write programs that solve it for *some* P's and Q's, but for each such program there will always be inputs for which the program gives a wrong answer or fails to give any answer. Rather than proving exactly this result, we will prove a simpler result that conveys the essential character of such arguments:

> **Theorem 13.2.** There is no TM program that
> - Will always halt,
> - Will accept any input tape that contains a TM program P, such that P would halt if P were fed its own description as input, and
> - Will reject all other inputs. □

This is just a restricted version of the halting problem. It says that no program can decide whether any other program, *given itself as input*, will halt.

To prove Theorem 13.2, we assume the contrary and show that a contradiction results. Essentially, we construct a paradox resembling, "This sentence is false." The reasoning here is a bit subtle, but not deep. If you miss it the first time you may want to reread the next few paragraphs a few times.

Proof of Theorem 13.2. Assume that there is a program, H, that will accept any valid program P that halts when given P as input, and rejects all other inputs. That is, assume that H solves the halting problem. Now, change H as follows: replace every REJECT statement with ACCEPT and every ACCEPT statement with "**while** *true* **do**;" (or any infinite loop). This gives us a new program, H', with the property that H' *accepts* any program, P, that *does not halt* when given P as input, and *goes into an infinite loop* when given any program Q that *halts* when given Q as input.

Let's ask "What will H' do if given H' as input?" If H' were to halt when given H' as input, then by construction H' would not halt when given H' as input. But if H' were not to halt when given H' as input, then by construction, H' must halt, accepting H'.

This is a paradox. Surely H' either halts or it doesn't. But both assumptions cause a contradiction. Therefore the only possible explanation is that H' can't exist. But certainly H' can be effectively constructed from H; thus H can't exist either. □

Technically, we should have distinguished, throughout the above discussion, between the executing program, H', and the string of symbols that represents it. For an informal discussion, such technical considerations are unproductive.

13.2.3.2 Consequences of the Halting Problem

There are several implications of the unsolvability of the Halting Problem. First, we save ourselves a lot of work should we ever decide to write a program that determines if other programs will halt. More significantly, we cannot hope to have a computer completely debug our programs for us, since one possible bug is the failure to terminate properly.

Of greater interest to the computer scientist is the fact that a very large class of other unsolvable problems can be shown to be equivalent to the halting problem and therefore many long and arduous proofs can be avoided. For example, it is not possible to construct an algorithm that will determine whether an arbitrary formula in the first-order predicate calculus is a theorem. A significant branch of theoretical computer science often concerns itself with the discovery of unsolvable problems and their proofs.

On the other hand, we cannot make certain inferences from the knowledge that certain problems are unsolvable. The unsolvability of the halting problem, for instance, *implies* that it is not possible to write a program that can always tell whether an arbitrary program will always halt. However, it *is* possible to write a program that will accept exactly those programs that really do halt and reject *some* of the programs that don't. Programs with this property are called *partial decision procedures*.

13.3 Metalanguages and Productions

You will recall that in our study of FSM's we explored a class of languages defined by regular expressions. We found that this class of languages was precisely the class that could be recognized by an FSR. In this and the following sections we will explore several classes of languages that turn out to be precisely those that can be recognized by NDPDA's and TM's. We define these classes in terms of a set of restrictions on a general language-definition scheme called *production grammars*.

In order to talk about something, and in particular to define it, we must use some language. When the object we are defining is itself a language, the language employed in its definition is known as a *metalanguage.* For example, the language of regular expressions was the metalanguage for defining a number of languages in Part One.

One powerful class of metalanguages, called production grammars, was introduced by the linguist Noam Chomsky in the late 1950s. The statements in these grammars are called productions or rewrite rules. The symbols appearing in each production are of three kinds: terminal symbols (or terminals), metavariables (or nonterminals), and metasymbols. The terminal symbols are symbols from the language being defined. The metavariables represent linguistic concepts that are part of the language being defined. If we were trying to define English, for example, all of the words in the dictionary would be terminal symbols. The nonterminals would be such linguistic concepts as "subject," "predicate," or "noun." The metasymbols serve as punctuation to identify various parts of a production. The following example should illustrate how production grammars work.

Consider the problem of defining all strings of the form $a^n b^n c^n$ that we used in previous sections. The following set of productions defines that language:

1. <legal phrase> → a <b part> c
2. <b part> → b
3. a <b part> → *aa* <b part> b <c part>
4. <c part> c → *cc*
5. <c part> b → b <c part>

Here, the symbols "a," "b," and "c" are terminals; <legal phrase>, <b part>, and <c part> are metavariables; the symbol → is a metasymbol that separates a phrase to be matched from the phrase that will replace it. In particular, we designate <legal phrase> to be the start symbol. A statement of the form $A \rightarrow B$ means, "You may replace an occurrence of A by B." We begin with the start symbol and apply these replacements until we have a string consisting of only terminal symbols. Each string produced along the way is called a *sentential form,* and the last one (with terminals only) is said to be a *sentence* in the language defined by the set of productions, or *grammar.* At any time, several productions may be applicable; we are allowed to pick *any one* of them. For example, the grammar above generates the string "aabbcc" as shown in Fig. 13.9.

The class of languages that can be described by this technique is known as the *type* 0 *languages.* Interestingly enough, we know the following:

Theorem 13.3. The type 0 languages are exactly those that can be recognized by a Turing machine. □

<legal phrase>	The start symbol
a <b part> c	Production 1.
aa <b part> b <c part> c	Production 3.
aabb <c part> c	Production 2.
aabbcc	Production 4.

Fig. 13.9 Derivation of "aabbcc"

Type 0 languages cannot, in general, be recognized by PDA's or FSR's. We will not prove this theorem here.

13.3.1 BACKUS-NAUR FORM (BNF)

In this section we study an important, but restricted form of type 0 grammar—the *type* 2 or *context-free* grammar. Type 2 grammars are often written in a notation known as BNF for Backus-Naur Form, after John Backus and Peter Naur, who adopted the notation in the late 1950's to describe the programming language ALGOL. It is also sometimes called the Backus Normal Form.

   The restriction is that the left-hand side of each production (everything left of the →) must be a single metavariable. By convention, we use the metasymbol ::= instead of → in writing BNF. For example, the following is another definition of unscaled real numbers (real numbers without a power of 10):

<unscaled real> ::= <integer> . <integer>

<unscaled real> ::= .<integer>

<unscaled real> ::= <integer>.

<integer> ::= <digit>

<integer> ::= <integer> <digit>

<digit> ::= 0

<digit> ::= 1

   . . .

<digit> ::= 9

Another convention allows us to merge productions that have identical left-hand sides. The metasymbol |, which is read "or," is used to concatenate the right-hand sides of these productions. Thus the vertical bar is not a part of the language being defined, but a punctuation mark that separates the possible right-hand sides of a merged production. The foregoing example becomes

<unscaled real> ::= <integer> . <integer> | . <integer> |

<integer> .

<integer> ::= <digit> | <integer> <digit>

<digit> ::= 0 | 1 | 2 | 3 | 4 | 5 | 6 | 7 | 8 | 9

To understand the definition better, read ::= as "is defined as"; and when two symbols follow each other on the right of the ::=, insert "followed by" between them. Also, read <x> as "something that satisfies the definition of x." Thus, the line

<integer> ::= <digit> | <integer> <digit>

should be read

An <integer> is defined as *either* something that satisfies the definition of a <digit> *or* something that satisfies the definition of an <integer> followed by something else that satisfies the definition of a <digit>.

The line

<unscaled real> ::= <integer> . <integer> | . <integer> |

<integer> .

should be read

An <unscaled real> is defined as *either* something that satisfies the definition of an <integer> followed by a decimal point followed by something else that satisfies the definition of an <integer>, *or* a decimal point followed by something that satisfies the definition of <integer>, *or* something that satisfies the definition of <integer> followed by a decimal point.

13.3.2 CLASSES OF LANGUAGES

The major reason for introducing BNF here is that it is the most common technique for defining programming languages. Since we already have claimed that regular expressions are too "weak" to define practical programming languages, it follows that BNF is significantly more "powerful." This is indeed the case; for example, consider the following language defined in BNF

<wff> ::= ε | (<wff> <wff>)

where ϵ denotes the null string. This defines the language of nested parentheses in which each matching pair of parentheses contains either nothing (ϵ) or two strings of parentheses, each subject to the same restriction. The strings in this language form a proper subset of the set of all strings of properly nested parentheses; we call it the language of "duplex nested parentheses." For example the following strings belong to the language

   ()

   (((((())))))

   ((()())())

but the following do not

   (()()())

   ((()

   ()(

We already know that this language cannot be accepted by a FSR, since the recognizer needs an unbounded amount of memory in order to keep track of the number of unmatched open parentheses.

   This kind of situation arises in the definition of practical programming languages all the time. Consider, for example, the definition of simple arithmetic expressions, <sae>s, in Fig. 13.10.

<sae>   ::= <t> | − <t | <sae> + <t> | <sae> − <t>

<t>   ::= <p> | <t> * <p> | <t>/ <p>

<p>   ::= X | (<sae>)

Fig. 13.10 A BNF grammar for simple arithmetic expressions

In Figure 13.10, the symbol X represents a simple operand. The reader should explore for himself what kinds of strings belong to a language defined by these rules.

   There is, in fact, a hierarchy of languages requiring more and more powerful grammars to describe them. Languages that can be described by regular expressions are called regular or type 3 languages. Those describable by BNF are known as context free or type 2 languages. In the next section, we show that

Theorem 13.4. The type 2 languages contain the type 3 languages. □

That is, all of the languages that can be described by a type 3 grammar can also be described by a type 2 grammar; moreover, there exist languages that can be described by a type 2 grammar that cannot be described by a type 3 grammar. Since BNF is a restricted form of type 0 grammar, we can also deduce

Theorem 13.5. The class of type 0 languages contains the class of type 2 languages. □

There is, in fact, a class of type 1 languages called context-sensitive languages, but we will not treat them in this textbook.

Just as type 0 languages correspond to TM's, type 2 languages correspond to NDPDA's. That is

Theorem 13.6.

a) For every NDPDA that recognizes a language L, there is a BNF grammar for L.

b) For any language, L', described in BNF, there is a NDPDA that accepts exactly L'. □

Exercise 13.30 is concerned with a proof of part (b) of this theorem. A proof for part (a) is beyond the level of this text.

Since much of the syntax of existing programming languages can be described using BNF, this theorem suggests a way to organize a program that *parses*—determines the grammatical structure of—programs. This is an important part of programming language translation. Actually, the problem is by no means as simple as Theorem 13.6 might lead you to believe. For one thing, it is difficult to write nondeterministic programs in most programming languages. Parsing is a large subject, however, the details of which are beyond the scope of this textbook.

13.3.3 BNF DESCRIPTION OF TYPE 3 LANGUAGES

Theorem 13.4 in the previous section asserted that BNF can describe type 3 languages; the proof is quite straightforward. Every regular expression, R, can be translated into the definition of a metavariable, $<L_R>$, describing the same language as follows:

1. If R is a single terminal symbol—say "a"—translate it to

   $<L_R> ::= a$

2. If R is $(R_1 R_2)$, translate it to

   $<L_R> ::= <L_{R_1}> <L_{R_2}>$

   where $<L_{R_1}>$ and $<L_{R_2}>$ are metavariables defining the languages described by R_1 and R_2.

3. If R is $(R_1 + R_2)$, translate it to

$$<L_R> ::= <L_{R_1}> \mid <L_{R_2}>$$

4. If R is $(R_1)^*$, translate it to

$$<L_R> ::= \epsilon \mid <L_{R_1}> <L_R>$$

Note that in each of the above cases, we used a production of the form

$$<L_R> ::= \beta$$

where β may contain $<L_R>$ and some other metavariables. Suppose that we put the production that defines those other metavariables *ahead* of the production that defines $<L_R>$ and eliminate any duplicate productions. Now we have a BNF grammar:

Definition 13.2. A BNF grammar has the *extended right-linearity property* if its productions can be ordered so that for any production

$$<A> ::= \beta_1 \mid \beta_2 \mid \cdots \mid \beta_n$$

each β_i contains only *previously defined* metavariables—those occurring on the left sides of preceding productions only—and $<A>$. Furthermore, if $<A>$ occurs in any of the β_i, it must occur as the rightmost (last) symbol in that β_i. $\square$

(We use the term "extended" to distinguish this definition from the standard definition of "right linearity" that appears in the literature.) As you might guess,

Theorem 13.7. Any extended right-linear BNF grammar describes a type 3 language. Any type 3 language has an extended right-linear BNF grammar. $\square$

The second part of Theorem 13.7 is simply a restatement of the preceding discussion. To prove the first part, do Exercise 13.31. We may also define *extended left-linearity* by substituting "leftmost (first)" for "rightmost (last)."

Theorem 13.8. Any extended left-linear BNF grammar describes a type 3 language. Any type 3 language has an extended left-linear BNF grammar. $\square$

We can prove this by straightforward modification of the proof of Theorem 13.7.

Let's consider an example of a linear grammar. The term "identifier" was defined as a regular expression in Section 1.3. What follows is an extended right-linear BNF grammar:

<letterordigit> ::= A I B I C I D ··· I 0 I ··· I 9

<postfix> ::= <letterordigit> I <letterordigit> <postfix>

<identifier> ::= <letter> I <letter> <postfix>

While all practical programming languages require the power of a type 2 grammar, often significant portions of their definitions are regular (that is, describable by a type 3 grammar). Hence the syntactic analysis of these languages is often split into two parts; *one* of these is a fast FSR (called a lexical analyzer) that accepts the regular parts of the language (see Chapter 19).

13.3.4 EMBEDDING

The finite-state restriction of FSA's implies that a particular FSA can distinguish between only a finite number of different input histories. Broad classes of input sequences must be treated as equivalent with respect to the future behavior of the machine. For example, the FSA that accepts the regular expression a*b must treat all strings of a's equivalently, no matter how many a's they contain. Languages defined by BNF are richer essentially because this restriction is relieved to some extent. Consider, for example, the definition

<px> ::= x I (<px>)

which defines the language {x, (x), ((x)), (((x))), ···}. It's clear that the proper matching of left and right parentheses in this language requires a potentially infinite memory; because the number of right parentheses most exactly match the number of left parentheses, an infinite number of different cases may arise. A careful analysis of BNF, one more careful than we'll pursue, will show that the increased power of BNF comes from a property called *self-embedding*, or just *embedding*.

We derive strings from the production rules of a BNF grammar by repeatedly applying the rules until no nonterminal symbols remain. Let's use α and β to denote arbitrary strings. Then, the notation <x> → α means that α, a string of terminal and nonterminal symbols, can be derived from the metavariable <x>. A BNF grammar is called self-embedding if there is at least one nonterminal of the language, say <x>, such that

<x> → α <x> β

where both α and β are nonempty strings. That is, starting with some metavariable, we can derive a sentential form in which that metavariable is surrounded by nonempty strings. Note that the form of the right- and left-linear grammars preclude embedding, which is why they are equivalent to regular expressions.

13.3.5 PARSE TREES

"Diagramming sentences" is sometimes taught in high school language classes. The diagram that results from this process is a pictorial representation of the grammatical structure of a sentence. For sentences described by BNF, we get similar diagrams, called parse trees. In this section, we explore them briefly.

Let's return for a moment to the grammar of Fig. 13.10 that defined simple arithmetic expressions, <sae>'s, in a variable X. Now, it happens that the string (X * (X + X))/X can be derived from <sae>. To see that this is so we need only apply a sequence of the productions to get the following sequence of sentential forms:

$$<sae> \rightarrow <t>$$
$$\rightarrow <t> / <p>$$
$$\rightarrow <p>/ <p>$$
$$\rightarrow (<sae>)/<p>$$
$$\rightarrow (<t>)/<p>$$
$$\rightarrow (<t> * <p>)/<p>$$
$$\rightarrow (<p> * <p>)/<p>$$
$$\rightarrow (X * <p>)/<p>$$
$$\rightarrow (X * (<sae>))/<p>$$
$$\rightarrow (X * (<sae> + <t>))/<p>$$
$$\rightarrow (X * (<t> + <t>))/<p>$$
$$\rightarrow (X * (<p> + <t>))/<p>$$
$$\rightarrow (X * (X + <t>))/<p>$$
$$\rightarrow (X * (X + <p>))/<p>$$
$$\rightarrow (X * (X + X))/<p>$$
$$\rightarrow (X * (X + X))/X$$

The same information can be displayed in a parse tree such as that in Fig. 13.11. The leaves of the tree are terminal symbols; indeed, the collection of leaves is the sentence derived above. The interior nodes of

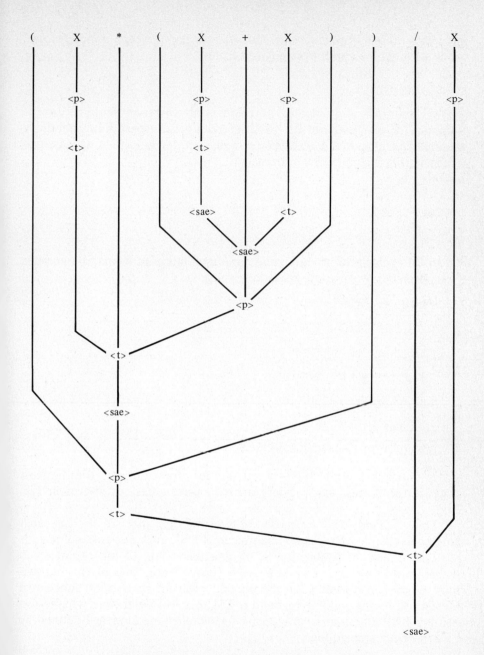

Fig. 13.11 A parse tree for $(X * (X + X))/X$

the tree are nonterminals, and the arcs ascending from a nonterminal connect the symbols that were derived from it.

13.3.6 AMBIGUITY

A BNF grammar is ambiguous if there is at least one sentence in the grammar for which there are two different parse trees. Equivalently, a grammer is ambiguous if there are two distinct derivations of at least one sentence. For example, the grammar

<f> ::= <g> | <h>

<g> ::= X

<h> ::= X

is a trivial example of an ambiguous grammar since an X may be either a <g> or an <h>. A more realistic example is

<x> ::= <y> | <z>

<y> ::= <bc> | a <y>

<z> ::= <ab> | <z> c

<bc> ::= bc | b <bc> c

<ab> ::= ab | a <ab> b

This grammar defines the language

$$L = \{a^i b^j c^k \mid i = j \text{ or } j = k\}$$

and is ambiguous for $i = j = k$. Consider, for example, the simplest string of this form, "abc." There are two parse trees, as shown in Fig. 13.12.

This language is especially interesting in that it is *inherently ambiguous*—that is, there is *no* unambiguous BNF grammar that defines it!

The notion of ambiguity is of practical interest, for example, to programming language designers. As a consequence, one of the implications of the unsolvability of the halting problem is also of substantial interest: It is not possible to build a TM that will determine whether an arbitrary BNF grammar is ambiguous (or whether the language defined by it is inherently ambiguous).

13.4 Further Reading

For those interested in the capabilities of the various automata, (PDA's, NDPDA's, and TM's), as well as those who would like to see proofs we

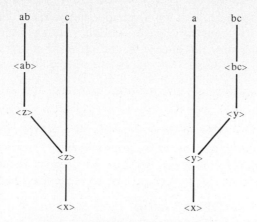

Fig. 13.12 Two parses for "abc"

omitted in this chapter, Hopcroft and Ullman (1969) is a definitive work. Minsky (1967) is also helpful, and it is not as formal.

BNF was first introduced in the syntactic definition of the programming language ALGOL 60, Naur (1963), whereupon it was noted that there were important connections between it and the context-free languages described earlier by Noam Chomsky (1956 and 1959).

A complete treatment of the subject of parse trees and parsing in general can be found in Aho and Ullman (1977).

Exercises

13.1 Which of the following are "effective procedures"? Why or why not?

    a) "Bake at 350 degrees for 50 minutes." (From a cook book.)

    b) "Repeat this procedure rhythmically until the dough becomes smooth, elastic, and shiny." (Instructions for kneading bread.)

    c) "First, catch a rabbit." (Old Welsh rabbit stew recipe.)

    d) "Perform takeoffs into strong crosswinds with the minimum flap setting necessary for the field length to minimize the drift angle immediately after takeoff." (From a pilot's operating handbook.)

    e) "Follow route 25 south to route 3A just past Polar Caves. Take route 3A south to Bristol, then take route 104 south toward Danbury." (Route directions.)

f) "A superior, man or woman, calling upon another employee may, of course, smoke without asking permission, but an outsider may not smoke in the office of someone else unless he is asked to do so." (From an etiquette guide.)

Find other examples of instructions that are (or are not) effective procedures.

13.2 Construct a PDA that accepts the set NMN, the set of strings containing N 1s followed by M 0s followed by N 1s for all nonnegative values of N and M. This is the set that we used in Section 1.5.1 to show that there were languages that could not be recognized by an FSM.

13.3 Consider the DPDA's in Fig. 13.13; as in the text, a question mark symbol (?) in a transition denotes several transitions, one for each possible replacement of the ?. For each of these DPDA's, (1) describe the set of strings it accepts, and (2) trace the execution of the DPDA for one "interesting," (or nontrivial) string.

(a)

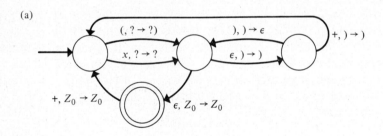

(b)

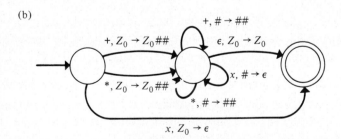

Figure 13.13

13.4 Consider the NDPDA's in Fig. 13.14. As in Exercise 13.3, (1) describe the set of strings that each accepts, and (2) trace the execution on one "interesting" string.

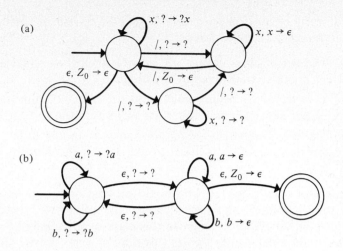

(a)

(b)

Figure 13.14

13.5 In Chapter 1 we formalized the definition of an FSM by specifying it to be a 4-tuple of states, an initial state, and NEXT and OUT functions. By analogy, define a general PDA. Apply your definition to write descriptions of the PDA's in Figs. 13.2, 13.4, and 13.6.

13.6 As we have defined them, PDA's have no output.

  a) Generalize the definition of PDA's in Exercise 13.5 to include an explicit output.

  b) Propose a notation for the labels of the diagrammatic description of PDA's.

  c) Show how you can use your generalized PDA's to output the reverse of an input string.

  d) Does your generalization add "power" to PDA's? Why?

13.7 Show that the following theorem is correct:

Theorem 13.9. Any language that can be recognized by a PDA, M, can be recognized by a PDA, M', that has no more than two states. □

The trick is to expand the stack alphabet of M' so that it can, in effect, keep track of the state of M as well as its stack contents.

13.8 Show that the following theorem is correct:

Theorem 13.10. Any language that can be recognized by a PDA, M, can be recognized by a PDA, M', whose stack alphabet has, at most, two symbols. □

13.9 For each of the following sets of strings, devise the simplest PDA you can to recognize exactly that set.

    a) The set of all strings of a's and b's having an equal number of a's and b's, intermixed in any order.

    b) The set of all palindromic numbers (that is, numbers such as 121 or 1441 that read the same backwards and forwards.)

    c) The set of all strings of a's, b's, and c's, in any order, such that the number of c's is equal to the sum of the number of a's and b's.

13.10 Consider the class of arithmetic expressions that involve just addition, multiplication, and parentheses. (Exercise 1.46 asked whether you could build an FSR to recognize this class of expressions.) Is it also possible to build a PDA to recognize this class of expressions? If so, do it. If not, why not?

13.11 Use Theorem 13.1 to show that no PDA can recognize the language

$$K = \{wxwyw \mid w, x, \text{ and } y \text{ are any strings}\}$$

[Hint: assume the contrary.] Extend your result to show why no PDA can recognize the set of all programs in which all identifiers are correctly declared before use.

13.12 Design Turing machines to compute the following functions, placing the final results on their work tapes in binary, with all other squares on the tape blank. The input tape will contain one or more numbers in binary representation, with the least significant (right-most) digits first and with each number terminated with a blank.

    a) The "monus" function, defined

        monus(x) = **if** $x > 0$ **then** $x - 1$ **else** 0,

        where x is the first number on the input tape.

    b) The sum of the first two numbers on the input tape.

    c) The product of the first two numbers on the input tape.

13.13 Design a TM that accepts (halts in an accepting state) any input consisting of a prime number of 1s followed by a blank and that rejects all other inputs. Can this be done by an FSM? By a PDA?

13.14 In Section 13.1.2 we argued informally that a PDA is limited by the fact that to reference an item in its stack, it must throw away the part of the stack above that item. A PDA can't save all of what it "throws away" because apart from its stack, it has only a finite memory. Suppose we defined PDA's to have *two* stacks that can be manipulated independently. Show that these improved PDA's would be as powerful as Turing machines.

13.15 What does the following Turing machine do when started with an input tape containing "0101101/101101;"? What does it do in the general case?

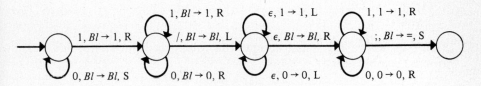

Figure 13.15

13.16 We said in Section 13.2.1 that adding nondeterminism to Turing machines would not increase the class of things they could do. Show why. Hint: If you were asked to work out the action of a nondeterministic Turing machine on paper, what would you do?

13.17 Write a BNF definition of the language defined by each of the following regular expressions

a) a^*b^*c

b) $b^*cb^* + c^*bc^*$

c) $a^*(bc + d)^*$.

d) $(a(c + d))^* + (b(e + f))^*$

13.18 Give a regular expression that defines the same language as defined by the following BNF productions with the start symbol $<s>$.

$<s> ::= <x> <y>$

$<x> ::= b \mid <c>$

$<c> ::= \epsilon \mid c <c>$

$<y> ::= \epsilon \mid d <y>$

13.19 The set of valid regular expressions is itself a context free language.

Give a BNF description of this language.

13.20 The set of valid BNF definitions is itself a context free language. Give a BNF definition of BNF. (Note that you will have to be careful with the symbols :: =, <, >, and I. You will have to adopt some convention to distinguish their use in the definition from their use in the language being defined. Carefully state your convention.)

13.21 Construct a parse tree for the following PASCAL program; use the grammar given in any PASCAL manual:

```
var i, j: integer;
begin
i := i + j * 2;
end.
```

13.22 Construct a PDA that accepts the following context free language, with start symbol <*S*>:

    <S>   ::= <NP> <VP>

    <NP> ::= the <NP1> I <NP1>

    <NP1> ::= <ADJ> <NP1> I <N>

    <N>   ::= boy I girl I cats I dog I people

    <VP> ::= <V> <NP> I <V>

    <V>   ::= walks I talks I likes I hit

    <ADJ> ::= small I big I furry I playful

Illustrate the operation of the PDA by showing its execution on the input "small furry people hit big cats."

13.23 The following context free grammar, with start symbol <*S*>, is ambiguous:

    <S> ::= <A> <S> I <A>

    <A> ::= <A> <S> I <B>

    <B> ::= b <S> a I ba

    a) Show that this grammar is ambiguous by giving a sentence that has at least two derivations.

    b) Is the language defined by this grammar inherently ambiguous? If so, why? If not, construct an unambiguous grammar to describe the language.

13.24 Consider the following BNF grammar, where <s> is the start symbol.

      <s> ::= X | (<a>)

      <a> ::= <s> | <a> <s>

Let S be the set of all strings formed from the symbols {X, (,), <s>, <a>} (not necessarily ones derivable by the grammar), such as X, (X, <s><a>, X<s>X. Define the relation $\rightarrow$ on S by the rule

> **Definition 13.3.** $x \rightarrow y$ for x and y in S iff by replacing one of the nonterminals in x according to one of the productions above, one can get y. If there are no nonterminals in x, then $x \rightarrow y$ is always false. □

Thus,

      ))<a><s> $\rightarrow$))<a>X,

but it is not true that

      ))<a><s> $\rightarrow$))<a>) or))<a><s> $\rightarrow$))<a>(<s>).

    a) Is $\rightarrow$ reflexive? symmetric? transitive? antisymmetric?   In each case, briefly defend you answer or give a counter-example.

    b) Let $*\rightarrow$ be the transitive closure of $\rightarrow$. Suppose y is some string in S containing only terminals. Suppose <s> $*\rightarrow$ y. What can you say about y? Suppose that it is *not* the case that <s> $*\rightarrow$ y. Now what can you say about y?

13.25 Consider the following BNF definition. The start symbol is <b>.

      <b> ::= b <ds> <ss> e | b <ss> e

      <ds> ::= Ds | <ds> Ds

      <ss> ::= <s> | <ss> s <s>

      <s> ::= S | <b> | I <s> T <s> E <s>

    a) Draw a parse tree for the string "bDsDsSsISTbSsSeESsSe."

    b) Is "bsSe" defined by this syntax? Why?

    c) Is "bDsSse" defined by this syntax? Why?

    d) Is the language defined by these rules describable by an RE? Why?

    e) Construct a PDA that accepts the strings in the language defined by its grammar. Illustrate its operation on a typical string.

13.26 The definition of $<$sae$>$ in Fig. 13.10 doesn't allow such function invocations as $SIN(X + X)$. Change the definition of $<p>$, and add any additional productions necessary to allow in expressions operands of the form $F(e_1)$, $F(e_1,\ e_2)$, $F(e_1,\ e_2,\ e_3)$, etc. Here, e_1, e_2, e_3, etc., are themselves $<$sae$>$'s. For example, the following should be legal under your extended definition.

$$X * F(X + X/X, X)$$
$$F(X)$$
$$F(X,\ X) + F(X)$$
$$F(X - F(X * (X + X - F(X, X * X, X + X - X, X/X))))$$

13.27 Construct a PDA to recognize the language defined by your extension of $<$sae$>$'s in Exercise 13.26.

13.28 Define a language that can be recognized by a PDA, but not by an FSR. Give the language definition in BNF and show the PDA.

13.29 Define a language that can be recognized by a TM, but not by a PDA. Give the language definition as a production grammar and show the TM.

13.30 Prove the second part of Theorem 13.6. That is, take the following steps to show that for every BNF grammar there corresponds a NDPDA that recognizes the same language.

- Suppose that you have a phrase containing nonterminals and terminals that has been generated from the start symbol by applying the BNF grammar production rules. Show that it does not matter which nonterminal you choose to subject to a production; you can always get the same final string. Thus you may always choose the left-most nonterminal for the next production without loss of generality. (Can you see now why BNF grammars are called "context-free"?)

- Now consider a NDPDA whose stack alphabet is the set of terminals and nonterminals. This NDPDA always does one of two things: *either* it compares the top of its stack with the next input, pops the stack if they are equal, and jams if they are not; *or* if the top of its stack is a metavariable, it pops off that metavariable and pushes on one of the things which that particular metavariable can produce. That is, if

$<$integer$>$ ::= $<$digit$>$ | $<$digit$>$ $<$integer$>$

is in the BNF grammar and $<$integer$>$ is on the top of the stack, the NDPDA can pop off $<$integer$>$ and push on $<$digit$>$, followed by <integer>. Show how this NDPDA can be made to recognize the language described by the BNF grammar. How many states will it need? What is the first thing it must push on the stack?

13.31 Prove that any extended right-linear BNF describes a type 3 language. Hint: essentially, this involves reversing the construction of Section 13.3.3. What properties of extended right-linear grammars make it valid to do so?

13.32 Write a program to convert regular expressions to the production grammar of Section 13.3.3.

13.33 Write a sentence generator. Set it up to read a BNF grammar from a file. Feel free to modify the standard form of BNF definitions to make them more convenient for your program to read. One way to make your program work is to make it simulate the following process:

Begin with the start symbol;
while there are any nonterminals left **do**
    Replace the left-most nonterminal symbol with one of
        the possibilities in the BNF grammar (chosen at
        random if there is more than one);
Print the resulting string of terminal symbols

Consider, for example, the grammar of Exercise 13.22. A possible computation sequence according to the above algorithm is

<S> → <NP><VP> →
the <NP1><VP> → the <N><VP> →
the boy <VP> → the boy <V><NP> →
the boy likes <NP> → the boy likes <NP1> →
the boy likes <adj> <NP1> →
the boy likes playful <NP1> →
the boy likes playful <N> →
the boy likes playful cats

13.34 Suppose that G is a context-free grammar. Show that there is another grammar, G', describing exactly the same language and, in which,

- The start symbol only appears to the left of ::= symbols, and
- the empty string ϵ, if it appears anywhere, appears only in a single production of the form

    <start symbol> ::= ϵ.

13.35 Using the definition of a PDA developed in Exercise 13.5 as a guide, construct a program that will

- a) Read in the definition of a *deterministic* PDA and check it for legality.
- b) Simulate the execution of the PDA that was read in, and print a meaningful trace of this execution.

You may assume that the stack will never get very deep; thus it is acceptable to use a fixed sized array to represent it.

13.36 Construct a program to read in the definition of a Turing machine, simulate its behavior, and print a trace of its execution. You may assume that the work tape will never get very long, so it is reasonable to represent it by an array. Illustrate the execution of your program with several of the TM's discussed in the text.

13.37 Suppose the assumption of "smallness" in both Exercises 13.35 and 13.36 were relaxed so that arrays would no longer be appropriate. Modify the program(s) to use the filing system of your computer to create a much larger memory capacity. Investigate what the practical limits of this scheme are on your computer, and report the results.

13.38 Both Exercises 13.35 and 13.36 concern the simulation of deterministic machines. Discuss the problems of, and design an implementation of programs that would simulate nondeterministic PDA's or TM's.

Chapter 14

Recursion and Related Topics

As illustrated in Section 2.2, an essentially equivalent FSM exists for any program that

a) Has only variables taking values from finite sets.

b) Uses only simple control structures: sequencing, conditional and unconditional transfers, and iteration.

We can add nonrecursive procedures (those that do not "call themselves" either directly or indirectly) and still maintain this equivalence. Nonrecursive procedures can be thought of as abbreviation mechanisms. In principle, each call could be replaced by a suitably modified version of the procedure body. Many programs satisfy these criteria. After all, scalar variables usually satisfy (a), as do arrays and records whose elements take values from finite sets.

In this chapter we explore features of programming languages that make the FSM model *inapplicable*. We may categorize these features according to which of the criteria, (a) or (b), they violate. We may relax (a) by introducing *dynamic data types* and relax (b) by introducing *recursive routines*.

After examining dynamic data types and recursion, we discuss two important programming techniques that use them: *divide-and-conquer* and *recursive operations on dynamic types*.

375

14.1 Dynamic Data Types

Section 13.1.1 showed that the addition of a stack to an FSM gave us an automaton of greater power. The reason, again, is that a stack can get arbitrarily large; the value of a stack—the sequence of values stacked within it—can be any of an infinite number of things. The type *stack* is an example of what we'll call a dynamic data type. Trees, graphs, linked lists, and any other type whose variables can "grow" in a similar fashion are also examples of dynamic data types.

> **Definition 14.1.** A *dynamic data type* is one whose domain of values is, in principle, infinite. □

We must hedge our definition with "in principle" because, of course, there are always limitations on machine resources. Note also that although the *domain* of possible values is infinite, any *particular* value is finite. For example, there are an infinite number of possible lists of integers, but any particular list has only a finite number of integers in it.

Lists and trees are examples of dynamic data types, but it is easy to come up with less exotic ones. Consider, for example, the type *integer*. As usually defined, there is a maximum and a minimum *integer* value; *integer* is not a dynamic type. One can imagine removing this restriction, so that the domain of the data type *integer* is, in fact, the standard domain of number-theoretic integers. Then *integer* would be a dynamic data type. The specifications change only in that some restrictions go away. That simple change in specifications, however, implies a large change in implementation. Since a single machine word, or any fixed number of words, cannot hold an arbitrarily large integer, we have to resort to a dynamically expandable representation (a linked list of digits, for example).

14.2 Recursion

Consider the language of nested parentheses used as an example of BNF in section 13.3.1:

$$\langle wff \rangle ::= \epsilon \mid (\langle wff \rangle \langle wff \rangle)$$

We may read this as "A $\langle wff \rangle$ is either the null string or two smaller $\langle wff \rangle$'s enclosed in parentheses." We refer to such definitions as *recursive* or *self-referential.*

Suppose we have the task of reading in valid $\langle wff \rangle$'s. That is, we want a procedure, *ReadWff,* specified

procedure *ReadWff*;
{Read in the longest possible valid <wff>. It is an error if
the input begins with an opening parenthesis that is not the start
of a valid <wff>.}

We specify "longest possible" to avoid ambiguity; otherwise, the definition
of <wff> always allows us to read in nothing (ϵ) and call it a <wff>.
The second sentence of the specification simplifies our program. It means
that, for example,

(()())

causes an error. That is, the longest valid <wff> for this input is ϵ, but
we only discover that fact after reading the first five characters of the
input. We therefore make this input illegal.

The definition of <wff> itself suggests a skeletal implementation of
the body of *Readwff*, namely

begin
if next character is '('
    then Read '(<wff> <wff>)'
    else Read nothing
end

Presumably, it is easy to "Read nothing"; the difficult part is "Read
'(<wff><wff>)'." Reading the two parentheses isn't difficult, but what
about those inner <wff>'s? The procedure *ReadWff* is supposed to read
in a <wff>, which is just what we want. So we can expand the skeleton
to

begin
if next character is '('
    then begin
        Read the '(';
        ReadWff; *ReadWff*;
        if the next character is ')'
            then Read the ')' **else** *ERROR*
        end
    else Read nothing
end

The natural objection to this routine is that it is supposed to be the
body of *ReadWff* and calls on *ReadWff* from within its own body appear to
be questionable—if not downright circular. Nonetheless, this program was
built by following the definition of <wff>, and we expect the program to
work if the definition "works."

The program does, in fact, work. To understand this a little better, consider a typical execution trace. Figure 14.1 shows the execution of *ReadWff* for the input (()) alongside a derivation of (()) from the BNF definition. The asterisks indicate the portion of the input that has been read at each point. The arrows (→) indicate the places where *ReadWff* has decided which of the two possible productions for <wff> it will use.

Call *ReadWff*	* <wff>
See '(' and read it;	→ (* <wff> <wff>)
Call *ReadWff*;	
See '(' and read it;	→ ((* <wff> <wff>) <wff>)
Call *ReadWff*;	
Don't see '(',	
so read nothing;	→ ((* <wff>) <wff>)
Leave *ReadWff*;	
Call *ReadWff*;	
Don't see '(',	
so read nothing;	→ ((*) <wff>)
Leave *ReadWff*;	
Read the ')';	→ (( ) * <wff>)
Leave *ReadWff*;	
Call *ReadWff*;	
Don't see '(',	
so read nothing;	→ (( ) *)
Leave *ReadWff*;	
Read the ')';	→ (( )) *
Leave *ReadWff*;	

Fig. 14.1 Execution trace of *ReadWff* on (())

The reason that *ReadWff* doesn't go into an infinite loop, as its apparent circularity might suggest, is that each time *ReadWff* is called, it has less input to work on. Sooner or later input runs out, and *ReadWff* must stop.

ReadWff is known as a *recursive routine*, just as the definition of <wff> is called a recursive definition. The structure of *ReadWff* is typical of many recursive routines. Its working part consists of a conditional statement of the form

> **if** condition
> > **then** some statements, possibly involving recursive calls
> > **else** some nonrecursive statements

The routine keeps calling itself until it finally takes an **else** branch and returns one level. Eventually, it works its way back to the original call.

Essentially, then, this example shows how a recursively defined problem leads naturally to a recursive procedure. Later in this chapter, and again in Chapter 22, we'll discuss how procedures that operate on recursively defined data structures are also naturally recursive.

14.3 The Power of Recursion

Section 14.2 defined a recursive procedure *ReadWff*, which reads strings in the language defined according to a particular BNF description. Theorem 13.6 says that there exists a NDPDA that does the same. In fact, for this case, there is a deterministic PDA that recognizes this language; it is shown in Fig. 14.2.

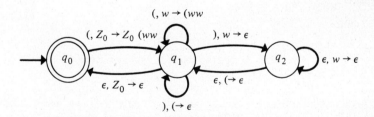

Fig. 14.2 A DPDA that recognizes a properly nested parentheses language

Essentially, a DPDA differs from an FSM only in the addition of a stack. An FSM, furthermore, corresponds to an ordinary, nonrecursive program that uses only nondynamic data. This suggests that we might be able to replace *ReadWff* with an equivalent nonrecursive program that uses a stack. In fact, we can. Using the stacks of Section 9.1.3, we can mimic the DPDA directly, as we show in Fig. 14.3.

```
procedure ReadWff2;
    type PlaceType = (Start, Place1, Place2, Stop);
    var S: stack;      {Stack of PlaceType}
        Place: PlaceType;
    begin
    Clear(S);  Place := Start;  Push(S, Stop);
    repeat
        case Place of
            begin
            Start:  if next character is '('
                        then begin
                            Read the '(';  Push(S, Place1);  Place := Start
                        end
                        else Pop(S, Place);
            Place1:begin
                        Push(S, Place2);  Place := Start
                    end;
            Place2:begin
                        if next character is ')'
                            then Read the ')'
                            else ERROR;
                        Pop(S, Place)
                    end
            end
        until Place = Stop
    end
```

Fig. 14.3 A nonrecursive version of *ReadWff*

Close examination of this program reveals that we have taken apart the original recursive *ReadWff* and put the pieces back within a repeated conditional statement. The variable "*Place*" tells "where to go next." It takes one of four values, corresponding to four locations in the *ReadWff* program:

- *Start*: The beginning of *ReadWff*.
- *Place1*: The point just after the first recursive call on *ReadWff*.
- *Place2*: The point just after the second recursive call on *ReadWff*.

- *Stop*: The point at which *ReadWff* is to return to its original calling program.

Instead of making recursive calls on *ReadWff*, *ReadWff2* pushes onto *S* a value for *Place* corresponding to the place where each recursive call is to return. It then sets *Place* to *Start*, since in *ReadWff* control then goes to the beginning of the recursively called *ReadWff* routine. In places where *ReadWff* returns, *ReadWff2* substitutes a "*Pop(S, Place)*," setting *Place* to the previously saved value where each recursive call is to return.

Note the striking similarity between the use of the stack here and in the representation of procedures (Section 4.3). In both cases, we used the stack to hold return addresses. In general, we may use the stack to hold all the data in the recursive routine that must be saved over the recursive call. For *ReadWff2*, this data consisted of the current position in the program. Section 14.5.3 includes an example where additional internal data is stored.

The construction we used to get *ReadWff2* from *ReadWff* can be made quite general. Indeed, in our attempt to make the relationship between the two programs clearer, we made *ReadWff2* rather more complicated than necessary. But we did so to illustrate the following:

> **Theorem 14.1.** Assume that f is a recursive routine all of whose parameters and local variables can be copied into a stack. Then there is an equivalent nonrecursive routine employing a single additional stack. Thus recursion adds to a program no more computational power than a stack. □

We won't try to prove this rigorously; basically, the proof involves nothing more than elaborating the conversion we made from *ReadWff* to *ReadWff2*.

Knowing that any recursive program may be converted to a nonrecursive one, we can always feel free to formulate algorithms recursively, because we can always convert them if necessary (if we are using FORTRAN, for instance). This flexibility can be advantageous, for there are many problems for which recursive formulations of the solutions are conceptually simpler than nonrecursive formulations.

14.4 Divide-and-Conquer

In this and the following section we explore the use of the two mechanisms discussed in this chapter—recursion and dynamic data types. One of the fundamental tricks of the trade in the design of algorithms is the technique of divide–and–conquer. A typical application of divide–and–conquer has the following recursive structure.

```
procedure Solve(x);
    begin
    if x is divisible into smaller problems
        then begin
        Divide x into two or more parts: x₁, · · · , xₖ;
        Solve(x₁); · · · ; Solve(xₖ);
        Combine the k partial solutions into a solution for x
        end
    else Solve x directly;
```

If combining small solutions to parts of a problem is substantially simpler than solving the whole problem directly, we can use this technique to produce rather efficient algorithms.

Consider, for example, the problem of finding the range of values in a given vector. (This example is taken from Aho, Hopcroft, and Ullman (1974).) That is, we need a procedure $range(A, i, j, low, high)$ that sets the output parameters low and $high$ to the minimum and maximum elements of the vector A between $A[i]$ and $A[j]$. An obvious implementation of $range$ is given in Fig. 14.4. The program simply scans the array, comparing each element with the largest and smallest seen thus far. Using the techniques of Chapter 6, it is easy to see that the cost of $range$ in terms of the number of comparisons between array elements will be $2N-2$ for a vector of N elements.

```
procedure range(A: array of integer;
                i, j: integer;
                var low, high: integer);
    var k: integer;
    begin
    low := A[i]; high := A[i];
    for k := i + 1 to j do
        begin
        if A[k] > high then high := A[k];
        if A[k] < low then low := A[k]
        end
    end
```

Fig. 14.4 Iterative range algorithm

Note that each element of A will be compared twice—once in finding the maximum and once in finding the minimum. The algorithm fails to take advantage of the fact that, for example, any element considered (from the successive values of *low*) to be a candidate for the minimum could never be also a candidate for the maximum except for the initialization condition. Thus our algorithm expends unnecessary effort examining each element twice. Suppose we want to avoid this unnecessary overhead. Consider the program of Fig. 14.5.

```
procedure range(A:  array of integer;
                i, j: integer;
                var low, high: integer);
   var mid, l1, u1, l2, u2: integer;
   begin
   if j ≤ i + 1 then
      begin
      if A[i] < A[j]
           then begin low := A[i]; high := A[j] end
           else begin low := A[j]; high := A[i] end
      end
   else
      begin
      mid := (i + j) div 2;
      range(A, i, mid, l1, u1);
      range(A, mid + 1, j, l2, u2);
      low := min(l1, l2);
      high := max(u1, u2)
      end
   end
```

Fig. 14.5 Recursive range algorithm

As you see, for short array segments (1 or 2 elements) the procedure includes a direct comparison. For longer segments, the procedure divides the interval in half and then compares only the largest and smallest results of these subtasks.

What is the cost (in number of comparisons) of this version of *range* for a vector of size N? We can sum the costs of the algorithm's components; but *range* calls itself and its cost is not yet known, so the resulting expression isn't *obviously* helpful. We present a general method for analyzing the cost of recursive routines in Chapter 18, but for the example of Fig. 14.5 the cost is $(3/2)N-2$ where N is a power of 2. For

other cases, that formula is a good approximation. A derivation of this
cost appears in Chapter 18, so we defer further discussion until then.

Let's take a second example of divide–and–conquer, but this time
let's develop the recursive algorithm directly instead of starting with an
iterative program. Suppose we want to sort a vector, V, in ascending
order. Following the divide–and–conquer "hint," we might divide V into
halves and sort as in Fig. 14.6.

```
procedure Sort(var V: vector; L, U: integer);
    {Sort the elements L through U of V in ascending order}
    var m: integer;
    begin
    if L < U then
        begin
        m := (L + U) div 2;
        Sort(V, L, m);  Sort(V, m + 1, U);
        {Combine the sorted halves of V}
        end
    end;
```

Fig. 14.6 Mergesort: a divide–and–conquer algorithm

Note the test for $L < U$: If $L \geq U$, we needn't do any work within
the segment $V[L..U]$, for that subarray contains at most one element.
Now we have to combine the sorted halves, a process called *merging*.
Because the two halves are already sorted, merging is a fast and
straightforward operation (although it is *not* easy to do in place.)

14.5 Recursive Operations on Dynamic Types: Tree Walking

Many of the algorithms used to manipulate trees have particularly elegant
recursive formulations. This is because *tree* is a *recursive type*, and again
the structure of Definition 9.4 suggests the structure of algorithms that
work on trees.

For example, one often has to *traverse* all the nodes in a tree,
performing some action on each one. If a tree is empty, there are no
nodes, and nothing needs to happen (part 1 of Definition 9.4). If the tree
is not empty, we must perform the desired action on its root node and
then traverse recursively each of its subtrees (part 2 of Definition 9.4).

Thus we have the abstract algorithm of Fig. 14.7.

procedure *Traverse*(*T*: *tree*);
   begin
   if not *IsEmpty*(*T*) **then**
      begin
         perform action on root of *T*;
         Traverse(*T*'s first subtree);

         .
         .
         .

         Traverse(*T*'s k^{th} subtree)
      end
   end

Fig. 14.7 A general tree traversal

Traversals of this sort are also called *tree walks*. We have included numerous examples of tree walks in the following sections, as well as the three chapter-length examples in Chapters 20, 22, and 23.

14.5.1 ALTERNATIVE ORDERS AND SOME NOTATION

Nothing in Definition 9.4 suggests that we must do things in precisely the order used in Fig. 14.7. The order in which we traverse the subtrees and the point(s) at which we perform the action(s) on the nodes depend on the particular problem we are solving. It is useful to have a notation for expressing the order used by a particular traversal algorithm.

    Since binary trees are so common, we'll start with them. For sample data, we'll use the tree in Fig. 14.8.

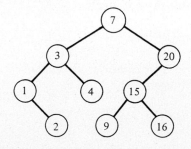

Fig. 14.8 The tree *T* (node data are integers)

Note that the nodes labeled 2, 4, 9, and 16 are leaves (both subtrees are empty), whereas node 1 has an empty left subtree and node 20 has an empty right subtree. We use *leftson* and *rightson* throughout this chapter to stand for *Subtree*$(x, 1)$ and *Subtree*$(x, 2)$.

We now apply the program *PreorderPrint* of Fig. 14.9 to *T*.

```
procedure PreorderPrint(x: tree);
  begin
  if not IsEmpty(x) then
    begin
    write(label(root(x)));
    PreorderPrint(leftson(x));  PreorderPrint(rightson(x))
    end
  end
```

Fig. 14.9 An NLR (preorder) traversal

The result is

7 3 1 2 4 20 15 9 16.

We call the traversal implemented by *PreorderPrint* an *NLR traversal* (or *NLR walk*). First, an action (in this case, writing) is performed on the root node (N); then the left (L) subtree is traversed; and finally, the right (R) subtree is traversed. This traversal is also called a *preorder walk* (the node action is prefixed to the subtree traversals).

Analogously, we have an LRN or *postorder* walk that first traverses the left (L) subtree, then the right (R) subtree, and finally performs an action on the node (N):

```
procedure PostorderPrint(x: tree);
  begin
  if not IsEmpty(x) then
    begin
    PostorderPrint(leftson(x));  PostorderPrint(rightson(x));
    write(label(root(x)))
    end
  end;
```

Fig. 14.10 An LRN (postorder) traversal

which for *T* yields

    2 1 4 3 9 16 15 20 7

Also, there is the LNR or *inorder* walk:

```
procedure InorderPrint(x: tree);
    begin
    if not IsEmpty(x) then
        begin
        InorderPrint(leftson(x));
        write(label(root(x)));
        InorderPrint(rightson(x))
        end
    end;
```

Fig. 14.11 An LNR (inorder) walk

which for *T* yields

    1 2 3 4 7 9 15 16 20,

a listing that is—not accidentally—in order.

In general, any combination of the letters N, L, and R defines a possible kind of traversal. For example, NRNLN describes a walk performing three actions on each node, traversing the right and then the left subtrees between actions.

Applying this technique to *k*-ary trees, we replace L and R with 1, 2, etc., for "first subtree," "second subtree," and so forth. Thus, on 3-ary trees, an N123 walk traverses the root and then the three subtrees in order (thus it can also be called a preorder walk).

14.5.2 SEARCH TREES

A principal use of trees is to organize data so that making *searches* through that data is easy. A tree may be organized so that a *discrimination function* can tell whether the object of any search is in the current node, and, if not, in which subtree it must be if it is present at all. The tree of Fig. 14.8 illustrates a tree so organized. Note that

- Every node in the left subtree of any node has a label value *less than* that of the node.
- Every node in the right subtree of any node has a label value *greater than* that of the node.

We call such a binary tree a *binary search tree.*

To find an object in a binary search tree, we could do a complete walk (in any order we please) until we find the object, but it is more efficient to avoid traversing subtrees that we know cannot contain the object. A routine for such a search appears in Figure 14.12.

```
function Find(T: tree;  x: Elt): tree;
    { Returns the subtree within T whose node label is x.
      Returns the empty tree if there is no such subtree. }
    begin
    if IsEmpty(T) then Find := EmptyTree
    else if label(root(T)) = x
            then Find := T
    else if x < label(root(T))
            then { x cannot be in the right subtree of T }
            Find := Find(leftson(T), x)
    else { x cannot be in the left subtree of T }
            Find := Find(rightson(T), x)
    end;
```

Fig. 14.12 Recursive binary tree search

The program of Fig. 14.12 has the property that after every recursive call, the function immediately exits, passing up the value of the recursive function call. That means that as soon as any call on the function exits, all recursive calls "above" it will exit without doing any further work. This is known as *tail recursion*; in such cases we may replace the recursion with an iteration without introducing a stack, as in Fig. 14.13.

We derived *Find2* from *Find* by

- Introducing a local variable, *R*, to hold the argument, *T*, to the recursive calls.
- Replacing all calls on *Find* with assignments to *R*.
- Replacing all returns from *Find* with **goto** statements that exit the entire *Find2* function.

Note that we did not need to introduce a variable to hold the argument *x*; it was the same for all recursive calls.

```
function Find2(T: tree; x: Elt): tree;
    var R: tree;
    begin
    R := T;
    repeat
        if IsEmpty(R) then
            begin Find := R; goto Stop end
        else if label(root(T)) = x
            then begin Find := R; goto Stop end
        else if x < label(root(T))
            then R := leftson(T)
        else R := rightson(T)
    until false;
Stop:
    end;
```

Fig. 14.13 An iterative tree search

14.5.3 OTHER ITERATIVE TREE WALKS

As illustrated in Section 14.3, we can write any recursive routine, and
specifically a tree walk, as a nonrecursive procedure using a stack. For
example, the NLR print routine can be rewritten as shown in Fig. 14.14.

```
procedure PreorderPrint2(x: tree);
    var S: stack;   {of trees}
        y: tree;
    begin
    Clear(S); Push(S, x);
    repeat
        Pop(S, y);
        if not IsEmpty(y) then
            begin
            write(label(root(y)));
            Push(S, rightson(y)); Push(S, leftson(y))
            end
    until Empty(S)
    end;
```

Fig. 14.14 An iterative NLR walk

Again, instead of calling itself recursively, *PreorderPrint2* places the subtrees to be printed on *S* in *reverse order.* The symmetry of the procedure eliminates the need to push a return address on the stack. The routine just keeps printing until there are no more subtrees that need to be printed (that is, the stack is empty).

Likewise, we can convert the LNR print of Figure 14.11 into the nonrecursive version in Fig. 14.15.

```
procedure InorderPrint2(T: tree);
    var S: stack;     {of trees}
        y: tree;
    begin
    Clear(S);  y := T;
    repeat
        if not IsEmpty(y) then
            begin
            Push(S, y);  y := leftson(y)
            end
        else if empty(S) then goto Stop
        else begin
            Pop(S, y);
            write(label(root(y)));
            y := rightson(y)
            end
    until false
Stop:
    end;
```

Fig. 14.15 An iterative LNR print

Here, we have replaced recursive calls with assignments to *y*. When going down the left subtree of a node, *InorderPrint2* saves the node on *S*, since it must eventually "return" to it. Since *InorderPrint* returns after traversing the right subtree, *InorderPrint2* need not save anything on *S* before setting *y* to the right subtree.

14.6 Further Reading

Recursion is seldom treated in isolation; however, see Horowitz and Sahni (1978) and Wickelgren (1974) for additional discussion.

Exercises

14.1 Which, if any, of the standard types *char*, *real*, and *boolean* are dynamic? What about enumerated types? If type *T* is not dynamic, is *set of T*? Is **array** [· ·] **of** *T*? What is the strongest characterization you can give of the set of all possible non-dynamic data structures in our language?

14.2 How would you implement the data type *FlexVec*, which is just like a *vector* except that there is no upper bound on the index set?

14.3 List the dynamic and nondynamic types in each language with which you are familiar.

14.4 How would the specifications of *stack* given in Section 9.1.3 change if *stack*s were a genuine dynamic data type in the sense of Secton 14.1?

14.5 Which types discussed in Chapter 9, and which of their implementations discussed in Chapter 4, are dynamic?

14.6 Section 14.1 indicates that unbounded integers form a dynamic data type. Suppose you were given a machine that implemented unbounded integers—but only supplied *one* such integer.

  a) Explain how to simulate a conventional, or finite, computer memory with a single unbounded integer.

  b) Explain how to implement an unbounded number of unbounded integers with a single unbounded integer.

14.7 Trace the execution of routine *ReadWff* for each of the following inputs:

  a) ((()())())
  b) (((())()))
  c)
  d) ()()

14.8 The *recursion depth* of a routine at some point in a computation is the number of calls on the routine that are active simultaneously. Thus, if *ReadWff* calls *ReadWff*, which then calls *ReadWff*, we say that the recursion depth of *ReadWff* at that point is three. What determines the maximum recursion depth attained by the routine *ReadWff* for some arbitrary input, *s*? If *s* has *n* characters, what is the maximum recursion depth that *ReadWff* can attain in processing *s*? What corresponds to recursion depth in the iterative version of *ReadWff*?

14.9 Prove Theorem 14.1 informally. That is, give a procedure whereby one can take almost any recursive procedure and convert it to a nonrecursive procedure that uses a stack and no other additional dynamic data structure. Why are the assumptions on Theorem 14.1 necessary?

14.10 Consider the routine *Pick* defined

> **function** *Pick*(*L*: ↑*listofreal*; *lower*, *upper*: *real*): ↑*listofreal*;
>   **begin**
>   **if** (*FirstItem*(*L*) ≧ *lower*) **and** (*FirstItem*(*L*) ≦ *upper*)
>   **then** *Pick* :=
>   *ConsList*(*FirstItem*(*L*), *Pick*(*Tail*(*L*), *lower*, *upper*))
>   **else** *Pick* := *Pick*(*Tail*(*L*), *lower*, *upper*)
>   **end**; { *Pick* }

Let *L* point to the list

$$[0.1] \rightarrow [6.5] \rightarrow [5.0] \rightarrow [8.7] \rightarrow \epsilon.$$

What is the result of *Pick* for each of the following calls?

   a) *Pick*(*L*, 1.0, 7.0)
   b) *Pick*(*L*, 0.0, 9.0)
   c) *Pick*(*L*, 3.0, 4.0)

What does *Pick* do in general (what are its pre- and postconditions)?

14.11 Complete and test the sort procedure of Figure 14.6 by implementing the necessary merge operation.

14.12 Convert the recursive version of *MergeSort*, from Fig. 14.6, into iterative form.

14.13 In the style of the recursive *ReadWff* developed in this chapter, write a set of mutually recursive procedures that will read valid simple arithmetic expressions, <sae>'s, as defined in the last chapter (see Fig. 13.10). You will need one procedure for each of the metavariables *ReadSae*, *ReadT*, *ReadF*, and *ReadP*.

14.14 Write a recursive procedure to evaluate postfix expressions as defined in Exercise 9.7.

14.15 Design a tree data structure to represent arithmetic expressions over the integers. Assume that a node may be either a literal value or a single-character variable name. Write an LRN tree-walk routine, in recursive form, that will find the value of an expression in the case that all the leaf nodes are literal values.

14.16 Consider the problem of testing set membership in the implementation of Section 10.5.2. In that implementation it was necessary for *IsMember* to examine all the elements in the set before returning *false*. Assume the implementation is changed to keep the vector sorted in ascending order of set values.

    a) Rewrite *IsMember* so it quits when it has scanned past the point where the element being tested would appear.

    b) Rewrite *IsMember* using a divide-and-conquer approach: divide the set or vector in half, then call *IsMember* for each half.

    c) Rewrite *IsMember* using a smart divide-and-conquer approach: divide the vector in half as above, but eliminate one of the halves and call *IsMember* recursively only on the half that must contain the element being tested (if it's there).

    d) Use the previous program to guide your design of an iterative program that uses the same search strategy.

    e) Analyze the execution costs of the original *IsMember* and each of the programs developed for this exercise.

14.17 A *game tree* is a data structure commonly used by game playing programs like chess, checkers, nim, etc. In a game tree, each node represents a position in a game. The sons of any node represent the positions resulting from the various possible moves. Thus, the root node represents a position in which the first player must move; the sons of this node are positions where his opponent, the second player, must move, and so forth. In order to choose which of the possible moves to take, we would like to *evaluate* the possible next positions. Suppose that we agree to assign positive integer values to the positions (nodes in the game tree): the higher the value assigned, the more favorable to the first player; hence, the lower the number assigned, the more favorable to the second player. Also suppose that we can somehow assign values to the leaves of a game tree, the leaves representing possible positions several moves ahead. To compute reasonable values for the other nodes, the *minimax* strategy is often used. It works as follows:

    ■ Since the first player wants to choose the best possible next move, it is reasonable to assign to any node representing a position that he faces the value of the best of its sons—that is, the *maximum* of the values of nodes immediately below it.

    ■ Since the second player wants to choose the best possible next move for himself, he wants to pick that move giving the *worst* position to the first player. Hence, it is reasonable to assign to any position faced by the second player the *minimum* of the values of the nodes immediately below it.

For example, Fig. 14.16 shows an assignment of values to the nodes in a game tree according to the minimax procedure (we assume that the leaf nodes were given values elsewhere.)

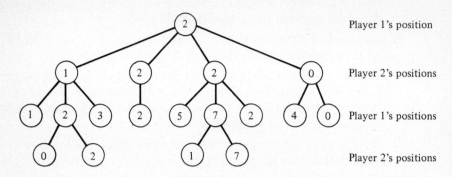

Figure 14.16

a) Write a recursive routine to evaluate the root node of a binary game tree using the minimax strategy. Invent any necessary notation (for the value of a game tree node, for instance.)

b) Many game-playing programs do not store the game tree explicitly, it being much too large. Instead, they provide routines that

- Compute a list of legal moves for a given position.
- Compute a new position, given a starting position and a move.
- Determine whether a position is a leaf for which a value can be computed, or *statically evaluated*, immediately.

Write a recursive routine that evaluates a position using these routines (give them any names and arguments you choose) and the minimax procedure.

14.18 If a tree has N nodes, of which K are terminal and L are of degree 1 (and therefore $N-K-L$ are of degree 2), how many times will the loop in the program of Fig. 14.15 be executed during a complete tree-walk?

14.19 Consider the tree shown in Fig. 14.17. What nodes are processed (and in what order) by each of the following traversals?

a) NRNLN

b) LNR

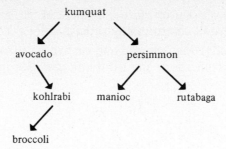

Figure 14.17

c) RRN

d) RL

e) LNRNRNL

14.20 Use the notion of a search tree from Section 14.5.2 to implement a "dictionary." The entries in the dictionary will be strings representing English words. Use the dictionary to implement a program that corrects spelling errors. The corrector should read a file and look up each word from the file in the dictionary. If a match is found the word is presumed correct; if no match is found, the word is either misspelled or not in the dictionary. In the latter case, your program should give the user two options:

- Add the word to the dictionary.
- Correct the spelling manually.

After you have the basic program working, you may wish to make it more sophisticated and *useful* in a number of ways. (By the way, the spelling in this book was checked using such a system at various stages of its development.)

a) Add an option so that when a potential error is detected, users may indicate that they wish to correct the spelling, and *automatically* correct the same misspelling on all of its recurrences. (Note that to do this you would really like to have *two* dictionaries, thus you'd be wise to define the concept of a dictionary as an abstract data type.)

b) Propose a method for correcting misspellings of words that aren't in the dictionary. To do this you will have to check for common spelling and typographical errors such as letter transpositions, incorrect vowels (for example, -able instead of -ible), common prefixes and suffixes, etc.

c) Include special user-specific dictionaries to hold corrections of a specific user's common misspellings.

d) Work out procedures for one or two of them. Think of more useful modifications and list them; use your imagination.

Chapter 15

The Interpretation of Identifiers

All high-level programming languages use identifiers to refer to data, as well as to parts of programs (routines and labeled statements, for instance). In this chapter we discuss some important issues concerning those identifiers that are used to name variables in order to emphasize both the differences (sometimes subtle ones) between languages and the crucial decisions about runtime representations. The most important issues fall into three categories—*scope*, *extent*, and *binding time*. Although these topics are not completely independent, we will treat them separately as far as possible.

15.1 Scope of Identifiers, Revisited

In most modern programming languages it is possible to use the same identifier to refer to distinct variables in different parts of a program. In PASCAL, for example, several procedures may each declare a local variable named x, and each of these variables will be distinct from the others. The notion of the *scope*, introduced in Section 8.4.3.1, is used to define which specific variable is named by a particular occurrence of the identifier, x.

The *scope* of an *identifier* is a unit of program text—specifically, it is that portion of the program text in which the identifier has a specific meaning. It is more accurate to refer to the scope of a *declaration*—the portion of program text in which a given declaration establishes the meaning of an identifier. However, we often use the phrase "scope of an identifier" to have both meanings. A *scope unit* is a syntactic unit in the language, such as a block or procedure body, within which an identifier can be assigned a new meaning. Both "scope of an identifier" and "scope unit" are conventionally shortened to "scope"; this usually causes no confusion.

The following three cases arise in determining the meaning associated with a particular occurence of an identifier:

1. *The identifier is declared locally.* That is, if the scope unit is the procedure, as in PASCAL, then all declarations within the procedure apply only to its body. If the unit of scope is the block, as in ALGOL, all declarations within the block apply only to the code contained in the block.

2. *The identifier is that of a formal parameter.* The identifiers of formal parameters to procedures are known within the procedures as though they had been locally declared.

3. *The identifier is "free."* A *free* identifier to a scope unit is one that is neither declared locally nor specified as a formal parameter to that unit.

Languages have *scope rules* that are used to decide which identifiers are known where. Virtually all languages treat the first two cases alike. In particular, locally declared identifiers and formal parameter identifiers are usually treated as they are in PASCAL, FORTRAN, ALGOL, etc.; they are known only in and throughout the scope unit that declares them.

The real issue concerning scope is the treatment of free identifiers. If an identifier is *not* declared local to a scope, what does it mean if that identifier is used within that scope? There are normally three possible treatments:

1. Treat the use of a free identifier as illegal and cause a compiler diagnostic.

2. See if the identifier is declared in a *lexically enclosing scope*. That is, see if the scope that uses the identifier is itself declared inside of another scope that declares the identifier. If the identifier is found in this manner (we keep searching outward until either we find it or we reach the outermost scope—the main program); then the identifier is indeed known in the inner scope. This search strategy is referred to as a *lexical scope* rule, and sometimes as a *static scope* rule, because the determination can be made statically by the compiler. In Fig. 15.1, under the lexical scope rule, a, b, c, and d are known at L_1; a and b at L_2; and e and f at L_3.

3. See if the identifier has been declared in a *dynamically enclosing scope*. That is, see if some procedure that has not yet completed execution declares the identifier. If so, the identifier is known. This is referred to as a *dynamic scope* rule, because the determination can be made during execution only. In Fig. 15.1 under the dynamic scope rule, e and f are known at L_3. Since P is called at L_3, e and f in addition to a and b are known at L_2. Since Q is called at L_2, all of a, b, c, d, e and f are known at L_1. However this determination cannot be made at compile time, since P may be called from other places outside of R, in which case e and f (at least *that* e and f) will not be known in either P or Q.

In the last two cases, it is possible that an identifier is declared in two distinct declarations in two lexically (or dynamically) nested scopes. Common practice is to consider the identifier as falling within the scope of the closest, or *innermost*, declaration.

> **procedure** P;
>   **var** a, b: t_1;
>
>   **procedure** Q;
>   **var** c, d: t_2;
>   **begin**
> $\cdots L_1 \cdots$
>   **end**;
>
>   **begin**
> $\cdots L_2$: Q; $\cdots$
>   **end**;
>
>   **procedure** R;
>   **var** e, f: t_3;
>   **begin**
> $\cdots L_3$: P; $\cdots$
>   **end**;

Fig. 15.1 Illustration of scope rules

Most compiled languages have lexical scope rules. These include PASCAL, ALGOL, PL/I, COBOL, etc. One reason for this design choice is representation efficiency. The next chapter shows how lexical scope is represented during execution, with most of the cost of determining the location of a variable absorbed at compile time. Dynamic scoping requires extra work at execution time. Those languages that have dynamic scope rules are, for the most part, the interpreted ones, including APL, LISP, and SNOBOL.

A second reason for lexical scoping is based on arguments about the understandability of programs. In a lexically scoped language, we can determine the meaning of an identifier by a simple, linear scan of the text. In a dynamically scoped language, we must examine all possible control-flow paths that might lead to the use of the identifier—a *much* harder thing to do. Interestingly, the first scope rule (free identifiers are illegal) is usually associated with older languages such as FORTRAN. However, these same arguments about understandability are even more valid about this scope rule: If there are no free variables everything must either be a local variable or a formal parameter. It is even easier to understand such programs. Consequently, this strict scoping rule has been adopted in some experimental languages, such as EUCLID and AL-PHARD.

One interesting effect of the second and third alternatives (lexical and dynamic scoping), is that of the "hole in scope." Consider Fig. 15.2.

```
procedure P;
    var x, y: t₁;
    procedure Q;
        var x, z: t₂;
        begin
            .
            .
            .
        end;
    begin
        .
        .
        .
    end;
```

Fig. 15.2 An illustration of a "hole" in a scope

Although the x and y declared in P would normally be known throughout the bodies of P and Q, the redeclaration of x in Q prevents the x of P from being seen in the body of Q. Thus the two x's are different variables, and anything done to x in Q will not be seen by P.

The same figure also illustrates the idea of a hole in dynamic scope, since, regardless of what procedure calls Q, the only x that Q can see is that declared locally within it.

15.2 Extent of Storage, Revisited

In Section 8.4.3.2, we introduced the concept of extent. The extent of a block of storage allocated as data was defined as the duration of that allocation during execution. In many FORTRAN implementations, the extent of all storage is the entire duration of program execution. Variables are allocated at the start of execution, and variables local to a subroutine have their values preserved across calls; on entry, a variable will have the same value that it had on exit from the last execution of the routine. In ALGOL 60, there are two possible extents for variables—local and own. Own variables have an extent equal to the entire time the program executes, just as in FORTRAN implementations. Local variables, on the other hand, are allocated when a scope is entered and then freed when the scope is exited. Thus local variables do not have their values preserved from one entry of the scope to the next; and, furthermore, no storage is allocated unless the scope is being executed. Because the local variables that are freed are always the last that were allocated (last in, first out), local extent is also called *stack allocation.*

A FORTRAN program that operates on a different amount of data for each run must either decide upon an absolute maximum amount or be recompiled each time it is used. The former decision may be grossly inefficient if larger amounts of data are rarely seen; the latter is inefficient if the compile time is significant relative to the execution time of the program. In ALGOL, the same program can be written to reserve judgment on the size of the data structure until execution. Of course there is a trade-off here, because the representation of ALGOL-like programs may be somewhat less efficient than that of FORTRAN programs, as we will see in the next chapter.

A third possible extent gives direct control of allocation to the programmer. This is known as dynamic extent, or *heap allocation.* The term "heap" seems to have been chosen to contrast with "stack." In normal, nontechnical use, "stack" connotes an orderly pile, whereas a "heap" suggests a disorderly one.

In PASCAL, for example, records may be dynamically created via the *new* operation. Storage is not tied to the creating scope, as is the case with local variables. Figure 15.3 illustrates the difference between stack and heap allocation. In Fig. 15.3(a), when A is called it allocates a new object of type r. In the body of $P1$ then, at point $L1$, this new r still exists even

```
type r = record · · · end;        type r = record · · · end;
  t = ↑r;
procedure P1;                      procedure P2;
  procedure A1;                      procedure A2;
    var s: t;                          var s: r;
    begin                              begin
    new(s); · · ·                      · · ·
    end;                               end;
  begin                              begin
  A1;                                A2;
L1: · · ·                          L2: · · ·
  end;                               end;
```

(a)   (b)

Fig. 15.3 Heap (a) versus stack (b) allocation

though A is "gone." In Fig. 15.3(b), on the other hand, A2 also allocates an r, but at point L2, it will have disappeared. Note that in Fig. 15.3(a), s will disappear too, but in this case s is only a local pointer, not the allocated object of type r.

When heap allocation is used, the issue arises of how long such allocation lasts; there are three possibilities:

- Forever.
- Until explicitly freed.
- Until there are no more references to it.

In some PASCAL implementations, the allocation lasts for the remainder of the program execution because there is no way to release the allocation explicitly. Some languages allow explicit freeing. For example, in other implementations of PASCAL, there is a function called *dispose* such that *dispose(q)* undoes the effect of *new(q)* and releases the previously allocated storage. This allows us to write more space-efficient programs, since we can return storage when we're done with it before the program finishes. In general this is beneficial, but it also creates a potential problem. In Fig. 15.4, storage is freed while there is still an active pointer to it. When q is used in P after A is called, an error will most likely result, since the object pointed at by q has been released. This can be deadly since that storage can now be reallocated somewhere else. This is known as the "dangling reference problem," since in a pictorial representation q will have an arrow leading from it to nowhere.

```
type r = record · · · end;
   t = ↑r;
      .
      .
      .
procedure P;
   var q: t;
   procedure A;
      var s: t;
      begin
      new(s);
      q := s;
      dispose(s)
      end;
   begin
   · · · A · · · q · · ·
   end;
```

Fig. 15.4 The dangling reference problem

PASCAL originally prevented dangling references by outlawing the *dispose* operation, but at the expense of wasting storage. (That expense was apparently sufficient that *dispose* is now part of standard PASCAL.) For example, in Fig. 15.5 the first *r* allocated is lost when *t* is overwritten with a pointer to the second *r* allocated. Since no accessible pointer now references that first allocation it has become garbage—a name given to allocated but inaccessible storage.

The third alternative is for the programming system to determine when there are no more references to an allocated area—that is, to determine when the area has become garbage, and to perform the *dispose* automatically. This alternative is both convenient and safe; the only problem is that it is expensive. The process of detection, which is called garbage collection, can require significant amounts of execution time.

15.3 Binding Identifiers to Values

The *binding time* of an identifier is the time at which the association is made between that identifier and its allocated storage. Thus, FORTRAN variables are bound at compile time, and PASCAL variables are bound during execution at procedure entry.

The concept of binding is closely related to those of scope and extent.

$$\textbf{type } r = \textbf{record} \cdots \textbf{end};$$
$$t = \uparrow r;$$
begin
$new(t);$
$new(t);$

.

.

.

end;

Fig. 15.5 Creating garbage

For example, binding in ALGOL takes place at each entry to a scope. From that we can infer that the ALGOL extent is the duration of a scope (except for **own** variables). More interesting variations in binding time can be found in the treatment of parameters. Parameters increase the versatility of routines by allowing them to be applied to different values at different times. Each time a routine with parameters is called, the formal parameter identifiers of the routine are bound to the actual parameter values given by the program that called the routine. There are a number of different ways to perform the binding—in other words, there are several different sets of rules that govern the relationships between a routine and its parameters. The following questions capture most of the differences among the parameter binding rules:

1. Can the routine use the input value of the parameter (that is, use the parameter to obtain data from the caller of the routine)?

2. Can the routine alter the parameter value (that is, use the parameter to communicate data back to the caller)?

3. How does the routine access the value of the parameter?

   a) Is the value copied to and from a local variable at the beginning and end of the routine?

   b) Is it accessed directly from the location it occupies outside the routine?

   c) Is it reevaluated each time it is used?

Several combinations of answers to these questions characterize rules in common use. All the names of these rules have the form "call-by-<binding rule>." This term is somewhat misleading; the name "pass-by-<binding rule>" would be more accurate, since a parameter binding rule is usually declared by the called routine rather than by the caller. We will, however, bow to tradition.

The rest of this chapter describes several common binding rules, using the following program to illustrate their consequences.

```
var i: integer;
    A: array[1..3] of integer;

procedure testbind(<binding> f, g: integer);
    begin
    g := g + 1;
    f := 5 * i;
    end;

begin
for i := 1 to 3 do A[i] := i;
i := 2;

testbind(A[i], i);

print(i, A[1], A[2], A[3]);
end
```

Note that the scope rules must be consulted to determine the meaning of the identifier *i* that appears in the procedure. It is in fact the *i* declared in the main program. This is *extremely* poor programming style; it is used in this example only to show the effects of the rules.

15.3.1 CALL-BY-VALUE

The binding rule *call-by-value* allows a parameter to be used for input but not for output. When a parameter is passed by value, the compiler creates a local variable in the routine. The formal parameter identifier names this local. When the routine is executed, the actual parameter supplied by the caller is evaluated, and its value is assigned to the local variable. Whenever the routine references the formal parameter, the local variable is used. Thus, the routine may not alter the actual parameter. In some languages, the routine is not even allowed to make assignments to the local copy of the parameter.

When we use this binding rule for our sample routine *testbind*, the result of the call is

$$i = 2$$
$$A = (1\ 2\ 3)$$

because neither the assignment to *f* nor the assignment to *g* may affect the actual parameter.

15.3.2 CALL-BY-ADDRESS

The binding rule *call-by-address* allows a parameter to be used for both input and output. When a parameter is passed by address, the address of the actual parameter is determined when the routine is called. If the actual parameter is an expression, the value of the expression is stored in a temporary local variable and the address of that variable is used. Whenever the routine references a formal parameter, the variable passed as an actual parameter is accessed directly.

This binding rule is often called *call-by-reference*. We have chosen to call it call-by-address to avoid confusion with reference types and variables as defined in Chapter 8. In PASCAL, parameters passed using this rule are called **var** parameters.

When we use this binding rule for *testbind*, the result of the call is

$$i \ = 3$$
$$A \ = (1 \ 15 \ 3)$$

because the assignment $g := g + 1$ changes i to 3 as soon as it is performed. Therefore the assignment $f := 5 * i$ uses the value 3 that i received from the assignment through the formal parameter g.

15.3.3 CALL-BY-VALUE/RESULT

The binding rule *call-by-value/ result* allows a parameter to be used for both input and output. The effect of this rule is much like that of call-by-value, except that the local temporary value in the procedure is modifiable, and its final value is copied back into the actual parameter at the end of the routine.

When we use this binding rule for *testbind*, the result of the call is

$$i \ = 3$$
$$A \ = (1 \ 10 \ 3)$$

because the assignment $g := g + 1$ does not change i to 3 until the value of the local g is copied back to i at the end of the routine. Therefore the assignment $f := 5 * i$ uses the value 2 that i had when the routine was entered.

15.3.4 CALL-BY-RESULT

The binding rule *call-by-result* allows the parameter to be used for output, but no value for the parameter may be assumed at the beginning of the procedure. The final value of the local variable is copied into the actual parameter at the end of the procedure. The use of call-by-result in our sample routine would result in an error when the increment of g was attempted, because g would have no defined value at that time.

15.3.5 CALL-BY-NAME

The binding rule *call-by-name* allows the parameter to be used for both input and output. The expression in the actual parameter position is reevaluated *each time* the formal parameter is used.

If the actual parameter happens to be a simple identifier, the effect is the same as that of call-by-address. However, if the actual parameter is an expression or a selector expression (for example, an array access), it is reevaluated for each use with whatever values the variables in the expression happen to have at the time. This means that if several parameters are passed by name and the actual parameters depend on each other, the routine may get a different value each time it accesses a parameter. This mechanism is very powerful, but it can lead to subtle bugs and can be relatively inefficient if used in a situation where one of the preceding mechanisms would serve.

When we use this binding rule for *testbind*, the result of the call is

$$i \quad = 3$$
$$A \quad = (1 \ 2 \ 15)$$

because the assignment $g := g + 1$ takes effect immediately and the actual parameter $(A[i] = A[3])$ that corresponds to f is reevaluated for the assignment $f := 5 * i$.

15.4 Further Reading

The subjects of scope, extent, and binding are usually discussed in texts that compare programming languages, for example Elson (1973) and Pratt (1975).

The parameter binding rules discussed in this chapter adequately cover rules in conventional languages. It is important for a programmer to have a good understanding of the issues presented here; programming in a new language without a complete knowledge of its parameter binding rules is one of the most subtle sources of bugs. The following languages provide examples of the scope, extent, and binding rules covered in this chapter: PASCAL has call-by-value and call-by-address. ALGOL has call-by-value and call-by-name. Most FORTRAN implementations have only call-by-address. ALGOL 68 has only call-by-value (yet can support the other mechanisms indirectly). The only language we know of with call-by-value/result is ALGOL W, a student programmer dialect of ALGOL; we include it only for completeness. All of these languages except FORTRAN support recursion, stack allocation, and nested (lexical) scopes. In addition, ALGOL 68, and to a lesser extent PASCAL and ALGOL W, support heap allocation. LISP and APL are good examples of languages with dynamic scoping.

Exercises

15.1 What do the following programs write under standard lexical scope rules and under dynamic scope rules?

a) **var** *i, k: integer,*
   procedure *P*(**var** *j: integer*);
       var *i: integer,*
           begin *i* := 1; *Q*; *j* := *i* **end**;
   procedure *Q*;
       begin *i* := *i* + 1 **end**;

   begin
   i := 3; *P*(*k*); *write*(*k*)
   end.

b) **function** *Glop*(**function** *Q: integer,   lower, upper: integer*): *integer,*
       var *i, S: integer,*
       begin
       S := 0;
       for *i* := *lower* **to** *upper* **do** *S* := *S* + *Q*;
       Glop := *S*
       end;   {*Glop*}
   function *A*;
       begin *A* := *i* * *i* **end**;

   begin *write*(*Glop*(*A*, 1, 5)) **end**.

15.2 Assume lexical scope rules and consider the following program:

   var *x, y: real*;
   function *f*(*a: real*): *real*;
       begin *x* := *a* + 1; *f* := *x* **end**
   begin
   x := 0;
   y := *x* * (*f*(*x*) − *f*(*x*)) / (*f*(*x*) * *f*(*x*));
   end

Note that each call on function f changes variable x. This is said to be a *side effect* of f on x.

    a) Suppose x is initially 0. What is its value after $f(f(x))$?

    b) Programs like the one above are sensitive to the order in which expressions are evaluated. Find evaluation orders for the assignment to y—that is, orderings for the evaluation of the four occurences of $f(x)$—that produce four different results.

    c) Discuss the effect of the use of side effects through free names on program clarity.

15.3 Do parts (a) and (b) of Exercise 15.2, substituting the following program:

```
var x, y: real;
function f(var a: real): real;
    begin a := a + 1; f := a end
begin
x := 0;
y := x * (f(x) − f(x)) / (f(x) * f(x));
end
```

15.4 Give a general algorithm in English for examining any line of a program in a language with lexical scope rules that states exactly what identifiers are declared at that point and which declaration of each identifier is currently "visible."

15.5 In a language without procedures, is there any difference between lexical scope and dynamic scope? Explain your answer.

15.6 Under the rules of lexical scope, is it true that the extent of the storage bound to an identifier is exactly the time during which the program is executing the text within the scope of that identifier? Explain.

15.7 Consider the procedure declaration

procedure G; $i := 2 * j$;

and the incomplete program given below. Assume that the program is written in a language with lexical scope rules. You are to fill in the blanks.

```
var i: integer;
begin
var j: integer;
```

```
procedure H( (a)  P: integer);
    var j: integer;
        begin j := 3; (b) end;

procedure W(P: integer);
    var j: integer;
        begin j := 5; (c) end;

begin j := 7; H( (d) ); write(i) end;
end.
```

[Note: (a) is a specification of a formal parameter's binding rule; (b) and (c) are statements; (d) is an actual parameter.]   Where would you insert the procedure declaration?   Fill in the blanks (a), (b), (c), and (d) with text that does *not reference* the variables *i* and *j* in order to cause the *write* statement to produce the values listed below.

a) 14

b) 6

c) 10

15.8 Can you add to the program of Exercise 15.7 a definition and call of a new procedure *E(P)* with all of the following properties?   (If you can, do so. If not, explain why not.)

a) The parameter *P* is of type **procedure**, and the procedure name *G* (the same *G* as defined in Exercise 15.7) is passed to *E* as the actual parameter.

b) The free variable *j* of *G* is bound to a *j* local to the procedure body of *E*.

c) The free variable *i* of *G* is bound to the same *i* as the one available to *H* and *W*.

d) *E* does not invoke procedures other than the one passed as parameter *P*.

15.9 Suppose you want to implement a version of a language without free names (FORTRAN, for instance) that supports recursion (nothing in the FORTRAN standard *prohibits* recursion). To be compatible with other versions of the language you would still have to implement the usual assumptions about the allocation and extent of variables. Given these assumptions, would the ability to recur be very useful?   Explain.   In addition, show how you could "fake" the effect of local variables.

15.10 Discuss the scope units and rules for all the languages you know.

15.11 How can one obtain the effect of **own** variables in PASCAL?

15.12 Own variables typically appear in applications such as the parameterless function *RAND*, defined

```
function RAND: integer;
    post  RESULT = a "random" integer;
  own var LastValue: integer;
    begin
    LastValue := (LastValue * c1 + c2) mod c3;
    RAND := LastValue
    end;  {RAND}
```

where $c1$, $c2$, and $c3$ are suitable constants. (Note that we intend the **own var** declaration to declare an **own** variable of *RAND*. Actually, PASCAL does not have such a declaration.) When a user of *RAND* writes

$x := RAND$; $y := RAND$;

the variables x and y are set to two different "random" numbers; *LastValue* keeps track of the last value returned by *RAND* so that the next call produces a different one.

Write a new version of *RAND* that does not use an **own** variable and contains no free identifiers (that is, all identifiers in the function must be standard PASCAL local variables or parameters to *RAND*). Calls on this new version of *RAND* may look different from calls on the original version. Can you think of any situation in which an **own** variable *must* be used to accomplish something?

15.13 A dynamic **own** array in the ALGOL 60 language is an array declared to be **own** in some routine (thus having an extent equal to the program's period of execution), but whose size changes with every entry to the routine. For example,

```
procedure Glitch(n: integer);
  own var A: array [1..n] of integer;
```

Each time *Glitch* is called, we are supposed to use the same array as the one that existed when the routine was last called, but we are to redimension it to hold exactly n elements, where n is a parameter to *Glitch*. What possible interpretations can you give to this specification, which is generally held to be a design flaw in ALGOL 60?

15.14 The results produced by the following program are influenced by the binding rules used for the parameters i and j of procedure F.

```
begin
var a: array a[1..5] of integer; i, j, k: integer;
procedure f(i, j: integer): integer;
{You are to determine the bindings for i and j}
    begin
    i := i + 1;
    j := j − 1;
    f := i − j;
    end;
for j := 1 to 5 do a[j] := j;
i := 1;
k := f(i, a[i]);
k := 2 * k − a[i];
write(k);
end.
```

Indicate what combination of parameter binding rules should be used for those parameters in order for the final value of k to be

      a) 1   b) 2   c) 3   d) 4

Briefly explain your answers. You should consider as candidate binding rules all those discussed in this chapter. In some cases, more that one rule will produce the indicated value.

15.15 Consider the procedure

```
integer procedure F(x, y: integer);
begin x := x + 1; y := y + 1; F := x − y end
```

    a) Show by one or more examples of calls on procedure F that call-by-name, call-by-value/result and call-by-address are different parameter passing methods. That is, show calls that produce different results for the different binding rules.

    b) Characterize the conditions under which the actual parameters of the following pairs give the same results independent of the procedure body.

       i) call-by-name, call-by-address

       ii) call-by-name, call-by-value/result

       iii) call-by-address, call-by-value/result

    Take into account the fact that a procedure may have several parameters. Explain your answer in each of the three cases.

15.16 Suppose you want to define

procedure *DoWhile*(*P*: *boolean*; **procedure** *G*);

so that *DoWhile*(*B*, *S*), where *B* is a boolean expression and *S* is a procedure, has the same effect as

while *B* **do** *S*;

The procedure declaration is legal Pascal, but the desired effect is not possible in PASCAL. Explain what you would do to make it possible.

15.17 Consider the following function definition (due to Donald Knuth and Jack Merner):

function *gps*(*A*, *B*, *i*, *n*: *integer*): *integer*;
    begin
    for *i* := 1 **to** *n* **do** *A* := *B*;
    gps := 1
    end; {*gps*}

The value of *gps* is totally irrelevant; it is the side effects that interest us. Suppose that the call-by-name parameter mechanism is used and assume the following additional variable declarations:

var *i*, *j*, *k*, *S*: *integer*; *A*, *B*: **array** [1..20] **of** *integer*;

and answer the following questions about *gps*.

a) What is the result of executing the following?

$S := 0$; $j := gps(S, A[i] * B[i] + S, 1, 20)$

b) What does the following statement do?

$j := gps(A[i], B[i], 1, 20)$

c) Suppose that the language supports conditional expressions such as

if $x > y$ **then** x **else** y,

whose value is the maximum of x and y. Write a single assignment statement of the form

$j := gps(\cdots)$

that sets *S* to the maximum of the values in *A*.

d) Assume the conditions of (c), and write a single statement that sorts the array *A* in ascending order.

e) The name *gps* stands for "general problem solver." Explain why the name is appropriate.

15.18 As the *gps* procedure in the last problem illustrates, call-by-name is a very powerful binding mechanism. It can cause problems, however. In a language like PASCAL with call-by-reference parameters, it is very easy to write a procedure that will exchange the contents of two variables. For example,

> **procedure** *swap* (**var** *a*, *b*: *integer*);
>   **begin**
>   **var** *t*: *integer*;
>   *t* := *a*; *a* := *b*; *b* := *t*
>   **end**

This procedure will not work, however, in a language with only call-by-name binding.

- a) Why does *swap* fail in a language with only call-by-name binding? (Hint: Consider calls of the form *swap*(*i*, *A*[*i*]).)
- b) Devise a body of *swap* that will work with call-by-name parameters. (Hint: You may need another function.)

15.19 Suppose you want a program to have the effects of all of the parameter bindings discussed in the text *except* call-by-name; furthermore, suppose that your programming language does not have reference variables. Which of the remaining binding mechanisms must be included, and which can be programmed in terms of the others? Specifically, design a program to get the effect of those that were omitted.

15.20 Consider a language (such as ALGOL 68) that has reference types whose values "point to" objects of any type. Assume that this language has a β operator that can return a reference to an arbitrary variable. Show that the effect of all the parameter-binding mechanisms, aside from call-by-name, can be achieved with only call-by-value in this language.

Chapter 16

Runtime Representation of High-Level Languages

Throughout this textbook we have constantly emphasized the difference between an abstraction and its representation. In the last few chapters we introduced several new abstract concepts, most notably recursion, scope, extent, and parameter binding. In the interest of completeness, and also to provide you with a firmer understanding of what goes on when you run a program, this chapter presents an overview of the runtime structure of programs—specifically those parts of the structure that are dictated by features of the programming language.

16.1 Languages with Static Storage Management

Although they are not strictly required by the FORTRAN standard, most FORTRAN implementations abide by the following rules:

1. The main program and each subroutine may declare local data, but all data are preserved across successive calls on subroutines.

2. Subroutines may not be recursive; it is illegal for a given subroutine to be called if there is an as yet unfinished call of that same routine. This forbids both direct recursion (P calls P) and indirect recursion (P_1 calls P_2, P_2 calls P_3, $\cdots P_n$ calls P_1). In addition, all FORTRAN implementations forbid dynamically sized arrays and equivalents of the PASCAL *new* operator. As a result, no FORTRAN program can create new data objects while a block or subprogram is executing.

415

Rule 1 implies that if recursion were possible, all instances of a routine would operate on the same data—that is, the storage assigned to variables local to a routine is associated with the routine definition, not a routine activation. Rule 2 prohibits even the limited recursion that Rule 1 permits. These two rules, in conjunction with the fact that a running FORTRAN program cannot create new data objects, imply that all storage for FORTRAN data can be allocated before execution begins, or *statically*.

In order to compare the storage management scheme for FORTRAN with those of other languages, we require a concept that is used in the next several sections: the *activation record*.

Definition 16.1. An *activation record* (AR) for a subprogram is the set of information necessary for the execution of that subprogram. □

The AR for a FORTRAN subroutine contains the storage allocated to the variables declared local to that routine. In Fig. 16.1, the AR for a simple subroutine is pictured. Note that the AR includes space for the actual parameters of P, X, and Y. For the moment, just note that space is reserved for them in the AR; more is said about parameters in Section 16.7.

SUBROUTINE P(X, Y)	X	Y	K
INTEGER K	A(1)	A(2)	A(3)
REAL A(10)	A(4)	A(5)	A(6)
K = 0	A(7)	A(8)	A(9)
.	A(10)	OTHER DATA	
.			
.			
RETURN			
END			

Fig. 16.1 The AR for a simple subroutine

In FORTRAN, each subroutine, the main program, and each **COMMON** block has an AR. In practice, the AR's may be grouped together in one contiguous area. When a subroutine executes, it always finds its AR in the same place, so that the addresses of variables, relative to the beginning of the program, are fixed for all time by the compiler.

All that remains to the runtime structure of FORTRAN is the representation of the state of subroutine-call nesting. That is, when a subroutine returns it must know *where* to return. This is quite easily accomplished by including space for a "return address" in the "other data" field of the AR. When P calls Q, Q stores in its AR the place to return to in P. When Q wants to return, it simply retrieves this value from its AR and branches to it. Figure 16.2 shows the state of the AR's when R is executing, after P calls Q and Q calls

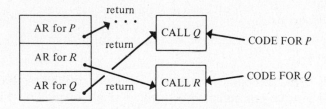

Fig. 16.2 State of AR's after P calls Q calls R

16.2 "Stack-Based" Languages

Most modern languages allow subroutines to be recursive. These same languages also support scope-entry declaration of local variables. In the presence of recursion, it is no longer sufficient to allocate a single area of fixed size for the AR of a routine. Since more than one activation of the routine can exist simultaneously, and since each may have a different return point and different values for local variables, we must make provision for multiple AR's for a single routine. Since we cannot predict at compile time how many activations of each routine can exist simultaneously, it follows that we can allocate space only when the routine is actually called.

Consider the implications of call-time allocation of the space for the AR's. First, we can see that allocations last for precisely the duration of a particular routine's execution. Thus we can allocate space when the procedure begins execution, and release that space when the procedure is finished. Second, we observe that if a routine P calls a routine Q, then P cannot complete before Q. This means that the extent of Q's AR is wholly contained in the extent of P's AR. Thus, the storage requirements of AR's are last-in-first-out, which suggests that a stack-like data structure will suffice to meet them. In Section 4.3, we saw a simple version of such a structure, which handled the bookkeeping of return points. Generalizing that simple version, a call on a routine P is represented as

> *rep*(P) =
> Push a new AR for P, containing #*Lnext* as its return field,
>   onto the stack;
> **goto** *Plabel*;
> *Lnext*: · · ·

The routine P itself is represented

> *rep*(**procedure** P; *BODY*) =
> *Plabel*:
>   *rep*(*BODY*);
>   Pop the current AR for P, saving the return point in T;
>   **goto** @T;

(where T is some temporary variable.) In some machines, the last two steps of this process can be combined into a single operation. Figure 16.3 shows what the situation would be like if P called Q, which then called P.

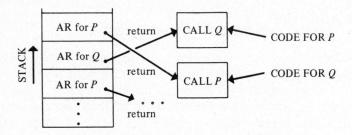

Fig. 16.3 State of AR's after P calls Q calls P

We used the term "stack-like" because although the operations described resemble those on a stack, they aren't entirely the same. For example, during its execution, *BODY* will change the values in its AR—the top item on the "stack"—an operation which we did not allow for in our definition of the type *stack* in Section 9.1.3.

16.3 Up-level Addressing and the Display

The representation discussed in the previous section works well in cases where procedures reference only formals and locally declared variables. Since the compiler is told precisely what the local variables are, it can quite easily lay out the AR for a given procedure and assign to each variable an "offset" address within the AR. All a procedure needs to know is where its AR starts, and it can find its local variables with no trouble.

The address of the current AR is usually kept in a designated register. When a procedure is called it allocates its AR on top of the stack, saving the caller's AR address in its own AR. When it returns, it simply restores the old AR address and branches to the return point.

What happens if a procedure can reference global variables? If the global variables are all declared in the outermost level of the program (that is, in the main program), there is no trouble. The main program's AR is always allocated at the bottom of the stack. Since the stack starts at a relative address known at compile time, each variable reference is made either

1. relative to the current AR for locals, or

2. relative to the stack base for globals.

However, in most block-structured languages there are *intermediate global variables*. The scope rules of ALGOL, PASCAL, etc. dictate that, for example, in the program segment of Fig. 16.4, x and y are accessible in P, Q, and R, and z is accessible in Q and R. Now we must ask whether R can tell where x, y, and z are located. It appears at first glance that R can tell where x, y, and z are located. After all, if AR_Q (the AR for Q) has a size known at compile time, and so does AR_R, then R can find z at a fixed offset from

$$\text{address of } AR_R - \text{size of } AR_Q = \text{address of } AR_Q,$$

and x and y can be found at

$$\text{address of } AR_R - \text{size of } AR_Q - \text{size of } AR_P = \text{address of } AR_P.$$

However, suppose that R calls itself recursively. If at some point in the midst of its recursion R chooses to reference z, it can no longer find AR_Q by subtracting the size of AR_Q from its own AR address. There will be some undetermined number of copies of AR_R in between the current one and AR_Q on the stack.

This problem is known as "up-level addressing." The name is derived from the concept of "lexical level." The lexical level of a procedure is an integer value one greater than the lexical level of the procedure in which it is declared. In Fig. 16.4, if P has level n, then Q has level $n + 1$, and R has level $n + 2$. We define the level of the main program to be zero. Up-level addressing is the term applied to the access to intermediate global variables—that is, to those at levels greater than 0 and less than the

```
procedure P;
begin
var x, y: t₁;

procedure Q;
begin
var z: t₂;

procedure R;
begin
var a, b: t₃;
.

.

.
end;
.

.

.
end;
.

.

.
end;
```

Fig. 16.4 Intermediate global variables

current level. How can we represent the execution history of a program to facilitate up-level addressing?

The basic problem in Fig. 16.4 is that in order for R to find x, y, and z, it must know where AR_P and AR_Q start. Since it cannot deduce this information from the size of the AR's alone, there must be another approach. Consider that the number of AR's accessible to R is precisely $level(R) + 1$. For example, the main program can "see" only the outermost declarations. Procedures declared in the main program can see both their own declarations and those of the main program. Procedures declared within these procedures can see three sets of variables, etc. This is a very important property, and one that is dictated by the static nature of scope in the languages under discussion.

The fact that ALGOL, PASCAL, and PL/I have lexical scope allows a relatively straightforward solution to the up-level addressing problem. Since each procedure P can access exactly $level(P) + 1$ AR's, we simply add *level*(*P*) + 1 locations to the AR for *P*. If we can arrange things so that each of these new locations contains the address of one of the accessible AR's, we should be in business.

The new locations are collectively called a *display*. The display of a given AR is a representation of the addressing environment (that is, of variables that can be "seen") dictated by the lexical scope rule. The display can be thought of as a vector of pointers to AR's. Suppose *display*[*k*] points to the accessible AR at level *k*. If *R* in Fig. 16.4 wants to reference *z*, it can do so via a fixed offset from *display*[*level*(*R*) − 1]. *R* can find *x* and *y* at fixed offsets from *display*[*level*(*R*) − 2]. Since *level*(*R*) is known at compile time, *R* has a compiled-in algorithm for getting at global variables.

The only problem is how to set up the display for a called procedure. If *P* calls *Q*, what should *Q*'s display look like? Assume that for *P* to call *Q*, *level*(*Q*) ≤ *level*(*P*).   (This assumption is valid only as long as procedure parameters are not allowed. See the next section.) If *P* and *Q* are at the same level, say *j*, (Fig. 16.5(a)) then the only difference between the two displays will be in the last element *display*[*j*]. Where *P*'s *display*[*j*] points to AR_P, *Q*'s *display*[*j*] should point to AR_Q. Thus we can simply copy *display*[0] through *display*[*j* − 1] from AR_P to AR_Q, and set *display*[*j*] to AR_Q.


```
    procedure Q;                        procedure Q;
        begin · · · end;                    begin · · · end;

    procedure P;                        procedure R;
        begin · · · Q · · · end;            begin
            ·                                   procedure P;
            ·                                       begin · · · Q · · · end;
            ·
                                            end;
```

          (a)   (b)

Fig. 16.5 (a) *level*(*P*) = *level*(*Q*) = *j*; (b) *level*(*P*) > *level*(*Q*) = *j*

If Q is up-level from P as in Fig. 16.5(b), then P and Q will have in common only the first $level(Q)$ display elements. These are $display[0]$ through $display[level(Q)]$. Just as before, $display[level(Q)]$ should be set to AR_Q.

Thus the algorithm is to copy $display[0..j-1]$ and set $display[j]$ to the new AR whenever the called procedure is at level j. When a procedure returns the act of restoring the old AR value to the designated register will implicitly restore the old display. The representation of a procedure, P, has now become

$rep(\textbf{procedure } P; BODY) =$
Plabel:
    set $display[0]$ to $display[level(P) - 1]$ in the current AR
        from the preceding AR;
    set $display[level(P)]$ to the address of the current AR;
    $rep(BODY)$;
    Pop the current AR for P, saving the return point in T;
    goto @ T;

Figure 16.6 shows the stack configuration that results from the program in Fig. 16.4 if R has called itself recursively twice. Note that, as

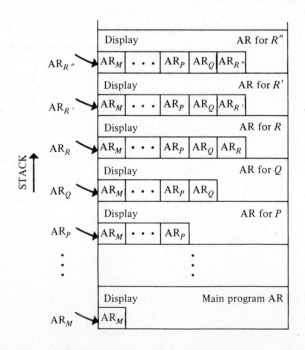

Fig. 16.6 Sample display configuration

should be the case, the current activation of R can reference the main program AR, AR_P, AR_Q, and its own AR, but not those of the other R activations. The up-level addressing problem is thus effectively solved.

As it happens, the implementation of displays described here has pedagogical advantages, but it is far from the most efficient implementation that can be devised. All that is really necessary is a single vector and one word for each AR. Each routine call must save and restore only one display entry. We'll leave the details of this scheme to an exercise.

16.4 Array Allocation

In FORTRAN and PASCAL, the size of arrays is fixed at compile time. Thus the AR of a routine that declares a local array can include space to store the array, and the size of an AR remains constant.

In other languages, including ALGOL and PL/I, the bounds of an array can be determined at execution time. The expressions whose values specify the bounds must be evaluated in the context of the procedure that declares the array. In order for the bounds to be evaluated, the AR must already be allocated, since the expressions can, in general, be arbitrary and thus may include function calls that alter the stack themselves. The array cannot then be part of the AR. Where does it go?

Actually this is not a tough problem. We simply include a *descriptor* for the array in the AR. The descriptor's size is determinable at compile-time. After the AR has been allocated, but before the body of the procedure executes, the bounds are evaluated. Although this may trigger an arbitrarily complex series of routine calls, all of this action will take place on top of the newly created AR. As each bound is calculated, the stack is returned to the environment of the AR in question. The bound value is stored in the descriptor, and the next bound computed, etc. When all bounds are known, the size of the array is computed, and the AR is "extended" by allocating the array on top of the stack. This process is repeated until all local arrays have been allocated. When the procedure returns, the arrays will disappear along with the AR when the old stack pointer of the calling routine is restored.

16.5 Block Structure

The PASCAL scope rules provide a limited form of "block structure," in that the unit of scope is the procedure. The languages more closely related to ALGOL are somewhat more general in that blocks (**begin** and **end** groups) may also declare their own local data. This does not present much of an additional problem. We have two options: either we can choose to allocate the local data when the block is entered, or we can allocate all of the local data in all of the blocks of a given procedure as soon as the procedure is entered. The former choice is better if it is common to declare many different blocks inside a procedure but only to enter very few during execution. The latter choice is better if most blocks are, in fact, entered during execution. Making the choice involves another time/space trade-off.

Of course if a block declares a local array with variable bounds, we have no choice but to wait to allocate it until the block is entered. Still, the array may simply be placed on the top of the stack. The descriptor can be part of the procedure AR.

16.6 Procedure Parameters

The analysis of Section 16.3 showed that the display can be easily established at procedure entry by copying a known number of elements from the display of the caller. This is not always possible if procedure names can be passed as parameters to other procedures. Consider the program skeleton in Fig. 16.7.

procedure Q(**procedure** F); **begin** $\cdots$ $F(x)$ $\cdots$ **end**;

procedure P;
    var a, b: t;
    procedure $R(y$: $t)$; **begin** $\cdots$ a $\cdots$ **end**;
    begin
    $\cdots$ P; $\cdots$ $Q(R)$; $\cdots$
    end;

Fig. 16.7 Example of procedure parameters

If P is invoked, it will possibly call itself recursively, eventually calling Q and passing R as a parameter. When F is invoked inside of Q, it is R that actually executes. How is R's display to be established? It will *not* copy display locations from its caller, Q, as before. This can be seen clearly from the fact that R must have access to the a and b of P, yet Q does not. The solution lies in the representation of the procedure parameter R. P must not only pass the address of R to Q, but it must also pass the current environment in the form of the address of AR$_P$. When R is invoked from within Q via F, it gets its initial display elements not from Q but from P's display at the time Q was called. We will not present all the details of this process, but it should be apparent that this is both possible and the correct thing to do.

16.7 Parameter Mechanisms

In Chapter 15 we discussed several different parameter mechanisms commonly found in high-level languages. These were:

- Call-by-value
- Call-by-address
- Call-by-value/result
- Call-by-result
- Call-by-name

In this section we will sketch the runtime representations of these various mechanisms. The differences among the various parameter mechanisms can all be characterized by what is stored in the representation of the actual parameter list by the caller and what is done with it by the called routine.

The simplest representation of the actual parameter list is a contiguous area on top of the runtime stack at the very front of the called routine's AR. Usually, if more than one parameter is to be passed, the parameters are evaluated left to right—that is, first the first parameter, then the second, etc. Thus, as each actual parameter value is determined, it can be pushed on the stack. If computation is necessary to obtain a parameter value, even for evaluation of arbitrary functions, that computation can be performed at the stack top without disturbing parameter values already computed and residing on the stack. Thus, when the called routine receives control, it can assume that its actuals are on top of the stack, and can treat that area as part of its complete activation record. Of course this simple scheme relies on the assumption that the number of parameters to a given routine is fixed at compile time. If not, the called routine would have to have some way of knowing how many parameters it received, perhaps by way of an extra word on top of the stack.

16.7.1 CALL-BY-VALUE

Actual parameters whose corresponding formals are declared as value parameters are evaluated in the current context and their values are pushed onto the stack. The called routine can use the stack locations of value parameters as local variables and when they are changed, new values can overwrite the old ones in the parameter list. Thus local changes are reflected locally; but as the caller never retrieves value parameters after the call, they are not reflected outside the procedure.

16.7.2 CALL-BY-ADDRESS

For address parameters, instead of storing the actual's value in the parameter list, the caller stores its address. When the called routine wants to reference the parameter, it does so *indirectly* through the appropriate parameter list location. Typically, an actual is restricted to be a variable (expressions are not allowed), so its address is known to the caller. If the actual is an array element, such as $A[i, j]$, its address must be computed by the caller before the call is actually made. Note the implication that, for example, a change to i or j by the called routine will not change the address of the actual, since that address is computed prior to the call and remains fixed. However, because the called routine receives an address instead of a value, it can cause the intended side effects.

16.7.3 CALL-BY-VALUE/RESULT

Call-by-value/result can be implemented by a combination of the techniques used for call-by-address and call-by-value. The actual parameter is passed as in call-by-address. Then, the first action of the routine body is to copy the value of the formal into a local variable. The body of the routine accesses this local. Finally, just before returning, the value of the local is copied back into the actual parameter location.

Alternatively, the *caller* can pass the parameter as for call-by-value and copy its final value back after the called routine finishes.

16.7.4 CALL-BY-RESULT

Call-by-result is simply a restricted form of call-by-value/result. Its implementation is identical, except that the local variable is not initialized from the actual parameter location.

16.7.5 CALL-BY-NAME

Call-by-name parameters cause much more difficulty than any of the other mechanisms because, unlike the others, they require evaluation in the *environment of the caller* rather than that of the called routine. ALGOL 60, which has the call-by-name parameter mechanism, suffers additional overhead in representation and procedure invocation because of the necessity of changing the environment to that of the caller every time a name parameter is accessed. We won't go into all of the details here, but the basic idea is to pass, not a value or a variable, but a *function* as the actual parameter for a formal declared call-by-name. This function doesn't exist at the source level (that is, the user didn't write it), but it is generated by the compiler at the call site. Thus this new function inherits the environment of the caller as though the user had declared it local to the caller. This function, often called a *thunk* (an onomatopoetic reference, rumor has it, to the sound made by a pointer as it moves instantaneously up and down the stack), takes no parameters itself, and delivers as its value the address of its corresponding actual. Thus if the actual $A[i]$ is passed for a call-by-name formal, a thunk is created at the call-site that computes the appropriate offset into the array using the global value of i. The thunk is passed as a procedure parameter (see Section 16.6) to the called routine. *Each time* it needs that parameter, the called routine invokes the thunk, as it would any other procedure parameter, and uses the resulting value as the address of the name parameter.

16.8 Heap Allocation

Some languages, such as PASCAL, allow the dynamic allocation of data objects in other than stack-based fashion. Consider the program fragment of Fig. 16.8.

> **type** *ref* = ↑*t*;
> **var** *x*: *ref*;
> **procedure** *P*; **begin** · · · *new*(*x*) · · · **end**;

Fig. 16.8 An example of non-stack-based extent

When P causes the allocation of an object of type t via the *new* operation, it assigns the address of that object to x. After P returns, x still points to that object. Hence it is possible for a procedure to cause the allocation of storage that outlives its own activation. If a new object were allocated on the stack, it would appear on top of AR_P. If AR_P "went away," it would leave a neat little hole in the stack under the newly created object. This would not do at all.

Storage whose extent is not tied to a particular scope cannot be allocated on the stack. It must be taken from a separate pool. The pool is usually referred to as the "heap." In order to avoid wasting storage, the

Fig. 16.9 Total memory allocation

stack and heap in implementation of many languages occupy the same block of memory as in Fig. 16.9.
The stack and heap occupy opposite ends of the block of memory. The stack grows toward the heap. When heap allocation is required, the search for free space is made in the direction of the stack.

If the total memory is insufficient, the stack and heap will eventually collide, causing the program to terminate with some such error as "Stack overruns heap."

16.9 Further Reading

The implicit runtime structure of programs is a subject properly included in the study of compilers. Complementary and more detailed discussions than the one we present here can be found in Aho and Ullman (1977) and Gries (1971). ALGOL 60 was the first language to require a stack-based allocation strategy, and is interesting in its own right. A good book devoted to the implementation of ALGOL 60 is Randell and Russell (1964).

Exercises

16.1 Assuming the representation in this chapter for FORTRAN subroutine calls, describe the effect of accidentally attempting recursion in FORTRAN.

16.2 Design a scheme by which a FORTRAN implementation can detect recursion when it occurs and print an error message.

16.3 Provide an example that shows why it is not always possible to anticipate the storage requirement of a recursive procedure at compile time.

16.4 Consider a graph representation of a program that abstracts the procedure-call relationships of the program. The routines and the main body of the program are represented by nodes; arcs represent the relation "calls."

    a) What restriction on allowable graphs corresponds to the restriction of no recursive calls in the program?

    b) What restriction on the uses of routine calls in a program corresponds to the restriction that the graphs must always be trees?

16.5 Why do you think that a language that requires a heap should bother with a stack at all? After all, if the heap must exist and be managed independently of the stack, why not just use the heap as a source of activation records also and forget the stack? Carefully specify an implementation that does exactly this before you answer the first question.

16.6 Exercise 4.16 introduced the concept of a coroutine. In the most general case, storage management for coroutines cannot be done conveniently using a single stack. Why not? Describe a good way to do storage management for coroutines.

16.7 It is often more efficient to access a variable if it is statically allocated (as in FORTRAN) than if it is allocated relative to a stack pointer. Can you devise a representation in which each identifier is always bound to the *same* storage location throughout execution, but which retains the effect of a block-structured language where any given identifier may refer to a large number of variables during program execution? If so do so; if not explain why.

16.8 In Section 16.6 we discussed briefly what must be done to handle routine names passed as parameters. Fill in the details. Exactly what must happen with each call of Q and each call of F within Q?

16.9 Assuming that dynamic arrays are permitted within procedures, trace all events that would occur in a call on the following procedure, using the representation discussed in this chapter.

    procedure $G(n$: *integer*; **var** y: *real*);
      var A: **array** $[1..n]$ **of** *real*;
      begin *body of G* **end**;

16.10 Which of the following language features would *force* you to use a stack implementation instead of static allocation? Which would force you to provide a heap?

    a) Block structure with static array sizes

    b) Block structure with dynamic array sizes

c) Local variables in routines

d) **Own** variables in routines

e) Pointers

f) The operator *new*

g) Nonrecursive procedures

h) Recursive procedures

i) Coroutines

16.11 Give an example showing why a heap and stack must be separate. Why can't the operator *new* just put objects on top of the stack where dynamic arrays go?

16.12 Write the SMAL code for the following program.   Assume the system's stack is called *SysStack*.

```
var x, y: integer;
function F(var w: integer): integer;
    var z: integer;
        begin z := w; P(w); F := z + w; end;

procedure P(value-result p: integer)
    var z: integer;
        begin p := p + 1 end;

begin
x := 1;
y := F(x);
x := F(y);
end.
```

16.13 Write the SMAL code for the following program.   Assume the system stack is *SysStack*.

```
var V: array [1..4] of integer; i: integer;
procedure P(var a, b: integer);
    begin a := a + 1; b := b + 1 end;

begin
for i := 1 to 4 do V[i] := i * 3;
i := 2;
P(i, V[i]);
P(V[i], i);
end.
```

16.14 Write the SMAL code for the following program using each of the five parameter binding rules and assuming static allocation and no stack. Which binding doesn't really make sense for this program? Explain.

```
var x, y: integer;
function Funny(W: integer): integer;
    begin W := W + 1; Funny := W end;


begin
x := 0;
y := Funny(x);
end
```

16.15 Write the SMAL code for the following program using each of the five parameter binding rules and assuming static allocation and no stack. Which binding doesn't really make sense for this program? Explain.

```
var V: array [1..3] of integer; i: integer;
function Funny(W: integer): integer;
    begin W := W + 1; Funny := W end;


begin
for i := 1 to 3 do V[i] := i;
i := 2;
y := Funny(V[i]);
end
```

16.16 Write the SMAL code for recursive and nonrecursive functions to compute Fibonacci numbers (see Exercise 3.15.) Trace the execution of both routines for $F(5)$.

16.17 Consider the routine declared

procedure F(**value-result** x: *integer*);

Is it correct to implement the call $F(A[i])$ as follows?

```
Create a new variable, x;
Execute x := A[i];
Execute the body of F;
Execute A[i] := x
```

If not, why not?

16.18 One way that we might implement a call such as $G(a, b)$ would be to replace that call with the body of G, with the actual parameters a and b replacing the formal parameter identifiers of G. Under what circumstances (with what parameter mechanisms, and what assumptions about the body of *G*) would this work? In cases where it would work, what advantages and disadvantages can you see to implementing such calls on *G*?

16.19 Exercise 15.13 introduced dynamic **own** arrays. How might you implement them? Your answer, of course, depends on how you answered Exercise 15.13. Suggest implementations for all the reasonable answers you gave to that exercise.

16.20 The implementation of displays discussed in the text requires $k + 1$ words per activation record, where k is the lexical level of the procedure. It also requires that k display entries be copied on each procedure entry. Neither of these is necessary. In fact, a display can be implemented using a single $m+1$-word vector, where m is the maximum lexical level of any procedure in a program. When this scheme is used, only one word needs to be copied on each procedure entry and return. Design this scheme, prove that it works, and show how it can be implemented in the SMAL representation of procedures.

16.21 Suppose you wanted to allow "partial parameterization" of routines. For example, suppose you had the definitions

```
type Vector = array [1 .. n] of integer;
procedure ForAll(A: var Vector; function F: integer);
    var i: integer;
    begin
    for i := 1 to n do  A[i] := F(A[i]);
    end;
function Sum(x,y: integer): integer;
    begin  Sum := x + y end;
```

and wanted to be able to say *ForAll*(*B*, *Sum*(?,1)), meaning "add 1 to each element of *B*." The expression *Sum*(?,1) denotes a function with one argument (indicated by the single "?" slot) obtained by taking *Sum* and fixing its second parameter at 1. Suggest an implementation for this construct.

Chapter 17

Reasoning about Recursion

Recursive program and recursive data structures admit of what might be called "recursive proofs." For each kind of recursion there is a corresponding form of inductive reasoning: induction on the *depth* of recursive calls, and induction on the *structure* of recursively defined data objects. In this chapter we illustrate both kinds of inductive proofs of recursive programs. We will do so informally, leaving the details as exercises.

17.1 Proving Recursive Programs

Induction on the depth of recursion is perhaps the most direct way of proving the correctness of a recursive procedure. Consider the time-worn factorial function:

$$n! = n \cdot (n - 1)! = 1 \cdot 2 \cdot 3 \cdot \cdots \cdot (n - 1) \cdot n$$

which can be coded directly (if somewhat inefficiently) as:

```
function factorial (n: integer): integer;
    begin
    if n < 2 then factorial := 1 else factorial := n * factorial(n - 1)
    end;
```

Obviously, the specifications for *factorial* are

> **pre** $n \geq 1$;
>
> **post** factorial $= n!$;

We can prove that *factorial* satisfies these specifications very easily by induction on n:

- Basis: For $n = 1$, we must prove

  $n = 1 \wedge n \geq 1$
  $\{$**if** $n < 2$ **then** *factorial* $:= 1$ **else** *factorial* $:= n * factorial(n - 1)\}$
  factorial $= n!$

  which is immediate. Note that since only an **if** statement and an assignment are involved, the program always terminates for $n = 1$.

- Inductive step: Suppose that *factorial*$(n) = n!$ for $1 \leq n < k$, where $k > 1$. By convention, we generally mean by this that *factorial*(n) is also defined—that the program terminates. Now consider the case $n = k$. We must prove

  $n = k \wedge k > 1 \wedge n \geq 1$
  $\{$**if** $n < 2$ **then** *factorial* $:= 1$ **else** *factorial* $:= n * factorial(n - 1)\}$
  factorial $= n!$

  Noting that $k > 1$ implies that $n \geq 2$, we eventually get the simplified verification condition

  $n = k \wedge k > 1 \supset n! = n \cdot (n - 1)!$

  which is obvious. Note also that our inductive hypothesis also allows us to conclude that *factorial*(n) terminates when $n = k$. $\square$

One interesting property of the foregoing proof is that it not only establishes the weak correctness, but also the total correctness of *factorial*. That is, we have proved that *factorial* always terminates. Recursive procedures need not terminate—even in the absence of loops. For example, if the body of *factorial* were

> **begin**
> *factorial* $:= factorial(n + 1)$ **div** $(n + 1)$
> **end**

the procedure would not terminate, even though it is certainly the case that

> $n! = (n + 1)!/(n + 1)$.

17.2 Structural Induction

Sometimes the proof of a program is dictated by the structure of the data it manipulates; this form of induction is known as *structural induction*. For example, consider programs that perform recursive tree traversals. A nonempty binary tree, by one definition, is a node with two subtrees. A tree is defined *recursively* to be composed of smaller trees, except for the empty tree.

The basis step of this induction is to show that whatever property you are trying to prove is true for the empty tree (or, alternatively, that it is true for all possible leaves). The inductive step is to show that, if the property holds for all subtrees of a tree, it must hold for the tree itself. The following example should serve to explain the technique. Consider the recursive *Find* program in Fig. 14.12 of Section 14.5.2.

```
function Find(T: tree; x: Elt): tree;
    {Returns the subtree within T whose node label is x.
    Returns the empty tree if there is no such subtree.}
    begin
    if IsEmpty(T) then Find := EmptyTree
    else if label(root(T)) = x then Find := T
    else if x < label(root(T)) then {x cannot be in the right son of T}
            Find := Find(leftson(T), x)
    else {x cannot be in the left son of T}
            Find := Find(rightson(T), x)
    end;
```

To prove that it works, we use the following proof

- Basis step: The program works if T is empty. This is left to the reader, but it should be clear intuitively.

- Induction step: Assume that *Find* works for all subtrees of T.

    a) If $x = label(root(T))$, *Find* works, because T is returned, as it should be.

    b) If $x < label(root(T))$, then if x is present in T, x must be in its left son. Since $leftson(T)$ is a subtree of T, the value of $Find(T, x)$ must then be $Find(leftson(T), x)$, by the inductive hypothesis.

    c) If $x > label(root(T))$, the argument is as in (b), with the substitution of "left" for "right."

Structural induction may be applied to any recursively defined data structure—not just trees. For example, a list (as defined in Chapter 9) is either empty or consists of an item of data (the head) and a list (the tail). Therefore, we have a recursive structure on which to apply structural induction. For example, consider the function *ReverseAppend*, defined as

> **function** *ReverseAppend*(*x, y*: ↑ *listofT*): ↑ *listofT*;
> **post** *RESULT* = *reverse*(*x*) ⁓ *y*;

That is, *ReverseAppend*(*x, y*) returns a pointer to a list consisting of the list *x* in reverse order concatenated to the list *y*. Consider the alleged implementation

> **function** *ReverseAppend*(*x, y*: ↑ *listofT*): ↑ *listofT*;
> **begin**
> **if** *IsEmpty*(*x*) **then** *ReverseAppend* := *y*
> **else** *ReverseAppend* :=
>   *ReverseAppend*(*Tail*(*x*), *ConsList*(*FirstItem*(*x*), *y*))
> **end**;

Is this correct? Consider the following proof:

- Basis: *ReverseAppend* works for *x* = *EmptyList* (trivial).
- Induction step: Suppose *ReverseAppend* works for *Tail*(*x*) and any *y*. Then

$$RESULT = reverse(Tail(x)) \sim FirstItem(x) \sim y$$

by the inductive hypothesis. The result is immediate, because

$$reverse(Tail(x)) \sim FirstItem(x) = reverse(x).$$

Of course, to make either of these proofs completely rigorous, we would need to fill in the details—using the *wp*'s from Chapter 5. Our purpose here is to illustrate the techniques used to verify recursive programs rather than to burden you with detail.

17.3 Further Reading

Mathematical reasoning about recursive programs can be difficult at first; in general it requires the use of mathematics slightly different from what one usually encounters as an undergraduate. A good place to start is Manna (1974).

Exercises

17.1 The function *Comb* of Section 3.3.4 is supposed to compute the number of combinations of *n* objects taken *k* at a time, given $n \geq k \geq 0$.

This number can be shown to be

$$\frac{n!}{(n-k)!\,k!}$$

Prove that *Comb* computes this number.

17.2 Fill in the steps showing that *Find* is correct.

17.3 In Section 14.4 we defined an iterative and a recursive version of procedure *Range*—a procedure to find the range of values in a given vector. Construct proofs of both versions, using ordinary *wp*'s and loop invariants for the iterative version, and induction on the depth of recursion for the recursive version.

17.4 Exercise 3.15 defined Fibonacci numbers and asked for iterative and recursive programs to compute them. Construct proofs of both versions, using ordinary *wp*'s and loop invariants for the iterative version and induction on the depth of recursion for the recursive version.

17.5 Chapter 14 develops four recursive programs. Write specifications for each of the following programs and prove that they work (including termination).

   a) Range of element values, Section 14.4, Fig. 14.5.
   b) Sorting a vector, Section 14.4, Fig. 14.6.
   c) One of the recursive tree walks of Section 14.5.1.
   d) Finding an element in a binary search tree, Section 14.5.2, Fig. 14.12.

17.6 Can you apply structural induction to programs manipulating graphs? Why, or why not, or under what circumstances?

17.7 Where would the inductive proof of *factorial* break down if we tried to prove the nonterminating version

$factorial := factorial(n + 1)$ **div** $(n + 1)$?

17.8 The following program is supposed to count the number of leaves (nodes without nonempty sons) of a binary tree. Prove that it does.

```
function Tips(T: tree): integer;
   begin
   if T = Empty then Tips := 0
   else if (LeftSon = Empty) and (RightSon(T) = Empty)
      then Tips := 1
      else Tips := Tips(LeftSon(T)) + Tips(RightSon(T))
   end;   {Tips}
```

Chapter 18

Analysis of Recursive Algorithms

The analysis techniques of Chapter 6 can work only as long as the steps of an algorithm are all "simpler" than the algorithm itself. When that is true, the cost of the algorithm can be determined by adding the costs of its component steps. When an algorithm is recursive, however, the component steps cannot all be expressed directly in primitive terms. When we introduced divide-and-conquer techniques in Section 14.4, we noted that the expression describing the execution time of such an algorithm is usually expressed in terms of the execution time for a smaller problem. The formulas that result are called *recurrence relations*.

In this chapter we examine the costs of several of the recursive programs developed in Chapter 14 and study ways to find closed-form solutions for the recurrence relations that express their cost. We assume that the "problem size"—for example, the length of vector or size of tree—is a power of 2. This simplifies the analysis and provides easy bounds. This assumption can be relaxed, but the computations become rather more complex.

In Sections 18.1, and 18.2 we derive the cost functions for several recursive programs and discuss some of the mathematical properties of the resulting formulas. In Sections 18.3, and 18.4, we show some general ways to solve problems of this kind.

18.1 Examples of Cost Functions for Recursive Programs

In Section 14.4, Fig. 14.5 is a recursive program for determining the range of element values in a segment of an array. The control flow is dominated by the recursion, and the cost is approximated by the number of comparisons (*max*, *min*) of array elements. The number of comparisons required is given by the formulas

$$Rng(N) = 2\ Rng(N/2) + 2$$

$$Rng(2) = 1$$

where $Rng(N)$ denotes the number of comparisons for computing *range* on a section of an array containing N elements.

Fig. 14.6 showed a recursive program for sorting a vector, based primarily on merge operations. Each merge combines two sorted segments of length $k/2$ into a sorted segment of length k. The cost of sorting a segment of length N, in terms of the number of comparisons required, is

$$MrgSrt(N) = 2\ MrgSrt(N/2) + Merge(N)$$

$$MrgSrt(1) = 0$$

where $MrgSrt(N)$ represents the cost of running *Sort* on a segment of length N and $Merge(N)$ represents the cost merging two vectors into a single vector of length N.

Section 14.5.1 contains three algorithms for traversing a binary tree and printing the values stored in the nodes. They differ only in the order of the operations in the procedure bodies, so a single analysis will suffice for all three. We assume not only that the number of nodes in the tree is $2^N - 1$ for some N, but also that the tree is balanced (that is, we assume that all terminal nodes are approximately the same distance from the root). Under these assumptions, the number of *Write* operations required is

$$TreePrint(N) = 2\ TreePrint(N/2) + 1$$

$$TreePrint(0) = 0$$

where $TreePrint(N)$ denotes the number of *Write* operations required to print a tree with N nodes.

Figure 14.12 shows a technique for finding an element in a binary search tree. The precise cost of a search depends on whether the key is in the tree and on how far it is from the root. However, if we assume that the tree is balanced but that the key is at the worst place (a leaf), the number of comparisons involving the key, $Fnd(N)$, is given by

$Fnd(N) = Fnd(N/2) + 1$

$Fnd(1) = 1$

for a tree with N nodes.

18.2 Familiar Recurrence Relations of Mathematics

The formulae in Section 18.1 are all examples of recurrence relations: that is, they express the value of a function at a point N in terms of its value at one or more points $M_i < N$. You are probably familiar with a few recurrences. For example, the factorial function is defined by one:

$Factorial(N) = N \cdot Factorial(N - 1)$

$Factorial(0) = 1$

Fibonacci numbers are also defined by a recurrence; this one refers to two preceding values:

$Fibonacci(N) = Fibonacci(N - 1) + Fibonacci(N - 2)$

$Fibonacci(1) = 1$

$Fibonacci(0) = 0$

Note that in both cases the definition involves a formula for the general case and enough formulas for special cases to be certain that the function is always defined.

    Recurrence relations are not especially convenient mathematical objects. Generally, we like to transform, or "solve," the relation and produce more familiar, closed-form, formulas. Unfortunately, not all recurrence relations can be summarized that way; but when they can, computations are greatly simplified. In Section 18.3 we present some rules that solve certain special kinds of recurrences that arise frequently in the solution of divide-and-conquer problems. In Section 18.4 we present a somewhat more general solution technique.

18.3 Solutions to Example Cost Functions

Recurrences are often solved by looking up the recurrence in a table to find the closed-form answer. In this section, we will supply three such solutions and show how they apply to the examples of Section 18.1.

Consider the tree-printing algorithm. The general rule that applies in this case is

if $T(N) = 2T(N/2) + k$

then $T(N) = N[T(1) + O(1)]$

where the O notation is as described in Section 6.2. For tree-printing, writing has constant cost (which we take to be 1,) so the rule applies. The cost is therefore

$$TreePrint(N) = N[TreePrint(1) + O(1)]$$
$$= O(N)$$

This corresponds with your intuitive assumption that writing each of N nodes once should require N *writes*. For the binary tree search, the applicable rule is

if $T(N) = T(N/2) + 1$

then $T(N) = T(1) + \log_2 N + O(1)$

This corresponds precisely to the formula for *Find*, so we obtain the solution

$$Fnd(N) = Fnd(1) + \log_2 N + O(1)$$
$$= O(\log_2 N)$$

We assumed a balanced tree with $2^N - 1$ nodes; the height of such a tree is $O(\log_2 N)$, so this is the result you would expect.

For merge sorting, we assume that the cost of a merge is $O(N)$. Then the useful recurrence solution rule is

if $T(N) = 2\ T(N/2) + O(N)$

then $T(N) = N[T(1) + \log_2 N + O(1)]$

We can apply this directly to obtain

$$MrgSrt(N) = N[\log_2 N + O(1)]$$
$$= O(N \log_2 N)$$

This result may not be so familiar, but most good sorting programs have times of $O(N \log_2 N)$.

18.4 A General Rule for Solving Divide-and-Conquer Recurrences

All the solution rules in the preceding sections of this chapter are special cases of a single rule. In this section, we present that rule without proof and show how to derive the previous results.

In all cases, we deal with recurrences of the form

$$T(N) = k\,T(N/2) + f(N)$$

$T(1)$ given

The recurrence is defined only when N is a power of two. To solve the recurrence we rewrite it in the following form:

$$T(N) = 2^p T(N/2) + N^p\, g(N)$$

$T(1)$ given

This rewrite is always possible because $p = \log_2 k$ and $g(N) = f(N)/N^p$. This new recurrence has the unique solution

$$T(N) = N^p\,[T(1) + h(N)]$$

where

$$h(N) = \Sigma_{i=1}^{\log_2 N}\, g(2^i)$$

We can simplify matters even further by tabulating some values of $h(N)$ in relation to $g(N)$, as in the following table:

$g(N)$	$h(N)$
$O(N^q)$ for $q < 0$	$O(1)$
$(\log_2 N)^j$ for $j \geq 0$	$(\log_2 N)^{j+1}/(j+1) + O((\log_2 N)^j)$

The derivations of this remain as exercises.

18.5 Further Reading

Analyzing the computational cost of recursive algorithms is not fundamentally harder than analyzing nonrecursive algorithms. It just requires careful thought and induction. Aho, Hopcroft, and Ullman (1974) and Horowitz and Sahni (1978) contain many more examples of this kind of analysis.

The technique for solving recurrence relations that is presented in this chapter was developed by Bentley, Haken, and Saxe (1978).

You may be interested to know that $O(N \log N)$ comparisons is the best that can be done for the sorting problem. An important advantage in knowing about such "lower bounds" is the effort you can save trying to find better ways to solve your problem. Note the analogy to the unsolvability of the halting problem (Chapter 13). The study of what can and cannot be done, and how fast it can be done, if at all, is a very interesting and important part of the theoretical body of computer science knowledge. The texts referred to here discuss in detail many of the known lower bounds.

Exercises

18.1 Reanalyze both range algorithms discussed in Section 14.4, this time counting *all* comparisons, both explicit and implicit.

18.2 Using the methods discussed in this chapter, find closed-form solutions for each of the following recurrence relations:

a) The following relation arises in the analysis of an algorithm to find the convex hull of a set of points in 2-space:

$$T(n) = 2\ T(n/2) + O(n^{1/2})$$

$$T(1) = c$$

b) The following relation arises in the analysis of an algorithm to find the "nearest neighbors" of a point-set in 3-space:

$$T(n) = 2\ T(n/2) + n \log_2(n)$$

$$T(1) = c$$

c) The following relation arises in connection with "extrapolative recursion" in numerical algorithms:

$$T(n) = 1\ T(n/2) + n \log_2(n)$$

$$T(1) = c$$

d) The following relation arises in connection with determining the number of bit operations necessary to multiply two numbers:

$$T(n) = 3\ T(n/2) + n$$

$$T(1) = 1$$

e) The following relation arises in the analysis of Strassen's algorithm for matrix multiplication:

$$T(n) = 7\ T(n/2) + n^2$$

$$T(1) = 0$$

18.3 Prove that $n - 1$ array element comparisons are sufficient to merge two vectors of length $n/2$.

18.4 Exercise 3.15 required iterative and recursive algorithms for computing Fibonacci numbers. Formally analyze both of your solutions using the techniques of this chapter and Chapter 6. What do you learn about writing recursive routines for problems that can in fact be expressed as recurrences?

18.5 Suppose that the *PreorderPrint* routine of Chapter 14 were rewritten as

```
procedure PreorderPrint(x: tree);
    begin
    write(label(root(x)));
    if x not a leaf then
        begin
        PreorderPrint(leftson(x)); PreorderPrint(rightson(x))
        end
    end
```

What difference would this change make in the analysis?

18.6 Prove the following theorem from the text:

Theorem 18.1. If N is a positive power of 2, then the solution of the equations

$$T(N) = 2^p T(N/2) + N^p\ g(N)$$

$$T(1) \text{ given}$$

is given by

$$T(N) = N^p\ [T(1) + h(N)]$$

where

$$h(N) = \Sigma_{i=1}^{\log_2 N} g(2^i). \quad \square$$

18.7 Section 18.4 gives closed-form conversions for $H(N)$ for two interesting cases of $g(N)$. Derive these formally.

18.8 The template for divide-and-conquer recurrences can be generalized to allow division by any constant. This exercise addresses recurrences of the form

$$T(N) = k\,T(N/c) + f(N)$$

$T(1)$ given

The recurrence is defined only when N is a power of c. To solve the recurrence we rewrite it in the following form:

$$T(N) = c^p T(N/c) + N^p\,g(N)$$

$T(1)$ given

These rewrites are always possible: $p = \log_c k$ and $g(N) = f(N)/N^p$. This new recurrence has the unique solution

$$T(N) = N^p\,[T(1) + h(N)]$$

where

$$h(N) = \Sigma_{i=1}^{\log_c N}\,g(c^i)$$

Prove this generalized rule. As before, the proof will involve mathematical induction on the powers of c.

18.9 Most sorting problems have times of $O(N \log N)$, where N is the number of items to be sorted. Suppose you ran an experiment and found that on every array you tried, your algorithm took at most $2N$ time units, regardless of N. Would this be a contradiction? Why or why not?

Part Four

Case
Studies

Part Four contains a set of extended examples that apply the theories of Parts One through Three to problems that require several pages to develop. Each chapter of this part develops a problem from initial statement through design, implementation, and analysis. Not all parts of the program development paradigm are carried out completely, but all are represented in at least one of the examples.

Chapter 19

Using FSM's in Programming: An Example

In this chapter we use an extended example to tie together the ideas presented in Chapters 1 through 6. The example was chosen to show how the ideas of *state* and *state transition* can be used to organize a programming task, as well as to illustrate the use of control constructs. We also introduce the ideas of using a program to simulate a machine and of a program that acts as an interpreter for the language of another machine.

As you recall from Chapter 1, finite-state machines (FSM's) are very simple models intended for abstract studies of computation. We have argued that a large class of programs are equivalent to FSM's; actually, however, it is seldom profitable to view an entire program as an FSM of the sort we've discussed. Nonetheless, in many cases a finite state model can be useful as an *organizational* tool for programs. This chapter explores one example—the construction of a *lexical analyzer*.

19.1 Background

Many programs are designed to process input in some special notation developed specifically for the application at hand. The program of this kind with which you are most familiar is probably a *compiler*, a program whose input is a string of characters that form a legal program in some programming language. The output of the compiler is a sequence of words that can be interpreted directly (or executed) by the computer hardware. We say that the compiler *translates* the programming language to machine language.

449

Large, complex programs such as compilers can be simplified by subdividing the work into smaller tasks. It is beyond the scope of this textbook to consider all the tasks involved in a compiler, but the initial stage of compilation is simple, and the techniques are useful for many other problems.

Physically, the input to a program is a stream of characters, including ends-of-lines and an end-of-file. However, in describing legal inputs, we generally wish to speak in less primitive terms than characters. For example, we might describe a declaration of some integer variables as

> The keyword **var** followed by a list of identifiers, followed by a colon, followed by the identifier *integer*, followed by a semicolon.

Note that we have ignored all details such as placement of blanks, ends-of-lines, and comments. We describe these once and for all at some other point. ("Blanks delimit keywords and identifiers and otherwise have no effect. Comments are enclosed by { } (braces) and have the same effect as blanks," etc.) It would be extremely confusing to write instead,

> The letter 'v' followed by the letter 'a' followed by the letter 'r' followed by any number of blanks, comments, and ends-of-lines, followed by any letter, followed by zero or more letters and digits, followed by any number of blanks, comments, · · ·

All of these details of how the keyword **var** and the identifier *integer* are formed or how identifiers are formed (where blanks go, where new lines can start, etc.) are too detailed to be useful at this level of description. We prefer to treat the input as a stream of keywords, identifiers, numbers, and punctuation marks—significant *clusters* of characters—and ignore the details of how they are arranged on the input lines. These clusters of characters are called *lexemes*.

The process of scanning the input for significant clusters of characters and eliminating irrelevant information such as end-of-line marks can usually be separated from the main part of the program. This is convenient, because processing the input string usually requires different techniques from the rest of the program. Separating the lexical analysis simplifies the job of writing the program and makes the program itself easier to understand. In other words, by dealing with spelling details in the lexical analyzer, we are able to address the compilation task at a more abstract level in the remainder of the program.

19.2 The Problem

Let's consider the problem of building a lexical analysis subprogram, called LEX, for use in a simplified compiler. (This subprogram could be modified to do lexical analysis for other tasks, as well.) The table below gives a representative set of lexemes and the regular expressions that define them. In this table, L stands for any letter; D, for any digit; and C, for any character other than a single quote or end-of-file.

Lexeme	Regular Expression
identifier or keyword	$L (L + D)^*$
integer	D^+
real	$(D^+ . D^*) + (. D^+)$
string	$' \ (C + \ '' \ )^+$
punctuation	$- + * + / + + + . + ,$ $+ : + := + ; +) + ($
end-of-file	a special character, denoted $

Additional lexical rules are

1. Blanks delimit lexemes and are otherwise ignored outside strings.

2. Ends-of-lines are ignored in strings, and otherwise behave as blanks.

3. The sequence

    { < any string of characters except "}"> }

    is a comment outside of strings and is to be treated as a blank.

4. The next lexeme is always the longest possible one that fits any of the definitions above. Hence, if the next characters are 125.6, the next lexeme is 125.6, and not 1, 12, 125, or 125 even though all these are valid lexemes.

5. In a string lexeme, two adjacent single quotes are interpreted as one single quote.

Note that several subtle points must be captured. For example, a colon (:) may be either the lexeme colon or the first character of $:=$. Similarly, a period (.) may be either a decimal point or the individual lexeme period. Moreover, by rule 4, a string of digits followed by a period is a real number, not an integer followed by a period.

19.3 A Solution Based on FSM's

Note that we can always decide how to treat a character by looking at the character itself and a few characters on each side. This suggests that we must keep track of only a small amount of information or a state in order to decide how to break the input into lexemes. Indeed, we found that regular expressions provided a convenient notation for describing legal lexemes.

On the basis of this observation, we choose to use the ideas of FSM's to organize our lexical analyzer. Unlike a true FSM, which produces one output for every input, our program is intended to collect the input one character at a time, discard some of it, and deliver the rest in bursts. We therefore generalize our notion of FSM's to permit us to handle this synchronization. We define a number of operations on the input and output; instead of producing an output for each input character, we perform one or more of these specially defined operations. As usual, we recognize a number of states ("collecting a character string," "collecting a number," ":" or ":=," and so on). Each time we read a character, we perform an action (instead of writing output) and go to a new state. Thus the processing involved in lexical analysis can be modeled as an FSM.

It is very easy to write an extremely complex and messy program to solve the preceding problem. But by organizing LEX using the FSM model for the central control logic, we obtain a program that is much easier to construct and understand. In the following subsection, we discuss the organization of the solution and then develop the program itself in subsequent sections.

19.3.1 A GENERALIZATION OF FSM'S

If it were possible to decide what a lexeme will turn out to be by looking only at the first character, the organization as an FSM would be very simple. For example, if we only had to process identifiers, numbers, and strings, the organization could be as in Fig. 19.1.

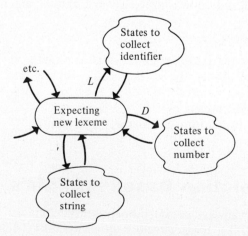

Fig. 19.1 Simplified lexical analysis

We could then build the FSM for each kind of lexeme and attach it to the starting state. The situation is somewhat complicated by the requirement that the output be synchronized with the input and by the fact that we can't classify a lexeme immediately upon seeing its first character (for example, a digit may begin an integer or a real). However, we use this general organization. In particular, the fact that the input is a sequence of independent lexemes will show up as a central, "neutral" state with the intuitive interpretation, "ready to start new lexeme."

The major difference between our program and our FSM involves coordinating the input (one character at a time) with the output (an entire lexeme each time). We use a collecting area, or *buffer*, to "save up" characters until we get a complete lexeme. The actions we take at each state transition will operate on this buffer and the current input character. The actions we allow for each transition can be built up from the following primitives:

- *AppendChar*: Append current input character to buffer.
- *Scan*: Read the next character from the input.
- *EmitLex(LexType)*: Output the contents of the buffer, labeling it as a lexeme of type *LexType* (*LexType* might indicate identifier, real, punctuation, etc.)
- *Fail*: Indicate that the input contains an error and cannot be analyzed.

19.3.2 ORGANIZING THE LEXICAL ANALYZER

We turn now to the problem of describing our lexical analyzer as a pseudo-FSM of the sort described in the previous section. By examining the problem statement we find

Input Alphabet = letters ∪ digits ∪

$$\{ \ ' \ . \ \$ \ \{ \ \} \ + \ - \ * \ / \ : \ = \ ; \ , \ (\ ) \ \}$$

Output Alphabet = actions (*AppendChar*, etc.)

Start State = NEW

State NEW will be the central neutral state; the machine will be in state NEW when it is ready to start scanning a new lexeme. When the machine is in this state, the buffer will always be empty and a fresh character will have been scanned. Now let's look at the things that might happen in state NEW.

The simplest case occurs when the machine reads a simple, one-character punctuation mark. The appropriate response is to put the character in the buffer (*AppendChar*), output the buffer tagged as punctuation (*EmitLex(P)*), scan a new character (*Scan*), and return to state NEW. Blanks are handled almost the same way; instead of emitting them as lexemes, we discard them. This part of the FSM looks like Figs. 19.2. (Note that in the figures we abbreviate the atomic actions by their first letters.)

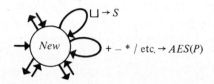

Fig. 19.2 Partial machine for punctuation and spaces

This figure is interpreted like the FSM diagrams of Chapter 1, except that action names replace outputs. Thus, if a space is encountered in state NEW, a new character is scanned; if a binary operator or something similar is scanned in state NEW, it is appended to the output stream and emitted as "punctuation," and then a new character is scanned.

Identifiers are almost as easy. If a letter is scanned in state NEW, we plant it in the buffer and continue scanning letters and digits in state ID. When some other character is scanned, we emit what was already collected (tagged as an identifier) and return to state NEW. The character that terminated the identifier scan must still be processed, so a new scan will *not* be made as we return to state NEW. Notice that this means we do not read a new character on every state transition. The identifier part of the FSM looks like Fig. 19.3. We use an "L" to stand for any letter and "Other" to stand for any other character.

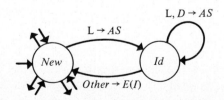

Fig. 19.3 Partial machine for identifiers

Comments are almost as easy as identifiers. They must be terminated explicitly by a brace (}), the characters are discarded, no lexeme is produced, and it is an error for the input to run out in the middle of a comment. Thus the comment part of the FSM is as in Fig. 19.4.

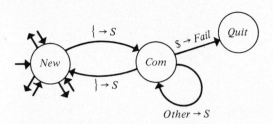

Fig. 19.4 Partial machine for comments

The interesting problem encountered with strings is getting rid of the quotation marks—both eliminating the single quotes at the beginning of the string and converting adjacent pairs of embedded quotes to single quotes. Figure 19.5 shows how selective use of the *AppendChar* operation saves only the characters that are actually part of the string. State STR collects everything except quotes. State STRQ processes a quote: If the next character is also a quote it puts only one quote in the buffer and resumes string scanning; otherwise state STRQ emits the string lexeme and returns to NEW.

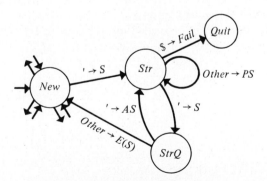

Fig. 19.5 Partial machine for strings

The colon and colon-equal (assignment) symbols present a slightly different problem. If the character after the : is =, then the pair is to be emitted as a punctuation lexeme. If the second character is anything else, the colon alone is emitted as a punctuation lexeme and the next character is saved to be reexamined in state NEW. This is captured in Fig. 19.6.

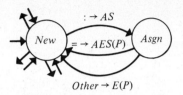

Fig. 19.6 Partial machine for : and :=

Integers, reals, and dots are related, so we treat them as a single unit. In state NEW, a dot may be a single-character lexeme or the beginning of a real, depending on the next character. Note the resemblance between states DOT and ASGN in this regard. In state NEW, a digit may begin an integer or a real. In this case, however, it may not be possible to decide until several characters have been scanned. Thus we remain in state NUM (meaning number—possibly integer, possibly real), collecting digits, until something other than a digit arrives. If this character is not a dot, we emit the integer and return to state NEW. If it is a dot, we go to state REAL. Note that the dot has a different interpretation in state NUM than in state NEW; however, neither state needs to be concerned with the possible interpretations of the other.   Note also that state REAL is entirely indifferent to whether it was reached through state NUM or state DOT. The diagram for this part of the machine is given in Fig. 19.8.

Finally, we need a way to stop the machine. Normally it will run until it encounters an end-of-file (denoted $). When an end-of-file is encountered, it is emitted with type $, but instead of returning to state NEW, the FSM halts by entering state QUIT. Unexpected characters also cause the lexer to quit. Figure 19.7 shows this part of the machine.

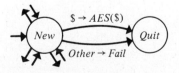

Fig. 19.7 Partial machine for EOF and bad characters

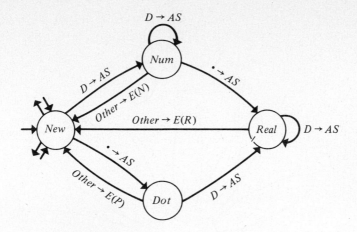

Fig. 19.8 Partial machine for numbers

The definition of the pseudo-FSM for the lexical analyzer is now complete. Note that we were almost always able to split off a lexeme type and define its submachine independently of all other lexemes. (The only exception arose when the definitions were not themselves independent.) Assembling the complete machine involves only collecting the definitions given above. They may be redrawn in a single diagram (Fig. 19.9), converted to a transition matrix (Section 19.5.4), or encoded directly as a program.

19.4 Abstract Program for the Solution

We split the solution of the lexical analyzer problem into two pieces. First, we devise a program that is capable of acting like one of the FSM's described in Section 19.3. This program will refer to stored transition matrices as it processes its input. Second, we decide what values to store in the matrices to make the program act like *our* FSM.

It may seem strange to solve a more general problem on the way to solving a specific one. Contrary to first appearances, doing so does not necessarily involve more work. Indeed, this approach may take much less work, because it divides the problem into the essential structure of a finite-state computation (which turns out to be the program) and the specific details of the current problem (which are encoded in the data). As a dividend, the division is still there after the program is finished. If you decide later that you should change the rules for forming lexemes (see

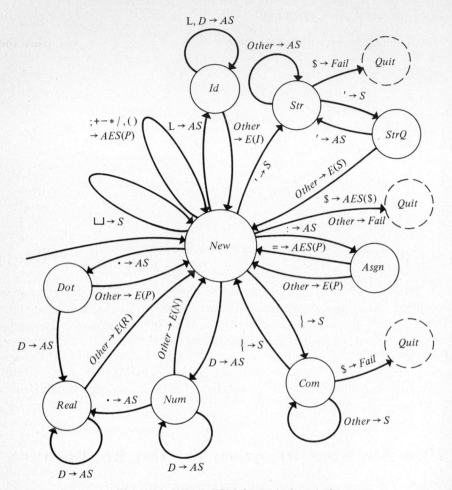

Fig. 19.9 Whole FSM for lexical analysis

Exercises 19.1 through 19.5), the particular language will still be explicitly defined.

When this program runs, it will refer to the stored tables in order to determine what actions to perform. Note the similarity to a computer, which refers to a stored program to determine the actions it will perform. Since our program is designed to behave like an FSM, we say it *simulates* the FSM. The tables play the same role as a stored program; we say the program *interprets* the tables, or that the program is *table-driven*. This sort of program organization can be very useful when many similar problems must be solved. It can be adapted to specific problems by loading different tables instead of by modifying the program itself.

19.4.1 AN FSM INTERPRETER

The main body of the lexical analysis program is a loop that uses the current character and state to choose an action and the next state. This program might look something like the program shown in Fig. 19.10.

```
var
    { Transition Matrices }
        F: array [State] of array [CharClass] of State;
        G: array [state] of array [CharClass] of Action;
    { Current state and character of FSM }
        CurState, NextState: State;
        CurCharClass: CharClass;
begin
Initialize FSM;
while CurState ≠ Quit do
    begin
    NextState := F[CurState, CurCharClass];
    Perform Action G[CurState, CurCharClass];
    CurState := NextState;
    end;
end.
```

Fig. 19.10 Abstract main body of lexical analyzer

Note that the next state is determined before the action is performed, but not assigned to the current state until after the action is complete. This is done because both *F* and *G* may depend on both the current state and the current character, and the action may scan a new character. We will return to this point in Section 19.6.2.

Note also that the program is described in terms of character classes rather than individual characters. (The characters themselves must, of course, be used to construct the output.) Our input alphabet contains 52 letters and 10 digits. The letters are always treated alike, as are the digits. If we were to store these as independent entries in the transition matrices, there would be enormous numbers of identical entries. Both the text that defines the matrices and the requirement for computer storage would be unreasonably large. We respond to this problem by defining a new type, *CharClass*, that we use to keep track of which *class* of character we are currently processing. These classes, not the actual literal characters as read from the input, are used for defining and controlling the FSM.

This program could serve as an interpreter for almost any FSM-like computation in which we want to treat some groups of inputs in the same ways. We can make it more useful for lexical analysis by building in the primitive operations *Scan*, *AppendChar*, *EmitLex*, and *Fail* and the combinations of these that we found useful in designing our FSM. In the following program, the primitive operations *Scan*, *AppendChar*, *EmitLex*, and *Fail* and the types *State*, *Action*, and *CharClass* are assumed to be defined elsewhere. Recall that primitive operation *EmitLex* requires a parameter to tag the lexeme type. This parameter is a function of the current state and character, which we will call *LexTag*. The program now expands to the text shown in Fig. 19.11.

19.4.2 TABLES TO DRIVE THE FSM INTERPRETER

In order to use this program to do lexical analysis, we need to provide values for the transition tables *F*, *G*, and *LT*. If there are k character classes and s states, each table will have $k \cdot s$ entries. For the machine we designed in section 19.3, $k = 19$ and $s = 10$, so we must provide 190 pieces of information for the transition tables.

One possible approach is to simply compute the tables "by hand" and to place them directly into the program text. Unfortunately, the relationship between the original transition tables and their translations into program text is often unclear. PASCAL, in particular, has no good provision for specifying constant tables. Therefore, we will write the transition tables in fairly natural notation and use an intialization routine in the lexical analyzer to generate the actual transition matrices used by the program.

Note that most states take special action for only a few characters. We can take advantage of this by arranging the input to the initialization routine so the special transitions for each state are collected as a group and followed by the "default" transition—the one used for all other characters. Then the input might consist of a sequence of groups such as:

State Name$_1$

    Char Class$_1$ *Action*$_1$ *LexType*$_1$ *Next State*$_1$

       .   .   .   .

       .   .   .   .

       .   .   .   .

    Other: *Action*$_n$ *LexType*$_n$ *Next State*$_n$

State Name$_2$ $\cdots$

```
var
    { Components of the state transition matrix }
        F:  array [State] of array [CharClass] of State;
        G:  array [State] of array [CharClass] of Action;

    { Current state and character of the FSM }
        CurState, NextState: State;
        CurCharClass:        CharClass;
begin
InitializeFSM;
while CurState ≠ Quit do
    begin
    NextState := F[CurState, CurCharClass];
    case G[CurState, CurCharClass] of
        S:   Scan;
        E:   EmitLex(T[CurState, CurCharClass]);
        AS:  begin AppendChar; Scan; end;
        AES: begin
                AppendChar;
                EmitLex(LexTag(CurState, CurCharClass));
                Scan;
                end;
        Fail: Failmsg;
        end;
    CurState := NextState;
    end;
end.
```

Fig. 19.11 Expansion of Fig. 19.10

which we could process with the following abstract program:

```
while Still States to Process do
    begin
    Get New State;
    Clear Chars— Done List;
    until Default Transition do
        Process Explicit Transition;
    Process Default Transition;
    end;
```

19.5 Concrete Program for the Solution

The abstract program of Section 19.4 is useful for organizing the solution and showing that it has certain properties. However, in order to actually perform lexical analysis on real input, we must provide enough additional information to compile and execute the program on a real computer. This information includes precise formats for the tables of transition functions, definitions of the *Scan*, *AppendChar*, *EmitLex*, and *Fail* operations that were previously treated as primitives, and further specification of such matters as the nature of the abstract types (for instance, *State*) and the buffer operations.

The program text in the following subsections is taken from an actual PASCAL program. Low-level details of input and file manipulation have been left out, but the rest of the program is presented in full.

19.5.1 DECLARATIONS FOR CONCRETE SOLUTION

The declarations for the complete program contain three major refinements of the abstract version. First, they define the particular sets of states, actions, character classes, and *LexType*s to be used. Second, they implement the transition functions and the *LexTag* function as the vectors *F*, *G*, and *LT*. Finally, they define the representation of the buffer.

The buffer for collecting the characters of a lexeme is implemented as an array of characters, *Buffer*, and the integer index of the last character inserted in that array, *BufPtr*. If *BufPtr* = 0, the buffer is empty. It is a policy of the design that the buffer should be empty whenever control returns to state NEW.

```
type
    State =        (NEW, ID, ASGN, COM, DOT, STR, STRQ, NUM,
                   REAL, QUIT);
    Action =       (S, AS, E, AES, FAIL);
    CharClass =    (Let, Dig, LBrace, RBrace, Col, Eql, Point, Sem,
                   Plus, Minus, Mult, Divd, Comma, LPar, RPar,
                   EndFl, Quote, Space, Othr);
    Word = array [1..4] of char;
    LexType = char;

var
{ The three components of the state transition matrix }
    F: array [State] of array [CharClass] of State;
    G: array [State] of array [CharClass] of Action;
    LT: array [State] of array [CharClass] of LexType;
```

{ Current state and character of the FSM }
    CurState, *NextState*: *State*;
    CurCharClass: *CharClass*;
    CurCharItself: *char*;

{ The Buffer }
    Buffer: **array** [1..100] **of** *char*;
    BufPtr: *integer*;

{ Used by SetUpFSM }
    CharsDone: **array** [*CharClass*] **of** *Boolean*;

19.5.2 MAIN PROGRAM

The main program for the running version, is very similar to the abstract
program in Fig. 19.11:

```
begin
SetUpFSM;
StartFSM;
while CurState ≠ Quit do
   begin
   NextState := F[CurState, CurCharClass];
   case G[CurState, CurCharClass] of
       S:    Scan;
       E:    EmitLex(LT[CurState, CurCharClass]);
       AS:   begin AppendChar; Scan; end;
       AES:  begin
                 AppendChar;
                 EmitLex(LT[CurState, CurCharClass]);
                 Scan;
             end;
       Fail: FailMsg(CurState, CurCharItself);
       end;
   CurState := NextState;
   end;
end.
```

This program assumes the existence of procedures to implement the
abstract operations *SetUpFSM*, *StartFSM*, *Scan*, *EmitLex*, *AppendChar*, and
Failmsg. Procedure *SetUpFSM* is assumed to initialize the transition
matrix; *StartFSM* scans a character and enters state NEW. *StartFSM* is
given below; the rest are discussed in the next few sections.

```
procedure StartFSM;
{ Set up FSM in starting state }
  begin   CurState := NEW; Scan; BufPtr := 0;  end;
```

19.5.3 IMPLEMENTATION OF ACTION PRIMITIVES

The four primitive operations maintain the current character, *CurCharItself* (together with its class, *CurCharClass*), and the buffer (array *Buffer* and integer *BufPtr*). No other part of the program should directly modify these variables. Some of the routines use a vector, *CharClassLit*, that is indexed by characters and preloaded with the corresponding literals of type *CharClass*. The code for these procedures is given in Fig. 19.12.

Function *FindClass* is used by procedure *Scan* to convert the character just read to its *CharClass*. Note that we have not completely filled in some details such as how *FindClass* is to determine that a character is a letter. To a certain extent, the best way to perform this test depends on the details of the compiler (and character set) being used. (See Exercise 19.7.)

19.5.4 INPUT FORMAT FOR TABLES

The overall organization of the input data was discussed in Section 19.4.2. This organization is carried into the concrete program using a very rigid input format, which may seem inflexible; but we are, after all, constructing the program that will help us cope with a variable input format.

Each line begins with a 4-character flag that indicates what kind of information is on the line. The fixed-format information on each line may be followed by other characters; the latter will be ignored by the program and may therefore be used for comments.

The first three lines of the input give, respectively, the spellings (external representations) of *State*, *Action*, and *CharClass* names that will be used in the input file. The names must be given in the same order as the program declarations. *State* and *Action* names are exactly four characters long and may include blanks; *CharClass* names are single characters. The names for each type must follow the flag **** immediately. No intervening blanks are permitted. For the FSM designed in Section 19.3 these lines are

```
****New Id  AsgnCOM DOT STR STRQNUM REALQUIT.
****S   AS E   AES FAIL.
****LD {}:=.;+-*/,()$'.
```

The remainder of the input file consists of a sequence of state descriptions. Each state description consists of a header line, zero or more explicit transitions, and a default transition:

```
function FindClass(Ch: char): CharClass;
  { Return character class corresponding to input character }
  var ChCl: CharClass;
  begin
  FindClass := Othr;
  if Ch is a letter then FindClass := Let else
  if Ch is a digit then FindClass := Dig else
  for ChCl := succ(Dig) to pred(Othr) do
      if CharClassLit[ChCl] = Ch then FindClass := ChCl;
  end;

procedure Scan;
  { Sets CurCharItself to next input character and CurCharClass
  to the class of that character. }
  begin
  read(CurCharItself);
  if eof(input) then CurCharItself := CharClassLit[EndFl];
  CurCharClass := FindClass(CurCharItself);
  end;

procedure AppendChar;
  { Appends current input character to buffer }
  begin BufPtr := BufPtr + 1; Buffer[BufPtr] := CurCharItself; end;

procedure EmitLex(Tg: LexType);
  { Outputs lexeme now in buffer (tagged Tg) and clears buffer }
  var i: integer;
  begin
  write('!', Tg, '! ');
  for i := 1 to BufPtr do write(Buffer[i]);
  BufPtr := 0;
  writeln(' ');
  end;

procedure FailMsg(CurSt: state; CurCh: char);
  { Error message }
  begin
  writeln('Failed scanning ', CurCh,
          ' in state', StateLit[CurState]);
  end;
```

Fig. 19.12 Primitive operations on input and output files and on buffer

- The header line must have the flag $>>>>$ followed immediately by a 4-character state name.
- The explicit transitions have the form

  bbbbCbAAAAbTbSSSS.

  where C is a *CharClass*, AAAA is an *Action*, T is a *LexType*, SSSS is the next *State*, and b denotes blank.
- The default transition has the form

  bb<<bbAAAAbTbSSSS.

The input file is terminated by a line with flag $<<<<$ when a new state is expected. The definitions for the FSM of Section 19.3 are as follows:

```
>>>>New ··· Neutral state.           >>>>Com ··· Comment.
    L  AS   - Id    .                    }  S    - New  .
    D  AS   - Num  .                     $  FAIL - QUIT.
    {  S    - Com  .                  <<   S    - Com  .
    :  AS   - Asgn.                  >>>>Str ··· Body of string.
    .  AS   - Dot  .                     '  S    - StrQ  .
    ;  AES  P New  .                     $  FAIL - QUIT.
    +  AES  P New  .                  <<   AS   - Str  .
    -  AES  P New  .                 >>>>StrQ ··· Quote in string.
    *  AES  P New  .                     '  AS   - Str  .
    /  AES  P New  .                  <<   E    S New  .
    ,  AES  P New  .                 >>>>Num ··· Number.
    (  AES  P New  .                     D  AS   - Num  .
    )  AES  P New  .                     .  AS   - Real  .
    $  AES  $ QUIT.                    <<   E    N New  .
    '  S    - Str   .                >>>>Dot ....Period or decimal point.
       S    - New   .                    D  AS   - Real  .
 <<   FAIL - QUIT.                     <<   E    P New  .
>>>>Id  ··· Identifier.               >>>>Real ··· Real number.
    L  AS   - Id    .                    D  AS   - Real  .
    D  AS   - Id    .                  <<   E    R New  .
 <<   E    I New  .                   <<<<
>>>>Asgn ··· Colon or colon-equal.
    =  AES  P New  .
 <<   E    P New  .
```

19.5.5 INITIALIZING THE FSM

Initialization has two stages: reading the external names of the values of the enumerated types *State*, *Action*, and *CharClass*; and reading the transition tables.

```
procedure SetUpFSM;
  { Load state transition tables }
  begin
  SetUpLiteralConversion;
  SetUpTransitionTable;
  end;
```

We omit the definition of procedure *SetUpLiteralConversion*; it simply reads and stores values from the input file. The procedure *SetUpTransitionTable* is similar to the abstract program of Section 19.4.2:

```
procedure SetUpTransitionTable;
  { Fill in State Transition Table }
  var ThisSt: State; ThisFlag: char;
  begin
  readln; ThisFlag := GetFlag;
  while ThisFlag = '>' do
    begin
    ThisSt := GetState;
    ClearCharsDoneList;
    readln; ThisFlag := GetFlag;
    while ThisFlag = ' ' do
      begin
      ProcessExplicitTransition(ThisSt);
      readln; ThisFlag := GetFlag;
      end;
    ProcessDefaultTransition(ThisSt);
    readln; ThisFlag := GetFlag;
    end;
  end;
```

This procedure relies on the four procedures whose bodies are given in Fig. 19.13 to read and store the actual transitions and to maintain the list of *CharClasses* that have been processed for a given state.

19.5.6 PROCEDURES FOR INPUT AND FILE MANIPULATION

The preceding sections have defined all but six short functions and procedures. We have omitted the concrete code for the following procedures because nothing would be gained by going into the details.

```
function GetFlag:char;
  { Read 4 characters from input, leaving last char in ch }
function GetState: State;
  { Read 4-char state literal and convert to type State }
```

```
procedure ProcessExplicitTransition(ThisSt: State);
  { Get CharClass, record it as defined, move values to tables }
  var ThisCh: CharClass; ch: char;
    begin
    ThisCh                := GetCharClass; read(ch);
    G[ThisSt, ThisCh]     := GetAction;       read(ch);
    LT[ThisSt, ThisCh]    := GetLexType;      read(ch);
    F[ThisSt, ThisCh]     := GetState;        read(ch);
    RecordCharDone(ThisCh);
    end;

procedure ProcessDefaultTransition(ThisSt: State);
  { Read default transition and use for all chars not yet defined }
  var TSt: State; TAct: Action; TCh: CharClass; TType: LexType;
      ch: char;
    begin
    read(ch); read(ch);
    TAct  := GetAction;    read(ch);
    TType := GetLexType; read(ch);
    TSt   := GetState;     read(ch);
    for TCh := Let to Othr do
       if not CharsDone[TCh] then
       begin
       G[ThisSt, TCh]    := TAct;
       LT[ThisSt, TCh]   := TType;
       F[ThisSt, TCh]    := TSt;
       end;
    end;

procedure ClearCharsDoneList;
  { Clear list of characters processed for current state }
  var ThisCh: CharClass;
    begin
    for ThisCh := Let to Othr do CharsDone[ThisCh] := false;
    end;

procedure RecordCharDone(ThisCh: CharClass);
  { Record current character as having been processed }
    begin
    CharsDone[ThisCh] := true;
    end;
```

Fig. 19.13 Support routines for *SetUpTransitionTable*

function *GetAction: Action*;
  { Read 4-char action literal and convert to type *Action* }
function *GetCharClass: CharClass*;
  { Read 1-char name for character and convert to type *CharClass* }
function *GetLexType: LexType*;
  { Read 1-char *LexType* for labeling lexeme. }
procedure *SetUpLiteralConversion*;
  { Sets up character equivalents for enumerated types. }

19.5.7 SAMPLE RUN

The result of running the program of Fig. 19.10 on the input of Fig. 19.14
is given in Fig. 19.15.

```
{ Test file for lexical analyzer }
var
   x1, x2: real;
   j,k,l: integer;
   AA,BB: char;
   ST: string;
begin
   x1 := 3.2;
   x2 := x1-.45 * x1 / 34.;
   j := 1;  for k := j to 12 do l := k+j;
   AA := 'c'; BB := "";
   ST := 'more than
one line';
   write(x1,x2,j,k,l,AA,BB,ST);
end.
```

Fig. 19.14 Sample input for LEX

19.6 Correctness of the Lexical Analyzer

19.6.1 VERIFYING PROPERTIES OF INDIVIDUAL COMPONENTS

The development of the lexical analyzer in this chapter included several
descriptions in different notations of various parts of the solution. Before
attempting to "verify" the lexical analyzer, we need to consider these
descriptions and the interesting properties that we might verify about each
of them. A convincing argument about the "correctness" of the final
program must deal with each of the descriptions and with what can go
wrong at each stage.

!I! var	!I! x2	!P! ;
!I! x1	!P! :=	!I! BB
!P! ,	!I! x1	!P! :=
!I! x2	!P! -	!S! '
!P! :	!R! .45	!P! ;
!I! real	!P! *	!I! ST
!P! ;	!I! x1	!P! :=
!I! j	!P! /	!S! more than one line
!P! ,	!R! 34.	!P! ;
!I! k	!P! ;	!I! write
!P! ,	!I! j	!P! (
!I! l	!P! :=	!I! x1
!P! :	!N! 1	!P! ,
!I! integer	!P! ;	!I! x2
!P! ;	!I! for	!P! ,
!I! AA	!I! k	!I! j
!P! ,	!P! :=	!P! ,
!I! BB	!I! j	!I! k
!P! :	!I! to	!P! ,
!I! char	!N! 12	!I! l
!P! ;	!I! do	!P! ,
!I! ST	!I! l	!I! AA
!P! :	!P! :=	!P! ,
!I! string	!I! k	!I! BB
!P! ;	!P! +	!P! ,
!I! begin	!I! j	!I! ST
!I! x1	!P! ;	!P!)
!P! :=	!I! AA	!P! ;
!R! 3.2	!P! :=	!I! end
!P! ;	!S! c	!P! .
!$! $		

Fig. 19.15 Sample output for LEX

We have given the following descriptions of various parts of the lexical analyzer:

Informal problem specification	Section 19.2
Design of the FSM	Section 19.3
Abstract FSM interpreter	Section 19.4
Program LEX	Section 19.5
Encoding of FSM in table	Section 19.5.4

Each of these descriptions was written for a different purpose. Each contains some important information and has its own set of properties to check. Let's discuss each in turn.

The informal problem specification is an attempt to write down the definition of "lexeme" that was in the mind of the language designer. There is no way to demonstrate that it is correct. Indeed, this is a weak link in all schemes for "proving correctness" of programs—we are never able to do more than show that the implementation is consistent with the specifications.

Although we did not go through rigorous derivations, we designed the FSM by examining the RE's for the various lexemes and writing down the submachines. We knew that the submachines were independent, so combining them was easy. Since the synchronization of input and output to the state transitions is not strictly like an FSM, we had to argue separately that the output had the desired relation to the input. This argument is most appropriately carried out for the FSM design; once we are confident the FSM does the right thing, we need only show that the remaining stages, in fact, implement or simulate the FSM correctly.

The abstract FSM interpreter is essentially a loop that performs transitions. We formally verify some aspects of the correctness of that loop in Section 19.6.2.

To argue about the implementations of the primitive actions *Scan*, *AppendChar*, *EmitLex*, and *Fail*, we must appeal to properties of the input and output streams and the implementation of the buffer. We addressed this problem in Section 19.5.

Finally, we must be certain that the transition matrices loaded for the program do, in fact, correspond to the FSM design. This is reasonably easy. The transformation from the state-transition diagrams of Section 19.3 to the table of Section 19.5.4 is essentially clerical, since each line of the table corresponds directly to an arc or a node of the FSM.

19.6.2 VERIFYING PROPERTIES OF THE FSM INTERPRETER

The core of this program is the "main loop," which interprets an FSM definition stored in tables F and G. In this section we show how to specify formally what it means to "interpret an FSM definition." We will limit the discussion to that property alone; we will not attempt to say anything about, for example, the output stream of lexemes.

19.6.2.1 The Main Loop
The fundamental characteristic of an FSM is that the history of the machine is irrelevant. That is, the current state and current input character and unused input are sufficient to determine what the machine

will do in the future. We need only show that *a single execution* of the loop correctly advances the state and input.

We will add assertions at the beginning and end of the loop to express this correctness. The annotated program is

assert *CurCharClass* = *FindClass*(*CurCharItself*);
NextState := *F*[*CurState*, *CurCharClass*];
case *G*[*CurState*, *CurCharClass*] **of**
    S: *Scan*;
    E: *EmitLex*(*LT*[*CurState*, *CurCharClass*]);
    AS: **begin** *AppendChar*; *Scan*; **end**;
    AES: **begin**
            AppendChar; *EmitLex*(*LT*[*CurState*, *CurCharClass*]); *Scan*
        end;
    Fail: *Failmsg*(*CurState*, *CurCharItself*);
    end;
CurState := *NextState*;
assert *CurState* = *F*[*CurState'*, *CurCharClass'*]
      $\land$ *CurCharClass* = *FindClass*(*CurCharItself*)
      $\land$ *G*[*CurState'*, *CurCharClass'*] is in {*S*, *AS*, *AES*} $\supset$
            Advanced(*input'*, *input*, *CurCharItself*)
      $\land$ *G*[*CurState'*, *CurCharClass'*] is in {*E*, *Fail*} $\supset$
            (*input* = *input'* $\land$ *CurCharItself* = *CurCharItself'*);

where we define

Advanced(*input'*, *input*, *CurCharItself*) $\equiv$
    (*input* = *input'* = <> $\land$ *FindClass*(*CurCharItself*) = *EndFl*)
    $\lor$ (*input'* $\neq$ <> $\land$ *input'* = <*CurCharItself*> $\sim$ *input*).

Informally, the initial assertion simply says that *CurCharClass* is consistent with *CurCharItself.* The final assertion says that

- *CurState* has changed correctly based on the values of the FSM state and input character class at the *initial* assertion,

- *CurCharClass* is still consistent,

- If the input was supposed to advance (the FSM output was *S*, *AS*, or *AES*), then the input advanced one character. Otherwise, the input did not advance.

Primed variables all refer to the values that variables had at the top of the loop (at the first assertion). The assertions treat the file variable *input*, which contains the FSM's input tape, as a sequence of characters. Except for details such as the behavior of the *eoln* predicate (which we don't use), this is an accurate abstract model.

For this program fragment to be correct (with respect to the two assertions, that is), we have to assume that *Scan* "advances the input." Symbolically,

    procedure *Scan*;
      post *CurCharClass = FindClass(CurCharItself)*
                ∧ *Advanced(input', input, CurCharItself)*;

We also assume that none of the other procedures affect *CurState*, *CurCharItself*, or *CurCharClass*.

The assertions we've provided are considerably more detailed than we would normally provide for such a program fragment. We went to the trouble of writing them mostly as an exercise in writing a rigorous specification. It should be quite clear that the program conforms to the specification. Thanks to our assumptions about *Scan* (which we should check, of course), the correctness of the whole fragment reduces to proving that

- *CurState* gets updated correctly, and
- *Scan* gets called exactly when the action corresponding to the original state and input class is *S*, *AS*, or *AES*.

Both specifications are clear.

If we were to verify the fragment formally, we would first treat the **case** statement as the equivalent **if–then–else** statements. Then we would note that because routines other than *Scan* do not alter the variables mentioned in our assertions, and because they have no preconditions, all the assertions in this proof will "pass through them" unaltered when the procedure call rule is applied.

19.6.2.2 Procedure *Scan*

We assumed above that *Scan* advanced the input properly. That amounts to showing

    read(CurCharItself);
    if *eof(input)* **then** *CurCharItself := CharClassLit[EndFl]*;
    CurCharClass := FindClass(CurCharItself);
    assert *CurCharClass = FindClass(CurCharItself)*
                ∧ *Advanced(input', input, CurCharItself)*;

The proof follows from the specifications for the *text* data type provided by PASCAL. Formally, the important parts of this specification for our purposes indicate that

    axiom *input* = < > ≡ *eof(input)*;

procedure *Read*(*input: text*; *ch: char*);
  pre **not** *eof*(*input*);
  post input = *trailer*(*input'*) $\wedge$ *ch* = *first*(*input*);

19.6.2.3 Other Action Procedures

In this section, we considered only the question of whether the loop correctly maintained the state and current character of the model FSM. If the specifications were extended to address the stream of lexemes produced by the program, precise specifications of *AppendChar*, *EmitLex*, and *FailMsg* would also have been required. As it is, we need only ensure that these routines do not, in fact, alter the variables *CurCharItself*, etc.

19.6.3 VERIFYING PROPERTIES OF THE ENTIRE SYSTEM

We can look at the verification problem from another standpoint, as well. In the previous pages we discussed the problem of verifying certain properties of each of the components of the solution. We can also describe what we would like to know about the entire solution, then look at as many components as necessary to establish the proof.

One such property is *termination*; that is, we would like to know that if the input stream is finite the program will always finish analyzing it. Note that at the outset the program itself is not guaranteed to terminate. If the tables *F*, *G*, and *LT* are loaded with a description of an FSM that reads no input and has a transition graph with a cycle, the simulation of that FSM will not terminate. Our termination argument must therefore address the system composed of the program and the values loaded in the tables. In this section we address only the question of whether the program terminates when loaded with the tables of Section 19.5.4.

Intuitively, we understand that the argument will show that every cycle in the graph of the FSM performs at least one *Scan* operation. The program (or, the FSM simulator) stops (or, enters state QUIT) when it reaches the end of the input. Therefore there is no way the program loaded with this data can loop indefinitely. More formally, we must find an expression that

- Decreases by a positive integral amount with each iteration, and
- Can never become smaller than some fixed quantity.

(See Section 5.3.7.3.) If every transition performed a *Scan*, we could use the length of the remaining input stream for the termination function: one transition is performed by each iteration, so the expression would decrease toward zero as required. Unfortunately, some transitions do not do *Scan*s, so we must look for a more complex function.

Note that the minimum path length from any state to NEW is never greater than two and that any transition that increases your distance from NEW performs a *Scan*. Using these two facts, we propose the termination function

$$3 \cdot (\text{length of remaining input}) + (\text{minimum distance to NEW})$$

The transitions that enter state NEW without scanning do not shorten the input stream but do shorten the distance from NEW. The transitions that leave NEW shorten the input by one and increase the distance to NEW by one or two, giving a net decrease in the termination function of two or one. You can check the effects of the other transitions on this function yourself.

19.7 Performance of the Lexical Analyzer

Now that we have an operating lexical analyzer, we need to know something about its performance. In this section we concentrate on speed; we want to know how the running time depends on the number of states, actions, and character classes, and on the length of the input itself. Our analysis will reveal a design decision that, if made differently, could significantly affect the speed of the program.

Since we are interested only in the order of magnitude of the time required, it is not necessary to carry much detail in the analysis. It is sufficient to count program statements. (True, nested statements sometimes make it hard to be precise about exact counts, but we are not interested in exact counts.) We begin with the main program. This program has two major phases: The calls on *SetUpFSM* and *StartFSM* initialize the FSM to be simulated; they are distinctly different from the loop that interprets the tables and simulates the FSM. We must analyze the initialization calls and the loop separately, then sum the results.

We cannot assume that one phase is more important than the other. If we set up a large definition and process a 3-character input, the initialization clearly dominates the cost. On the other hand, if we set up a modest definition and process input for hours, the interpretation phase will dominate.

The problem, then, can be divided into two parts: finding the time required by the initialization section and the time required by the simulation section. The costs are expressed in terms of S, the number of states; A, the number of actions; C, the number of character classes; and N, the number of characters in the input. In order to simplify the analysis, we count only assignment operators and condition tests (for example, **if**'s and tests for ends of loops).

19.7.1 COST OF INITIALIZATION

To get a crude idea of a lower bound on the time required, note that all the elements of three $C \times S$ matrices must be initialized. It may take more than a constant number of operations for each, but since only one element can be initialized with a single statement, we can expect a cost of at least $O(C \cdot S)$.

We perform the analysis in somewhat more detail. Initialization consists of setting up the literals, setting up the transition table, and starting the machine. (The first two operations are in *SetUpFSM*, the last in *StartFSM*.)

The procedure for setting up the literals is not given here, but it must simply read the literals corresponding to values of enumerated types and store them in vectors. The cost will be $O(S + A + C)$.

The body of *SetUpTransitionTable*, annotated with operation costs, is

```
procedure SetUpTransitionTable;
  { Fill in State Transition Table }
  var ThisSt: State; ThisFlag: char;
  begin
  readln; ThisFlag := GetFlag;                          1
  while ThisFlag = '>' do                               S · [ 1
     begin
     ThisSt := GetState;                                   + 1
     ClearCharsDoneList;                                   + CCDL
     readln; ThisFlag := GetFlag;                          + 1
     while ThisFlag = ' ' do                               + k · [ 1
        begin
        ProcessExplicitTransition(ThisSt);                    + PET
        readln; ThisFlag := GetFlag;                          + 1 ]
        end;
     ProcessDefaultTransition(ThisSt);                     + PDT
     readln; ThisFlag := GetFlag;                          + 1 ]
     end;
  end;
```

The procedure for initializing the transition matrices thus consists mainly of a loop that is executed approximately once per state, so the number of executions of the loop body is about S. Within the loop, we find two procedure calls, another loop, and five other statements. The first called procedure, *ClearCharsDoneList*, consists of a loop that iterates through the character classes, so its cost is $CCDL = O(C)$. The inner loop is executed for each nondefault transition of the state. Let k denote the number of such transitions for any state. Procedure *ProcessExplicitTransition* has a fixed cost, so the cost of the inner loop is $PET = O(k)$. Procedure *ProcessDefaultTransition* consists of several initialization statements and a loop whose body is executed for all states that were not explicitly initialized. Thus the cost of *ProcessDefaultTransition* is $PDT = O(C - k)$. We leave the detailed derivations of the cost of these subprocedures as exercises. Taking the costs from the annotated program (and taking advantage of the fact that we know approximately how many times each loop is executed) we determine the cost of *SetUpTransitionTable* to be

$$1 + S \cdot (CCDL + k \cdot (PET + 2) + PDT + 4)$$

Substituting values of $CCDL$, PET, and PDT, we reduce this to

$$1 + S \cdot (O(C) + k \cdot (O(1) + 2) + O(C - k) + 4)$$

and then to

$$\text{Cost of } SetUpTransitionTable = S \cdot (O(C) + O(k) + O(C - k))$$
$$= O(S \cdot C)$$

The cost of *StartFSM* depends on the cost of *Scan*, which is $O(C)$:

$$\text{Cost of } StartFSM = O(C) + 2 = O(C)$$

Combining the three components of the initialization cost, we obtain a total cost of $O(S + A + C) + O(C \cdot S) + O(C)$, or $O(C \cdot S)$ as expected.

19.7.2 COST OF SIMULATION

Now let us turn to the processing phase itself. This section of the main program is simply a loop, so we need to determine the number of times the loop is executed and an estimate of the cost of each loop body.

We will need estimates of the costs of procedures *Scan*, *AppendChar*, *EmitLex*, and *FailMsg*. The formal derivation of these costs is left as an exercise, but the results are:

Scan	$O(C)$	because *FindClass* is $O(C)$
AppendChar	$O(1)$	that is, constant
EmitLex	$O(len_i)$	where len_i is length of the lexeme output on the ith transition
FailMsg	$O(1)$	

We can now do the accounting for the main loop:

```
while CurState <> Quit do                          M * [ 1
begin
NextState := F[CurState, CurCharClass];             + 1
case G[CurState, CurCharClass] of
    S:   Scan;                                       + if S then  O(C)
    E:   EmitLex(LT[CurState, CurCharClass]);        + if E then O(len_j)
    AS:  begin                                       + if AS then
         AppendChar;                                   (O(1)
         Scan;                                        + O(C))
         end;
    AES: begin                                       + if AES then
         AppendChar;                                   (O(1)
         EmitLex(LT[CurState, CurCharClass]);         + O(len_j)
         Scan;                                        + O(C))
         end;
    Fail: Failmsg(CurState, CurCharItself);          + if Fail then  O(1)
    end;
CurState := NextState;                               +1]
end;
```

The cost of the main loop body depends on the state of the FSM. If we let $K(x)$ be the number of times during the execution of the program that the machine is in state x and let M be the number of state transitions required, we see that the cost of the loop can be estimated by

$$[3 \cdot M + K(S) \cdot O(C) + K(E) \cdot O(len_j) + K(AS) \cdot [O(1) + O(C)]$$
$$+ K(AES) \cdot [O(1) + O(len_j) + O(C)] + K(Fail) \cdot O(1)]$$

$$= [3 \cdot M + [K(S) + K(AS) + K(AES)] \cdot O(C)$$
$$+ [K(E) + K(AES)] \cdot O(len_j)$$
$$+ [K(AS) + K(AES) + K(Fail)] \cdot O(1)]$$

because

$$K(S) + K(E) + K(AS) + K(AES)) + K(Fail) = M.$$

The loop itself is executed once for each of the M transitions in the simulation of the FSM. (If we were doing a probabilistic analysis, we would use the probability of being in a state rather than a count.) Unfortunately, FSM transitions do not correspond directly to operations on the input string. We can, however, make a crucial observation by inspecting the definition of the FSM in Fig. 19.9. Notice that the longest path in the FSM without a scan operation is of length one. (The really important thing is that there's no cycle without a scan, but we can make the stronger statement here.) This means that a new character MUST be read on at least every other transition, or every other time through the main program loop. Thus that loop can be executed at most $2N$ times for an input of length N. We also observe that no transition ever scans more than one character. Thus M is less than or equal to $2N$, and it follows that the number of loop executions is between N and $2N$, or $O(N)$.

The cost of *EmitLex* depends on the number of characters in the lexeme emitted, but no character of the input is emitted in more than one lexeme, so the cost of *EmitLex* over the entire program must be no worse than proportional to the length of the input. In other words,

$$[K(E) + K(AES)] \cdot O(len_i) \leq \sum_{i=1}^{M} O(len_i) = O(N).$$

If *Scan* had unit cost, we could declare the cost of the case statement to be a weighted average of a set of constant costs, hence constant. The result would be that the simulation section of the program would have cost $O(\text{const} \cdot N)$ or $O(N)$.

Unfortunately, this *Scan* contains a loop that searches through the character classes. Hence its cost is $O(C)$. Moreover, it is called for somewhere between half and all of the transitions, because N characters are scanned and the number of loop executions is between N and $2N$. It follows that the cost of the case statement is $O(C)$. The cost of the simulation loop is therefore actually $O(N \cdot C)$. N is usually much larger than C, so the cost penalty is not enormous. However, this analysis points out that seemingly minor design decisions can have significant impact on program cost (see Exercise 19.7).

19.7.3 TOTAL COST

We can now return to the total cost, which depends on both initialization and actual simulation. It is $O(C \cdot S) + O(N \cdot C)$. As we remarked at the outset, there are no grounds for arguing that one or the other of these terms dominates the cost.

Exercises

19.1 Write a version of the lexical analyzer given in this chapter that (a) works and (b) is as faithful as possible to the design given here. You may use this program in the following exercises.

19.2 Change the lexical analyzer given in this chapter to also accept an equal sign (=) as an individual lexeme. You should be able to do this by changing only the data that define the machine, not the program.

19.3 Change the lexical analyzer given in this chapter so that the lexemes (* and *) delimit comments instead of { and }.

19.4 Change the lexical analyzer given in this chapter so that it also accepts reals defined by the regular expression

$$((D^+ . D^*) + (. D^+) + \epsilon) E (+ + - + \epsilon) D^+$$

Note that this is an extension of the current definition—the leading expression is identical to the current definition. The character ϵ (meaning "empty," or no character) is used to simplify the regular expression. You will probably find that you need to change the definition of type *CharClass* and the state definitions of several other lexemes. You must also decide what to do with inputs such as

    1324.56E491

Is this a real or a real followed by an identifier? Think about these complications next time you see a language with this syntax.

19.5 Change the lexical analyzer given in this chapter so that it also accepts two periods (..) as a lexeme. The other two uses of a period (.) must continue to work. The case

    23765..

is to be interpreted as an integer followed by the two-period lexeme (..). You will probably find that you need to add a new primitive action.

19.6 The program given in this chapter is *table-driven*; it uses an explicitly stored transition matrix. We could have written function *F* and *G* that used conditional statements to determine the next state and action. Discuss the advantages and disadvantages of encoding the transition matrix directly in the program logic.

19.7 As the program is written, function *FindClass* searches through all the possibilities in order to determine the class of its input. An alternative way to organize the program is to set up an array indexed by character, initialize it with the *CharClass* corresponding to each character, and rewrite function *FindClass* to fetch from the vector. Rewrite your program to do this and discuss the efficiency trade-off involved.

19.8 In function *Findclass*, suppose each **else** were replaced with a semicolon (;). What would be the effect on the execution time (a) of the function, (b) of the interpretation phase of the program?

19.9 The lexical analyzer in this chapter has a fixed and unalterable set of character classes. Change it to read in a list of character class names and the characters in each of the classes.

19.10 The program uses predetermined files for input and output. Change the program to ask the user where the input may be found and where the output should go. Your program should be able to accept a series of lexical analyzers and several input streams for each one. Note that these additions to the program could be useful for controlling other programs as well. You should be able to make them with only slight perturbations to the basic program.

19.11 Section 19.5.1 asserts that the buffer should be empty every time control of the FSM is in state NEW. Prove that this is true for the transition matrix of Section 19.5.4.

19.12 Change the lexical analysis program so that the transition matrix is an array of records rather than three arrays of scalars.

19.13 Discuss how scope rules affect your ability to enforce the policy that only the action primitives may directly modify the current character and the buffer.

19.14 The lexical analyzer given in this chapter is organized to write its outputs on a file. Most often, you will want to use such a program to process input for some more elaborate program. To do so, you must face the problem that certain state information must be retained from one call to the next. Compare the advantages and disadvantages of the following approaches to turning the given program into an input processor:

a) Store certain values (Which ones?) in variables that retain their values from one procedure call to the next.

b) Use variables external to the procedure to save these values. Pass the state values to and from the procedure as parameters

c) Use coroutines.

Be sure to consider which features of a language will affect your decision.

19.15 Choose one of the approaches described in Exercise 19.14 and modify your program to act as an input routine for some other program.

19.16 The lexical analyzer treats all ends of lines as blanks. That is, each time *Scan* reads an end-of-line, it sends back a blank in *CurCharItself*. Suppose that we allow *Scan* to return an end-of-line character (which we'll write as the symbol @). We also establish the following continuation convention:

> If the symbol @ is followed immediately by a plus sign (+), (in other words, if a plus sign (+) begins a line), ignore both the @ and the + (that is, pretend there is no end-of-line). If @ is not followed by a +, treat the @ as a blank (as before).

Thus,

Square

+oftheHypotenuse

is the same as

SquareoftheHypotenuse

whereas without the +, it is the same as

Square oftheHypotenuse

A continued line must have something on it besides a plus sign; the sequence @+@ is illegal. Thus, one may not write

TheBeginningand

+

theEnd

a) Show how to change the FSM for identifiers and keywords (Figure 19.3) to accommodate this new convention.
b) Change the RE describing identifiers and keywords to correspond to this new FSM. That is, the new RE should allow for embedded @+'s, but should not allow @+@.

19.17 Formally prove the main program of Section 19.6.2.1 with respect to the assertions provided.

19.18 Write an input file (FSM description) for which the program developed in this chapter will not terminate.

19.19 Determine formally the costs of procedures *ClearCharsDoneList*, *ProcessExplicitTransition*, and *ProcessDefaultTransition* as functions of the numbers of states, actions, and character classes.

19.20 Determine formally the costs of procedures *Scan*, *AppendChar*, *EmitLex*, and *FailMsg* as functions of the numbers of states, actions, and character classes.

Chapter 20

Fast Implementation of Sets: An Example

In this chapter we use an extended example to tie together the ideas of Chapters 7 through 12. Specifications for finite homogeneous sets were given in Section 9.2.3, and two distinct implementations to fit these specifications were given in Section 10.5. In this chapter we review the abstract requirements for the data type *set*, then develop a richer implementation than those already presented by using linked storage techniques to create a dynamic data structure and to relax the requirement for a fixed upper bound on the set size.

20.1 Background

In Section 10.5 we gave an implementation of sets based on bit-strings in which each position in the bit-string corresponds to one particular value that might be stored in the set. The amount of time required by each of the operations on such sets is essentially constant, because only one bit must be inspected to determine whether a value is present in the set.

In the same section, we gave an implementation based on a vector in which set values are stored contiguously. In this case, it was necessary to search the set to find values, and, as a result, insertion and deletion required time proportional to the number of elements in the set. Operations such as union and intersection that must search individually for all elements in a set therefore required time proportional to the square of the number of set elements.

483

Both the bit-string and vector-pointer representations allocate a fixed amount of storage for each set. The storage that is dedicated in this way is not available for any other use. The cost of dedicating this unused storage has the effect of limiting the maximum size of the sets: in the case of the vector representation, the user must be careful to select a suitable maximum size for each set. Further complications are introduced when different size limits are used for different sets (see Exercise 20.1).

Chapter 10 showed how flexibility and speed can often be gained by dedicating space to pointers. Indeed, it showed that when space is used for pointers, the total space requirement of a data structure may even be reduced. For example, in the case of sets, a linked representation can actually reduce the average space requirement if the *average* set size is sufficiently small relative to the *maximum* size requires (see Exercise 20.14).

When we design data structures, we usually find that the time required for the operations depends very heavily on the data structure. For example, in the bit-string representation of sets, the cost of each operation was essentially constant. However, in the vector-pointer representation, a search through the existing values was usually required to avoid introducing duplicate values. As a result, the time required depended on the number of elements in the set—for an N-element set, most operations required either $O(N)$ or $O(N^2)$ steps.

In the following sections we revisit the problem of implementing sets: We develop and analyze another representation, this time based on lists and trees. We relax the fixed-maximum-size restriction by using dynamically allocated storage, and the implementations we provide for all operations are significantly more efficient than the operations of the vector implementation.

20.2 The Problem

We will implement sets according to the following specifications. Our implementation will use a flexible amount of storage; the storage requirement for each set will depend on the number of elements inserted in that set. The design of this representation emphasizes efficient searches for elements; the primitive operation of finding an element is significantly faster than $O(n)$ for large sets.

20.2.1 OPERATIONS ON THE SET TYPE

The specifications for sets that we use here are essentially those of Section 9.2.3. We define sets of unlimited, but finite, size. We'll call the set type *SetofElts*. The elements of the sets are of some single type, *EltType*.

Figure 20.1 shows the specifications for *SetofElts*. Compare these operations on sets to those specified in Section 9.2.3. The operation *MaxSize* has been eliminated. Except for the elimination of restrictions on the set size, the specfications of the operations previously defined are unchanged.

> **procedure** *Clear*(**var** *S*: *SetofElts*);
>   **post** $S = \{\}$;
>
> **procedure** *Insert*(**var** *S*: *SetofElts*; *x*: *EltType*);
>   **post** $S = S' \cup \{x\}$;
>
> **procedure** *Remove*(**var** *S*: *SetofElts*; *x*: *EltType*);
>   **post** $S = S' \sim \{x\}$;
>
> **function** *IsMember*(*S*: *SetofElts*; *x*: *EltType*): *boolean*;
>   **post** $RESULT \equiv (x \in S)$;
>
> **function** *IsSubset*(*S*, *T*: *SetofElts*): *boolean*;
>   **post** $RESULT \equiv (S \subseteq T)$;
>
> **function** *IsEmpty*(*S*: *SetofElts*): *boolean*;
>   **post** $RESULT \equiv (S = \{\})$;
>
> **procedure** *Union* (**var** *S*: *SetofElts*; *T*: *SetofElts*);
>   **post** $S = S' \cup T$;
>
> **procedure** *Intersect*(**var** *S*: *SetofElts*; *T*: *SetofElts*);
>   **post** $S = S' \cap T$;
>
> **procedure** *Copy* (**var** *S*: *SetofElts*; *T*: *SetofElts*);
>   **post** $S = T' \wedge T = T'$;
>
> **procedure** *Compact*(**var** *S*: *SetofElts*);
>   **post** $S = S'$;

Fig. 20.1 Specifications for *SetofElts*

Two new operations have been added. The *Copy* operation is simply the assignment operator for sets. The *Compact* operation, according to its specifications, has no effect on the abstract value of *S*. *Compact* does have an effect, but the effect is not directly visible in the value of a set. Since the specifications only show such direct effects, they don't explain *Compact*. The effect of *Compact* is on the future efficiency of the set operations, not on the current value of the set. In the implementation we have chosen, obsolete information can accumulate in the representation of each set. This obsolete information takes up space and slightly slows down operations on the set.   *Compact* is a "clean-up" operation that flushes obsolete information out of the data structure. This will be discussed in more detail in Section 20.5.10.

20.3 An Implementation Using Lists and Trees

20.3.1 REPRESENTING SETS WITH TREES

Section 14.5.2 discussed the use of trees in organizing searches. Adapting the discussion there to our current purposes, we can represent sets using k-ary trees (that is, trees with k descendants at each node). Each node will contain one element of type *EltType*. Furthermore, if some node of one of these trees contains some *EltType* value, x, then no other node of that tree contains the value x. The motivation for this structure is that we want to speed up the operation of finding a particular element of the set—or of discovering that the element isn't in the set—while at the same time allowing for arbitrary expansion of the set. Roughly speaking, if the tree is organized so that at each node it is easy to choose which of the possible subtrees *must* contain the value we are looking for, we can search rapidly, since we reduce the size of the search by a factor of k with each choice.

Suppose we have a discrimination function, *Discrim*, such that if y is the *EltType* value contained in node n, then

- *Discrim*(y, x) has a value denoting node n if $y = x$, and
- *Discrim*(y, x) has a value—say i, where $1 \leq i \leq k$—if x must be in the ith subtree of n (or not in the tree at all).

For example, assume that $k = 2$ (that is, the trees are binary), that the values stored in the tree are strings, and that the value of *Discrim* depends on alphabetical order.   Then the set {*kohlrabi, rutabaga, persimmon, avocado, kumquat*} might be represented by any of the trees in Fig. 20.2. Note that these are all legal representations according to the rule above. Every value can be found reliably in each tree. Further, if new values are added at the places indicated by the rule, they too can be found again.

Fig. 20.2 Possible tree representations of a set

To see this, consider adding *broccoli* to the set. The representations of Fig. 20.2 become those of Fig. 20.3.

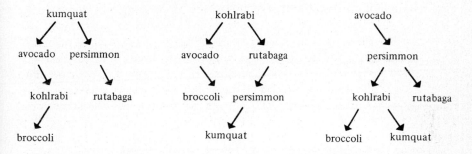

Fig. 20.3 Sets after addition of new element

The trees for two equal sets may have their values arranged differently because the values were inserted in different orders. The first value inserted in a tree becomes the value at the root. Moreover, if two values lie on the path to some leaf, the value closer to the root must have been inserted before the value closer to the leaf. (See Exercise 20.2.)

Finding values in this tree is easy; the discrimination function steers the search in the right direction at each node. If the tree contains n nodes, the cost is usually less than $O(n)$; indeed, the cost is about $O(\log_2 n)$. (The analysis to support this claim is in Section 20.7.1.) However, the structure is less than ideal for operations that need to access all of the values, as we will see in the next section.

20.3.2 THE PROBLEM OF DELETING ELEMENTS

In this tree implementation, removing an arbitrary element can sometimes be expensive. For example, if we remove an internal node from a tree, we have no place to reattach the two sons of the deleted node unless we also rearrange many other elements. We can avoid this problem by including with each element a flag indicating whether it's *really* present in the set we are representing or whether it's obsolete (that is, was formerly a member that has been deleted). This approach speeds up deletions, but it can allow obsolete information to accumulate in the data structure. It can also speed up insertions if the same elements are repeatedly inserted and deleted. We provided a special operation, *Compact*, to remove obsolete and invalid information from the data structure (see Section 20.5.10).

20.3.3 REPRESENTING SETS WITH LISTS

Occasionally, we may want to operate on all the elements of a set in some arbitrary sequence. For this purpose, a simple linked list, as in Fig. 20.4, is more convenient than the tree structure.

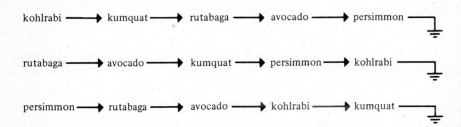

Fig. 20.4 Possible list representations of a set

Unfortunately, in most cases considerably more time is required to search a linear list for one particular element than is required in a tree structure.

20.3.4 A COMBINED REPRESENTATION

We would like to find a representation that provides the best attributes of the two organizations we just discussed. We can do so using a data structure in which the data in the set can be treated as *both* a tree and a linear list. We will use the "tree part" of the representation for some operations and the "list part" for others.

Figure 20.5 shows three possible representations of {*kumquat, rutabaga, avocado, kohlrabi, persimmon*} with the same tree structure as the first representation of Fig. 20.2. Solid lines represent the tree structure;

dashed lines, the linear list. Inserting broccoli into each of these will produce, respectively, the trees of Fig. 20.6.

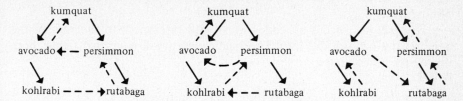

Fig. 20.5 Threaded trees representing sets

Fig. 20.6 Threaded trees after next insertion

Note that the new element was always inserted in the tree at the same place as in Fig. 20.3, but in each case the element is inserted at the *head* of the linear list—just after *ListThread*. Note also that, although the tree part of the structure is the same in all cases, the linear thread is different. All three structures are still correct; the differences may occur because the linear order depends on the sequence in which elements are inserted (see Section 20.5.5). This data structure is an example of a *threaded tree*.

To represent sets in this way, we need to store several pieces of information with each set element. Each element will occupy a node containing five items:

Data:	The value stored at the node.
Valid:	A bit indicating whether this data value is currently in the set.
LeftLink:	Reference to the left son subtree.
RightLink:	Reference to the right son subtree.
Thread:	Reference to the next node in the linear list.

The *Data* field will store the value of a set element. The *Valid* field will indicate whether the value is currently in the set or has been removed since it was last inserted. The *LeftLink* and *RightLink* fields are used to represent the set as a tree. The *Thread* field is used to represent the set as a linear list.

Each set also needs a header word with references to the root of the tree and the head of the list. The header word will have two fields; both are references to nodes:

> *TreeHead*:   Reference to the root of the tree
> *ListThread*:   Reference to the head of the linear list

The following sections develop an implementation that uses this combined representation, linking the elements of the set both as a tree and as a list.

20.4 Using the SetofElts Type

When the type *SetofElts* has been specified and implemented, we should be able to use it much like an ordinary data type. For example, suppose we want to know the sets of factors of the first N integers. One way is to define a vector of sets indexed, say, 2 to N, then construct the set of factors of K in the Kth vector element. The following program does this:

```
var Factor: array [2..N] of SetofElts; i, j: integer;
  begin
  for i := 2 to N do Clear(Factor[i]);
  for i := 2 to N do
    begin
    if IsEmpty(Factor[i]) then Insert(Factor[i], i);
    j := 2 * i;
    while j <= N do
        begin Union(Factor[j], Factor[i]); j := j + i; end;
    end;
  end.
```

First, all sets are initialized. Then the sets are examined in order. If *Factor[K]* is empty when its turn comes, it has no smaller factors and therefore must be prime. We therefore insert K as a factor of *Factor[K]*. In either case, we add (using *Union*) the set of factors of K to the set of factors of all multiples of K.

20.5 Programs for the Solution

We are now ready to implement sets using the threaded trees of Section 20.3.4. To simplify the choice of a discrimination function, we consider only the case where *EltType* is the type *integer*. For illustrative purposes, we will use the tree in Fig. 20.7 as a base-line example. This figure shows one representation of the set {85, 25, 42}. The data structure also contains cells with invalid elements; 47, 65, and 39 have been in this set previously, but they have been removed.

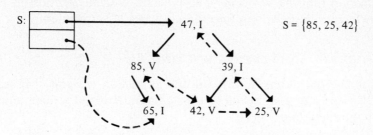

Fig. 20.7 Standard example tree

As before, tree links are depicted as solid lines and linear-thread links as broken lines. The status of the *Valid* field is indicated by a *V* for valid or *I* for invalid at each node.

20.5.1 THE DATA STRUCTURE

The fields of each node and the purposes of the reference fields were explained in Section 20.3.4. The actual declarations are

```
type
    EltType = integer;

    RefSetElt = ↑SetElt;
    SetElt = record
        Data: EltType;
        Valid: boolean;
        LeftLink, RightLink, Thread: RefSetElt;
        end;

    SetofElts = record
        TreeHead, ListThread: RefSetElt;
        end;

    DiscrimClass = (SameClass, LeftClass, RightClass);
```

EltType is declared as a type so the type of the set elements can be changed easily. If other declarations (procedure headers, for example) specify type *EltType* instead of naming *integers* explicitly, then changing this line to, say, *real* (and possibly adjusting the discrimination function) will provide sets of reals instead of sets of integers without further ado.

The type declarations for *SetElt* and *SetofElts* exactly parallel the description in Section 20.3.4. *SetElt* is the node type; *SetofElts* is the set type, that is, the header record itself.

DiscrimClass specifies the values that can be returned by the discrimination function, *Discrim*. The function *Discrim* depends on *EltType* and may have to be rewritten for different types of set elements.

20.5.2 FINDELT: AN INTERNAL ROUTINE

Several of the operations on sets will use a function, *FindElt*, that searches for an element x in a set S and returns a boolean B and a node pointer N with the interpretation

$\quad$ B $\quad \supset N\uparrow.Data = x$ $\,(N\uparrow.Valid$ may be *true* or *false*).

$\quad \sim$B $\,\supset x$ is not in S. If it is to be added, it should be

$\qquad$ attached to $N\uparrow$.

Understanding this routine—what it does, why it does it, and why we choose to make it do the things it does—is the key to understanding the set implementation.

FindElt will be used both by routines that simply look for values in the structure and by routines that will alter the structure. These routines may do several things with the information returned in N and B:

- Discover that x is not in S by checking for $B = $ *false*
- Determine whether x is currently in S by checking $N\uparrow.Valid$.
- Remove x from S by setting $N\uparrow.Valid$ to *false*.
- Reinsert x in S (after finding that it has previously been inserted and deleted) by setting $N\uparrow.Valid$ to *true*.
- Insert x in S (after finding that no node contains x) by adding a new node as a left or right son of $N\uparrow$.

Note first that *FindElt* does not decide whether the value x is currently in set S, but only whether some node contains the value. In other words, *FindElt* operates on the tree rather than on the set represented by the tree. *FindElt* also returns a pointer to the node where the value is, so it's easy for the calling program to test $N\uparrow.Valid$. Some programs (*Insert* and *Remove*, for example) don't need to know.

In order to use *FindElt* in this way, it is necessary to know the details of the data structure. It is appropriate for routines such as *Insert* and *Union* to use these details, but it is *not* appropriate for a user of the set type to do so—it might be necessary to change the implementation some day. For this reason, we do not include *FindElt* in the specification of the type set.

Note also that the interpretation of N depends on B. If B indicates that the value x is in the tree, then N is a pointer to the node that contains x. If x is not in the tree, however, N is a pointer to the node that will be the *parent* of the node containing x if the caller of *FindElt* chooses to add x to the data structure. *FindElt* does not actually add the new node because the caller may, for example, be testing membership rather than inserting a value.

Suppose we execute *FindElt* according to the standard example (Fig. 20.7) and the values of x given below. The results would be as indicated.

x	B	Node referenced by N
85	*true*	[85, V]
65	*true*	[65, I]
40	*false*	[42, V]
45	*false*	[42, V]
47	*true*	[47, I]

Let's turn now to the actual code for *FindElt*, given in Fig. 20.8.

First, note that the function uses S both to name the set and to return the result. As a consequence, the input parameter (of type $\uparrow SetNode$) is the root of the tree part of the structure or a subtree (of type $\uparrow SetNode$) rather than a full set header. Thus S and x correspond directly to the abstract description; N is the final value of S; and B is the functional result. *FindElt* is not a pure function; it alters its input parameter.

Next, note that two pointers are used to traverse the tree. Parameter S is used to develop the result. Local variable *Next* is used to investigate the next element in the tree; if x is not in the tree, *Next* will eventually be advanced off the end of a branch (that is, become *nil*) and S will retain a reference to the node we last processed.

FindElt searches the data structure using the tree part of the structure. At each node it updates S, quits if the current node contains x (the "equal" return from *Discrim*), and otherwise chooses a subtree for further search.

```
function FindElt(var S: RefSetElt; x: EltType): boolean;
post (RESULT ⊃ S↑.Data = x) ∧
      (~ RESULT ⊃
                x not in S'↑ and if it is to be added, it should
                be a son of S↑; S = nil denotes the set header);
label 1;
var Next: RefSetElt;
  begin
  Next := S; FindElt := false;
  while Next ≠ nil do
      begin
      S := Next;
      case Discrim(Next↑.Data, x) of
          SameClass:  begin FindElt := true; goto 1; end;
          LeftClass:   Next := Next↑.LeftLink;
          RightClass:  Next := Next↑.RightLink
          end;
      end;
1: end;  {FindElt}
```

Fig. 20.8 Routine for finding nodes in the tree

The programmer must define a discrimination function to go with
EltType. *Discrim*(x, y) yields one of the three of type *DiscrimClass*:

$$Discrim(x, y) = SameClass \text{ iff } x \text{ and } y \text{ are equivalent } EltTypes,$$

$$= LeftClass \text{ iff } x \text{ is less than } y,$$

$$= RightClass \text{ iff } x \text{ is greater than } y.$$

What "less than" and "greater than" mean depends on what makes sense
for *EltType*. When *EltType* is *integer*, we can use $<$ and $>$ on integers. If
EltType were some sort of string, we could use alphabetic order. For this
example, however, we have chosen to assume that the values in the set
are integers. We can thus use the discrimination function shown in Fig.
20.9.

Describing *Discrim* in this way has led us to introduce an enumerated
type and an explicit function. If the type of the set elements is a design
decision that is not likely to be changed, the discrimination function may
be coded directly in the two places it appears. (See Exercise 20.27.)

We can now turn to the operations themselves.

```
function Discrim(x, y: EltType): DiscrimClass;
post (RESULT = SameClass ≡ x = y)
      ∧ (RESULT = LeftClass ≡ x < y)
      ∧ (RESULT = RightClass ≡ x > y);
    begin
    if x = y
        then Discrim := SameClass
        else if x < y
            then Discrim := LeftClass
            else Discrim := RightClass
    end;  {Discrim}
```

Fig. 20.9 One possible discrimination function

20.5.3 THE OPERATION CLEAR

The operation *Clear* is easy. It simply sets the pointers in the header to *nil*.

```
procedure Clear(var S: SetofElts);
  post S = { };
    begin
    S.TreeHead := nil; S.ListThread := nil;
    end;   {Clear}
```

20.5.4 THE OPERATION REMOVE

For the *Remove* operation, we use *FindElt* to locate the element. If it's not there, fine. If it is there, we set the *Valid* bit had to *false*.

```
procedure Remove(var S: SetofElts; x: EltType);
  post S = S' ∼ {x};
  var R: RefSetElt;
    begin
    R := S.TreeHead;
    if FindElt(R, x) then R↑.Valid := false;
    end; {Remove}
```

Note that if *FindElt* returns *true*, it does not matter what value the *Valid* bit in the beginning. The precondition of *Remove* doesn't demand that *x* be in *S*; but the postcondition does guarantee that *x* isn't in *S* *after* the operation.

Figure 20.10 shows the state of the standard example of Fig. 20.7 after the sequence of operations

Remove(*S*, 85);
Remove(*S*, 75);
Remove(*S*, 65);

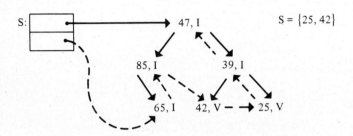

Fig. 20.10 Set after some elements are removed

20.5.5 THE OPERATION INSERT

Insert, shown in Fig. 20.11, is somewhat more complex than *Remove*, because a new node in the data structure may be required. If *FindElt* locates the value being inserted, we set the *Valid* bit to *true*. If the value isn't anywhere in the structure, *FindElt* tells us which node should be its parent. We then create a new node, initialize its fields, and link it into the tree and the linear list at the appropriate places.

The procedure begins by selecting the root of the tree from the set header for *S*. It then calls *FindElt*. If *x* is anywhere in the structure, *Insert* sets the *Valid* bit to *true* and quits. (Just as for *Remove*, it doesn't matter whether *x* is currently in *S*.) If *x* isn't in the structure, *Insert* does the following:

1. Construct a new node. *MakeCell* creates the node and initializes the reference fields to *nil* and the *Data* and *Valid* fields to the values given as parameters.

2. Insert new node at the head of the linear thread. Note that pointer-swing rather than node-overwrite techniques are used, because the "tree part" of the data structure "assumes" that data values remain in the nodes where they were originally placed.

3. Insert the new node in the tree. It can go in one of three places. If there are no elements (either valid or invalid) in the set, the new node must be attached to the *TreeHead* field of the set header. If

there are elements in the set, the new node is attached to either the
LeftLink or *RightLink* of the node indicated by *FindElt*. The *Discrim*
function is used to determine the proper subtree.

```
procedure Insert(var S: SetofElts; x: EltType);
   post S = S' ∪ {x};
   var R, T: RefSetElt;
      begin
      R := S.TreeHead;
      if FindElt(R, x)
         then R↑.Valid := true
         else
            begin
            MakeCell(T, true, x);
            T↑.Thread := S.ListThread; S.ListThread := T;
            if R = nil
               then S.TreeHead := T
               else if Discrim(R↑.Data, x) = LeftClass
                  then R↑.Leftlink := T
                  else R↑.RightLink := T;
            end;
      end; {Insert}
```

Fig. 20.11 Implementation of *Insert*

Insert calls *MakeCell* to cause a new node to be allocated and
initialized. *MakeCell*, like *FindElt*, is a private operation of the set
implementation; it is *not* to be called directly by users of the type. The
code for *MakeCell* is straightforward:

```
procedure MakeCell(var S: RefSetElt; Val: boolean; Dat: EltType);
   post S↑ Valid = Val ∧ S↑.Data = Dat ∧ S↑ has no sons;
      begin
      new(S); S↑.Valid := Val; S↑.Data := Dat;
      S↑.LeftLink := nil; S↑.RightLink := nil; S↑.Thread := nil;
      end; {MakeCell}
```

Figure 20.12 shows the state of the standard example of Figure 20.7
after the sequence of operations

Insert(S, 85);
Insert(S, 75);
Insert(S, 65);

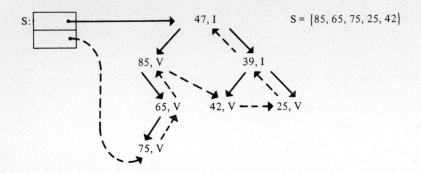

Fig. 20.12 Set after some elements are inserted

IsMember simply uses *FindElt* to see if *x* is in *S*.

> **function** *IsMember*(*S*: *SetofElts*; *x*: *EltType*): *boolean*;
> **post** *RESULT* ≡ (*x* is in *S*);
> **var** *R*: *RefSetElt*;
> **begin**
> *R* := *S.TreeHead*;
> *IsMember* := *FindElt*(*R*, *x*) **cand** *R*↑.*Valid*;
> **end**;   {*IsMember*}

The function begins by selecting the tree root, then calls *FindElt* and tests the results. In a language that does not provide the *cand* operation, the following test is expressed with the conditional statement,

>   **if** *FindElt*(*R*, *x*) **then** *IsMember* := *R*↑*Valid* **else** *IsMember* := *false*;

(See Exercise 20.17.)

20.5.6 THE OPERATION ISEMPTY

At first glance, it might appear that the operation *IsEmpty* could be implemented as a simple test for *nil* in, say, the *TreeHead* field. However, this would produce an incorrect result for a set that had a tree with one or more nodes in which all *Valid* bits were set to *false*. It is therefore necessary to search through the tree until a *Valid* bit of *true* is found.

Since we do not care about a *particular* value, we scan the structure by searching along the *ListThread*. If a *Valid* element is found, the search terminates successfully at that point, and otherwise terminates unsuccessfully at the end of the list. The implementation is as follows:

```
function IsEmpty (S: SetofElts): boolean;
  post RESULT ≡ (S = { });
  label 1;
  var R: RefSetElt;
    begin
    IsEmpty := true;
    R := S.ListThread;
    while R ≠ nil do
        begin
        if R↑.Valid then begin IsEmpty := false; goto 1; end;
        R := R↑.Thread;
        end;
1:  end; {IsEmpty}
```

20.5.7 THE OPERATION ISSUBSET

The function *IsSubset* returns the value *true* exactly when every element of its first parameter is a member of its second parameter. The implementation captures that exactly, using the *ListThread* to scan through the first parameter:

```
function IsSubset(S, T: SetofElts): boolean;
  post RESULT ≡ (S ⊂ T);
  label 1;
  var x: EltType; R: RefSetElt;
    begin
    IsSubset := false;
    R := S.ListThread;
    while R ≠ nil do
        begin
        if R↑.Valid cand not IsMember(T, R↑.Data) then goto 1;
        R := R↑.Thread;
        end;
    IsSubset := true;
1:  end; {IsSubset}
```

The functional result is initially set to *false*. It can be set to *true* only if the **while** loop exits normally—that is, if every element of *S* is a member of *T*.

20.5.8 THE OPERATION UNION

Union is implemented just as you might expect: Every element of the second parameter is inserted into the first. We use the list thread to simplify the problem of finding elements in the second set.

```
procedure Union(var S: SetofElts; T: SetofElts);
  post = S' ∪ T;
  var x: EltType; R: RefSetElt;
    begin
    R := T.ListThread;
        while R ≠ nil do
            begin
            if R↑.Valid then Insert(S, R↑.Data);
            R := R↑.Thread;
            end;
    end; {Union}
```

20.5.9 THE OPERATION INTERSECT

Intersect must remove from *S*, the first parameter, all the elements not also in *T*, the second parameter. The *Valid* bit from *T* is copied into *S*. As a result, if an element had been deleted from *T*, it would not be in the intersection *S* ∪ *T*. If a value from *S* is not found in *T*, it is deleted from the intersection.

```
procedure Intersect(var S: SetofElts; T: SetofElts);
  post S = S' ∩ T;
  var RS, RT: RefSetElt;
    begin
    RS := S.ListThread;
    while RS ≠ nil do
        begin
        RT := T.TreeHead;
        RS↑.Valid := RS↑.Valid cand FindElt(RT, RS↑.Data)
                        cand RT↑.Valid;
        RS:= RS↑.Thread;
        end;
    end; {Intersect}
```

RT must be reset for each loop because *FindElt* alters it. Again, the **cand** may need to be written as a conditional statement in some languages.

20.5.10 THE OPERATION COMPACT

As we have noted, obsolete elements can accumulate in this representation. We provide the *Compact* operation to clean up. This is not accomplished by manipulating the data structure of the set to be compacted—the data structure is simply too complex. Instead, we create a new, empty set and insert each valid element of the original. We then throw away the original, replacing it with the newly created version.

```
procedure Compact(var S: SetofElts);
  var B: SetofElts; x: EltType; R: RefSetElt;
  begin
  Clear(B);
  R := S.ListThread;
  while R ≠ nil do
    begin
    if R↑.Valid then Insert(B, R↑.Data);
    R := R↑.Thread;
    end;
  S := B;
  end; {Compact}
```

This can be a very slow operation (see Section 20.7.1). The result of executing *Compact* on the set of Fig. 20.7 is given in Fig. 20.13.

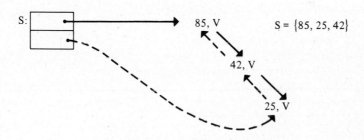

Fig. 20.13 Compacted set

20.5.11 THE OPERATION COPY

The *Copy* operation, shown in Fig. 20.14, copies the entire structure, *including* invalid elements. An internal function, *Icopy*, is used to construct a duplicate of the "tree" part of the data structure. Function *Icopy* is also recursive—it constructs each subtree by calling itself to copy the sub-subtrees, then connecting them to make the subtree. As *Icopy* runs, it develops a new linear thread in *R*.

20.6 Correctness of the SetofElts Type

We have developed an implementation for the abstract data type *SetofElts*. In order to show that this is a *correct* implementation, we must perform the steps described in Chapter 11: stating a representation mapping,

```
procedure Copy(var S: SetofElts; T: SetofElts);
  var R: RefSetElt;

  function Icopy(Q: RefSetElt): RefSetElt;
    {Returns pointer to root of copy of tree rooted at Q}
    var c: RefSetElt;
    begin
    if Q = nil
        then Icopy := nil
        else
            begin
            MakeCell(c, Q↑.Valid, Q↑.Data);
            c↑.LeftLink := Icopy(Q↑.LeftLink);
            c↑.RightLink := Icopy(Q↑.RightLink);
            c↑.Thread := R; R := c; Icopy := c;
            end;
    end;  {Recursive inner Icopy}

  begin  {Copy body}
  R := nil;
  S.TreeHead := Icopy(T.TreeHead);
  S.ListThread := R;
  end;  {Copy}
```

Fig. 20.14 Implementation of Copy

finding an abstract and a concrete invariant, verifying that initialization is done properly, and verifying that each operation preserves the concrete invariant and satisfies its own specification. In this section we perform these steps informally, but you should be able to see how to make them formally precise.

20.6.1 REPRESENTATION MAPPING

The set is represented as a linked data structure in which each node contains a value and an indication of whether the value is *currently* a member of the set. The abstract set thus depends on the *Data* and *Valid* fields of the nodes, but not on the way the nodes are linked. The set is

$$rep(S) = \{x \mid \text{There is a node, } c, \text{ in the data structure with}$$
$$c.Data = x \wedge c.Valid)\}$$

20.6.2 INVARIANTS

In this case we are implementing a data type that corresponds directly to one of the formal systems we use to write abstract specifications. That is, we are implementing sets, and we can (and do) choose to model them as sets. We do not have any limits on the sizes of these sets, so any mathematical set can, at least in principle, be represented. Thus the abstract invariant is

$A \equiv$ *true.*

The concrete invariant is more complex. We have chosen a sophisticated data structure and arranged the values stored in it so that certain operations, notably searches, can be performed efficiently. If a value is stored out of place, the search will fail. It is therefore extremely important for the necessary properties of the data structure to be stated precisely and preserved by all operations. These necessary properties constitute the concrete invariant. To write the invariant formally, we need a predicate that says that the values arc arranged in a satisfactory way. We will call this property of a binary tree *DiscrimProp*, and define it

$DiscrimProp(T) \equiv (T{\uparrow} = nil) \vee$

$\qquad (DiscrimProp(T{\uparrow}.LeftSon)$

$\qquad \wedge DiscrimProp(T{\uparrow}.RightSon)$

$\qquad \wedge (x$ is a *Data* field in $T{\uparrow}.Leftson$

$\qquad\qquad \supset (Discrim(x, T{\uparrow}.Data) = LeftClass))$

$\qquad \wedge (x$ is a *Data* field in $T{\uparrow}.Rightson$

$\qquad\qquad \supset (Discrim(x, T{\uparrow}.Data) = RightClass)))$

Note that this property is unrelated to the values of *Valid* fields. We can now state the concrete invariant for the representation of a set *S*.

$C \equiv$ (The path from *S.ListThread* through successive *Thread* fields

$\qquad$ passes through all nodes of the structure exactly once)

$\qquad \wedge DiscrimProp(S.TreeHead)$

20.6.3 INITIALIZATION

Sets are initialized by the operation *Clear*. *Clear* sets both the *TreeHead* and *ListThread* fields to *nil*. The specifications say that initialization of a set is required to make it empty. This requirement is satisfied because the representation mapping picks up all valid nodes in the data structure, and the data structure is empty.

In this case, the concrete invariant is also vacuously satisfied. There are no elements, so the list thread of *nil* links them all. The predicate *DiscrimProp* also holds because of its initial *nil* clause.

20.6.4 CORRECTNESS OF OPERATIONS

We need to argue for each of the operations that it does what it is supposed to do and that it leaves the concrete invariant intact. In the following subsections we make *informal* arguments for *Remove* and *Insert* and leave the rest of the proofs as exercises. Our arguments follow the same pattern as the precise, formal proofs we have made elsewhere, but we keep them informal by detailing only major steps and issues. You should be able to fill in the details. We emphasize the point that you can apply verification methods informally as well as formally.

20.6.4.1 *Remove*

The only way *Remove* can modify the data structure is by changing one *Valid* field to *false*. This does not affect the linking or the arrangements of values, so the concrete invariant is preserved.

The postcondition of *Remove* demands that x be removed from S. *FindElt*'s postcondition promises that if x is in the subtree named by the input value of R (which is the whole tree), then the output value of R will name the node with Data field x. If such a node is found, *Remove* sets its *Valid* field to *false*, thus removing it from the set.

20.6.4.2 *Insert*

We can separate the analysis of *Insert* into two cases: either x was previously a member of S (and may or may not have been deleted), or it was not (and therefore does not appear in any *Data* field).

The first case (x is a *Data* field somewhere in the data structure) is directly analogous to *Remove*, and the correctness argument for this case of *Insert* exactly is the same as the one for *Remove*.

The second case hinges on the **else** clause:

```
begin
MakeCell( T, true, x);
T↑. Thread := S. ListThread;  S. ListThread := T;
if R = nil
    then S. TreeHead := T
    else if Discrim( R↑. Data, x) = LeftClass
        then R↑. Leftlink := T
        else R↑. RightLink := T;
end;
```

We enter this program fragment when *FindElt* returns *false*. Let's start with the postcondition $S = S' \cup \{x\}$ and work backward. The **if** statement that tests for $R = nil$ puts T in the tree structure. In order for the program to work properly, R must indicate where T is to go and T must have the correct values. To achieve the postcondition, the precondition of the **if** statement must be, informally,

> (T is already in the list part of the structure)
>
> $\land$ ($T\uparrow.Data = x \land T\uparrow.Valid = true \land T\uparrow$ has no sons)
>
> $\land$ (x is not in S' and if it is to be added,
>
>   it should be a son of $S\uparrow$; $S = nil$ denotes the set header)

The two assignments involving *ListThread* put T into the list part of the structure. The precondition of this pair of statements thus reduces to

> ($T\uparrow.Data = x \land T\uparrow.Valid = true \land T\uparrow$ has no sons)
>
> $\land$ (x is not in S' and if it is to be added,
>
>   it should be a son of $S\uparrow$; $S = nil$ denotes the set header)

The remaining statement in this program fragment is a procedure call on *MakeCell*. The postcondition of that procedure assures us that T contains the right data and the procedure has no precondition, so the precondition for the program fragment is

> (x is not in S' and if it is to be added, it should be a son of $S\uparrow$;
>
> $S = nil$ denotes the set header)

But this is guaranteed by the "**if** *FindElt*" test, so proof of *Insert*'s postcondition is complete.

Note that the concrete invariant is preserved when an element is added: The thread is maintained because the new element is threaded exactly once; the *Discrim* property is maintained because the new *Value* is added as a leaf at a position determined by *FindElt* and *Discrim*.

20.7 Performance of the SetofElts Type

Because the size and organization of the data structure changes as elements are added, the analysis of its space and the time requirements is not as straightforward as for the other set implementations.

20.7.1 SPEED OF THE SETOFELTS TYPE

We approximate the execution cost of the routines that operate on threaded trees by counting the number of *SetElt* cells they access. Most of the routines of the set type use *FindElt* to locate elements in the tree. When we analyze these routines, we count cells in *FindElt*, as well as direct references to *SetElt* cells. We also estimate the cost of *FindElt* alone and reduce the costs of individual operations to primitive terms. One advantage of retaining *FindElt* terms in the cost estimates is that the analysis can easily be updated to reflect the costs of a new version.

20.7.1.1 Cost of *Intersect*

The analysis of *Intersect* is typical. (We will analyze *Intersect* in detail and leave the rest as exercises.) The program, annotated with operation counts, is

```
procedure Intersect(var S: SetofElts; T: SetofElts);
  post S = S' ∩ T;
  var RS, RT: RefSetElt;
    begin
    RS := S.ListThread;
    while RS ≠ nil do                    SS · [
      begin
      RT := T.TreeHead;
      RS↑.Valid := RS↑.Valid             1
        cand FindElt(RT, RS↑.Data)       + if RS↑.Data in S then FindElt
        cand RT↑.Valid;                  + if RS↑.Data also in T then 1
      RS := RS↑.Thread;
      end;                               ]
    end;  {Intersect}
```

The **while** loop follows the thread completely through *S*, so *RS* is the number of cells in the representation of *S*. However, the first **cand** prevents *FindElt* from being called for invalid elements. Thus *FindElt* is called once for each element actually in the set *S*. If the element in question has ever actually been in the set *T*, a further cell access is needed to determine whether it is currently a member. The cost of *Intersect* is thus

$$(\# \; Elts \; \text{in} \; S \; \text{tree}) + cardinality(S) \cdot (\text{cost of} \; FindElt \; \text{for} \; T) + k$$

where $k \leqq cardinality(S)$.

This reduces to $O(cardinality(S) \cdot (\text{cost of} \; FindElt))$, assuming that $(\# \; Elts \; \text{in} \; S \; \text{tree})$ is $O(cardinality(S))$.

20.7.1.2 Costs of Other Operations

Figure 20.15 tabulates the costs of the operations. We leave the derivations as exercises. For simplicity, we assume that both operands of binary set operators have cardinality N.

Operation	Cost
Clear	$O(1)$
Insert	$O(FindElt) = O(\log N)$
Remove	$O(FindElt) = O(\log N)$
IsMember	$O(FindElt) = O(\log N)$
IsSubset	$O(N \cdot IsMember) = O(N \cdot \log N)$
IsEmpty	$O(N)$
Union	$O(N \cdot Insert) = O(N \cdot \log N)$
Intersect	$O(N \cdot FindElt) = O(N \cdot \log N)$
Compact	$O(N \cdot Insert) = O(N \cdot \log N)$
Copy	$O(N)$
FindElt	$O(\log N)$, but see next section.

Fig. 20.15 Costs of operations on *SetofElts*

20.7.1.3 Cost of FindElt

FindElt locates elements in the tree by searching downward from the root. The time required to find any given value is determined by the length of the path from the root to the node containing that value. Thus the average cost of *FindElt* for the elements in a given representation of a set is determined by the average distance of the elements from the root. As a result, the representation of a given set is not unique. For example, the set of Section 20.3 might be represented as any of the structures in Fig. 20.16. We assume that the elements are usually well-distributed in the tree so that the distance from the root to an element is $O(\log N)$ for a tree containing N elements. We will, therefore, use $O(\log N)$ as the cost of *FindElt*.

There is a further complication in the analysis of *FindElt*. We would like to express the cost of the set operations in terms of the number of elements in the abstract set. Unfortunately, the threaded tree representation may be cluttered up with invalid elements. In an extreme case,

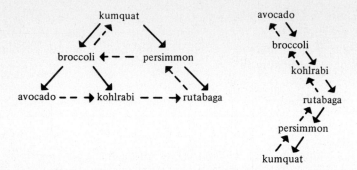

Fig. 20.16 Good and bad representations of a set

there might be several times as many deleted elements as valid ones. This does not affect an analysis that refers to the number of elements in the representation, but it does not satisfy our goal of a description that makes sense in terms of the abstract set.

One way to deal with this problem is to express the costs in terms of the history of the set—that is, to refer to "the number of elements that were ever in the set" rather than to "the number of elements now in the set." If we do this, we need to know details about the patterns of insertions and deletions in order to complete the analysis. We will instead make another simplifying assumption, namely that the *Compact* operation (which rebuilds the data structure to contain only valid elements) is used often enough to keep the proportion of invalid elements small. By making this assumption, we lose some precision, but we simplify the analysis. On this assumption we can say

Cost of *FindElt* $= O(\log N)$, where N is the set size.

20.7.2 SPACE OF THE SetofElts TYPE

Space costs can be determined from the declarations, so we will assume that integers, references, and booleans cost one word apiece. Then the annotated declarations are

type

$EltType = integer;$ $s_{EltType} = 1$

$RefSetElt = \uparrow SetElt;$ $s_{RefSetElt} = 1$

$$SetElt = \textbf{record}$$
  Data: EltType;
  Valid: boolean;
  LeftLink, RightLink, Thread: RefSetElt;
  end;

$s_{SetElt} =$
  $(\ s_{EltType}$
  $+\ s_{boolean}$
  $+\ 3\ s_{RefSetElt}$
  $)$

$$SetofElts = \textbf{record}$$
  TreeHead, ListThread: RefSetElt;
  end;

$s_{SetofElts} =$
$2\ s_{RefSetElt}$

$DiscrimClass = (SameClass, LeftClass, RightClass);$ $s_{DiscrimClass} = 1$

Every set requires one *SetofElts* for a header and one *SetElts* for every node in the data structure. Thus the storage cost for sets is $5M + 2$, where M is the number of elements in the structure. We can again make the simplifying assumption that *Compact* is run frequently enough to keep the number of invalid elements small. By making this assumption we obtain the result

Storage cost of sets $= O(N)$, where N is the set size.

20.8 Choosing a Representation

We have presented three implementations of sets, one in Chapter 20 and two in Section 10.5. Each implementation has different time and space costs. In addition, they have slightly different restrictions. In this section we discuss the constraints on and abstractly visible differences among the implementations, compare their time and space requirements, and discuss the criteria for choosing an implementation for a specific problem.

20.8.1 DIFFERENCES AND CONSTRAINTS

The three set implementations differ in two major ways. First, they provide different kinds of control over the sizes of sets; second, they have different restrictions on the values that can be set elements.

In a bit-string representation, each value corresponds to a bit in a machine word. As a result, the maximum size of a set represented as a bit-string is determined by the number of bits in a machine word. Furthermore, all such sets have the same maximum size. In the vector-pointer representation, a vector large enough to store all the set elements is allocated when the set is declared; thus the maximum size of a set is determined by its declaration. Finally, the size of a set represented as a threaded tree changes in response to individual operations. The maximum size of such a set is limited only by the pool of storage that is provided for dynamic allocation. In addition, the data structure grows and shrinks as the program runs. Thus the three representations differ in the time at which the maximum set size is determined: for bit-string representations it is fixed by the implementation; for vector-pointer representations it is fixed when the set is declared and allocated; for threaded trees it is flexible, even during execution.

The values that can be stored in sets in the vector-pointer and threaded tree representations can be any values of the declared type. In the bit-string representation, however, each bit is assigned to a specific value; the complete set of potential values must therefore be declared in advance. Bit-string representations are primarily useful for sets whose elements are drawn from enumerated types or small subranges of the integers.

These differences are significant when a set is used near one of the limits of its implementation. Programs should, in any case give the same results for all implementations unless one of the limits is reached—and if such is the case, they should give clear error indications.

20.8.2 TIME COMPARISON FOR SET IMPLEMENTATIONS

The cost of an individual operation on any of the representations can be determined as in Section 20.6 or Chapter 6. Figure 20.17 simply summarizes them in tabular form. We assume that all sets contain N elements.

Operation	Bit-String	Vector-Ptr	Threaded Tree
MaxSize	$O(1)$	$O(1)$	-
Clear	$O(1)$	$O(1)$	$O(1)$
Insert	$O(1)$	$O(N)$	$O(\log N)$
Remove	$O(1)$	$O(N)$	$O(\log N)$
IsMember	$O(1)$	$O(N)$	$O(\log N)$
IsSubset	$O(1)$	$O(N^2)$	$O(N \cdot \log N)$
IsEmpty	$O(1)$	$O(1)$	$O(N)$
Union	$O(1)$	$O(N^2)$	$O(N \cdot \log N)$
Intersect	$O(1)$	$O(N^2)$	$O(N \cdot \log N)$
Copy	-	-	$O(N)$
Compact	-	-	$O(N \cdot \log N)$
Make Cell	-	-	$O(1)$
FindElt	-	-	$O(\log N)$

Fig. 20.17 Summary of operation costs for N element sets

20.8.3 SPACE COMPARISON FOR SET IMPLEMENTATIONS

When evaluating the space consumption of a set representation, you may need to consider both the number of elements actually stored in a set, say, N and the maximum number that will be allowed, say M. The bit-string representation requires one word. The vector-pointer representation requires an array of M elements and an integer; if each element uses one word, then the size is $M + 4$ words (according to the rules and specific assumptions of Chapter 12). For a threaded tree representation, the size is $5N + 2$ words (according to the analysis in this chapter).

20.8.4 CRITERIA FOR CHOOSING AN IMPLEMENTATION

Suppose you are writing a program that involves sets of objects and need to choose an implementation. In order to select the representation that will be most suitable, you need to spend some time analyzing your problem. Your first concern must be to pick a representation with constraints that fit your problem. For example, if you will need to insert more distinct values than there are bits in a word, you rule out a bit-string implementation immediately. Your second concern is efficiency. Consider whether it is more important to save time or to save space. If you are choosing between the vector-pointer representation and threaded trees, you will often find that vectors are smaller, but trees are faster.

If your chief concern is with speed, you will compare the estimated execution times for the operations you expect to perform on the sets. In this case the ranking is the same for almost all the operations:

1. bit strings
2. threaded trees
3. vector-pointer

Note, however, that the ranking is not always the same for all operations: sometimes one representation supports fast execution of one group of operations, whereas a different representation expedites another group. Before you can make a decision in such cases, you must predict which operations will be most heavily used.

If your chief concern is with storage space, you will compare the estimated storage costs for the representations. In this case, you will find that the estimates are not directly comparable: One is expressed in terms of maximum set size, but another is expressed in terms of actual size. To resolve the problem, you must learn enough about your intended application to know whether

- You can set a reasonable upper limit on the size.

- The set will be more likely to be full or empty.
- The problem is so small that the costs do not really matter.

In the present case, we can make a ranking only by stating additional conditions.

- The bit-string representation will be preferred when the number of distinct values to be inserted is small enough.
- The vector-pointer representation will be preferred when the average number of elements in the set is a large enough percentage of the maximum size.
- The threaded tree representation will be preferred when the average number of active elements is a small enough percentage of the maximum size and when a maximum size cannot be determined.

Exercises

20.1 Compare the implementations of *Union* in the three implementations of sets that you have seen. Discuss the changes that would be required to handle sets of different size limits.

20.2 Find two possible orders in which each of the trees of Figure 20.2 could have been created.

20.3 Write an implementation of the set type that (a) works and (b) is as faithful as possible to the design given here. You may use this program in the following exercises.

20.4 Write an implementation of the set type using the bit-vector representation.

20.5 Write an implementation of the set type using the array representation.

20.6 Use the implementation of sets given in this chapter to solve the following problem:

a) Let a class record consist of a class number, a set of prerequisites, and the set of students requesting the course for next semester.
b) Let a student record consist of a student number, the set of courses completed, and the set of courses requested for next semester.

c) Read values for class records, student numbers and the "completed-course" portions of student records. Then fill in the "courses-requested" part of the student records. As you do this, check to see that each student has completed the prerequisites for all courses that he or she has requested.

d) Keep track of the students who have made illegal requests (in a set, of course—you will want to see each bad student once only).

e) When processing is complete, print the results.

20.7 Change the type of the set elements from *integer* to *real.*

20.8 Write a print routine that uses the thread to find elements. Print only the values actually in the set. Note that this will be a new operation of the type—it will need to refer directly to the fields of the data structure.

20.9 Write a print routine that uses the tree structure to scan through the elements in the set. Print only the values actually in the set. Note that this will be a new operation of the type—it will need to refer directly to the fields of the data structure.

20.10 Change one of the print routines of Exercises 20.8 and 20.9 to be useful as a debugging aid. Specifically, make your program print information about *all Data* fields and *Valid* bits in the structure, together with enough information to determine where all the other pointers point. Remembering that you can't print a pointer value directly, print enough information so that someone can easily draw the entire data structure.

20.11 Change the type of data element from integers to 4-character strings.

20.12 Compare the time requirements of the implementation of Exercise 20.11 with those for the two implementations of Section 10.5.

20.13 Using the sample tree of Fig. 20.7, show how to insert elements in such a way that the cost of *IsMember* is $O(n)$ instead of $O(\log_2 n)$.

20.14 Compare the space requirements of the set implementation given in this chapter to the requirements of the vector representation of Section 10.5.2. What assumptions do you have to make about sizes? Using those assumptions, let P be the fraction of *MaxSize* elements that are, on the average, in the set. Assume that obsolete elements are compacted out frequently. For what values of P does the threaded tree use space more efficiently? For what values of P does the vector representation use space more efficiently?

20.15 Assume cells are allocated between addresses 60000 and 70000. Produce a "memory map" that could represent Fig. 20.7. That is, show which words of memory are used in the data structure and what each contains. Remember that words of a record are stored contiguously, but individual records need not be contiguous. Assume the fields are packed so that references occupy 1/2 word, integers occupy 1 word, and booleans occupy 1 bit.

20.16 The tree portion of the data structure developed in this chapter is a binary tree. Change the implementation to use a ternary tree (that is, one in which each node has a left son, a middle son, and a right son). (Hint: Begin by thinking of a discrimination function that will produce equality and *three* other outcomes.)

20.17 Give two reasons for using **cand** instead of **and** in the *IsMember* test. (Hint: Consider the empty set.)

20.18 Run the program of Section 20.4 with each of the three representations. Compare the results.

20.19 Using the techniques of Section 6.5, measure the performance of the program of Section 20.4 with the bit-vector representation. Did you get the results you expected?

20.20 Using the techniques of Section 6.5, predict the performance of the program of Section 20.4 with the bit-vector representation. Compare this prediction of performance to the performance measured in Exercise 20.19.

20.21 Using the techniques of Section 6.4, measure the performance of the program of Section 20.4 with the array representation. Did you get the results you expected? Discuss.

20.22 Using the techniques of Section 6.5, predict the performance of the program of Section 20.4 with the array representation. Compare this prediction to the performance measured in Exercise 20.21.

20.23 Using the techniques of Section 20.4, measure the performance of the program in Section 20.4 with the representation of this chapter. Did you get the results you expected? Discuss.

20.24 Using the techniques of Section 6.5, predict the performance of the program of Section 20.4 with the representation of this chapter. Compare to the measured performance from Exercise 20.23.

20.25 One way to compute the set of primes between 2 and N is to use the method called the "Sieve of Eratosthenes." Figure 20.18 shows a program to do this for $N = 200$. Compare the space utilization of this

```
const N = 200;
var Sieve, Primes: SetofElts; next, j: integer;
begin
Clear(Sieve); Clear(Primes);
for j := 2 to N do Insert(Sieve, j);
next := 2;
repeat
    while (not IsMember(Sieve, next)) do next := next+1;
    Insert(Primes, next);
    j := next;
    while j ≦ N do begin Remove(Sieve, j); j := j + next; end;
until IsEmpty(Sieve);
end.
```

Fig. 20.18 The Sieve of Eratosthenes

program for the set representation of this chapter and the two in Section 10.5. Also compare any restrictions on the size of N.

20.26 Change *FindElt* to a recursive routine.

20.27 Rewrite routines *FindElt* and *Insert* to make the discrimination test directly rather than by calling *Discrim* and using a **case** statement. Compare the readability and efficiency of the two versions.

20.28 The discrimination function, *Discrim*, must have certain properties in order for the *SetofElts* implementation to function properly. For example, it is necessary that

$$Discrim(x, y) = SameClass \text{ iff } x = y.$$

Under what circumstances would the implementation fail to meet its specifications if this property did not hold? Give an example.

Now suppose that the property (1) does hold. What other properties must *Discrim* have in order for the implementation of *SetofElts* to work properly? Give an example of a *Discrim* function for which these properties do not hold and an example of where, as a result, the implementation fails.

20.29 How could we write the routines and data structures implementing *SetofElts* so that by changing *only* the definitions of *DiscrimClass* and *Discrim*, we could change the tree structure to have nodes of any fixed degree we choose? That is, how can we write the program so that going from binary to ternary trees, for example, is only a matter of changing the literals in *DiscrimClass* and the body of *Discrim*?

20.30 Suppose that the discrimination function had the property that

$$Discrim(x, y) = SameClass \text{ iff } x \text{ E } y$$

where E is some equivalence relation (not necessarily $=$). The set-manipulating routines of our implementation would still "almost" work, but their specifications would have to be changed as follows:

1. *Insert*(*S*, *x*) would insert *x* into *S* iff *S* does not already contain another value, *y*, such that $x \text{ E } y$. If *S* does contain such a *y* already, *Insert* would have no effect.
2. *Remove*(*S*, *x*) would remove any value, *y*, from *S* such that $x \text{ E } y$.
3. *IsMember*(*S*, *x*) would yield *true* iff there were any element, *y*, of *S* such that $x \text{ E } y$.
4. *IsSubset*(*S*, *T*) would yield *true* iff for every element, *x*, in *S*, there were an equivalent element, *y*, in *T* (that is, $x \text{ E } y$), and vice-versa.
5. *Union* and *Intersect* would change similarly.

In addition, the *Insert* routine would have to change (see Exercise 20.28). Write the revised specifications of *Insert, Remove, IsMember, IsSubset, Union,* and *Intersect* (that is, the new preconditions and postconditions); revise the *Insert* routine appropriately.

20.31 As a concrete example of the ideas described in Exercise 20.30, consider the problem of implementing a symbol table. A symbol table is a data type used for associating information with symbols. For example, a compiler uses a symbol table to keep track of what identifiers have been declared and what they mean. The identifiers are the symbols, and the information about each identifier (data type, place of declaration, etc.) is associated with the identifier in the symbol table.

In terms of sets, a symbol table is a set of ordered pairs, each pair having the form

<symbol, associated data>

Two pairs are termed equivalent iff their symbol parts match (this is the "E" relation).

Using the implementation of *SetofElts* as revised by Exercise 20.30, and changing *only* the definitions of *EltType* and *Discrim*, make *SetofElts* behave as a symbol table type. The symbols should be strings of four characters, and the associated data should be integers.

Symbol tables are useless without the operation *LookUp*, which is specified

    procedure *LookUp(S: SetofElts; x: EltType;* **var** *y: EltType)*;
      pre *IsMember(S, x)*;
      post $y \mapsto S_M \lor x \mathrel{E} y$;

(where, again, "E" is the equivalence relation between *EltTypes*—"symbols are equal" in the case of symbol tables). That is, *LookUp(S, x, y)* sets *y* to the element of *S* equivalent to *x* (that is, to the element whose symbol matches that of *x*). Implement *LookUp* two different ways:

a) Using only the abstract operations of Section 20.2.1 to manipulate *S*.

b) As another primitive operation, able to use the internal routines of the set implementation and able to manipulate the tree structure of *S* directly.

20.32 There are several problem of efficiency and convenience with using the data structure of Exercise 20.31 for symbol tables. If we wished to change the data associated with a symbol, we would first have to remove the symbol and its associated data and then insert the symbol with its revised data. If there were a large amount of data associated with each symbol, this could be less than optimal. Indeed, it is somewhat costly to reference the data associated with any node, since *LookUp* copies data into its second argument. How could you alleviate this problem without changing the *SetofElts* routines or data structures (aside from defining *EltType*)?

20.33 Suppose that *DiscrimClass* has only two elements (*SameClass* and one other). Give an example of such a function. What kind of tree structure would this choice imply? Is the structure really a tree at all? How else might you describe it?

20.34 Why do we need a *Copy* operation at all? That is, what is wrong with writing

    $S := T$

to copy set *T* into set *S*?

20.35 Argue convincingly that *Compact* in Section 20.5.10 is implemented correctly.

20.36 Argue convincingly that *Union* and *Intersect* (Sections 20.5.8 and 20.5.9) are implemented correctly.

20.37 Argue convincingly that *IsMember* is implemented correctly. Do you need to be concerned with the invariant?

20.38 Analyze the routines of the *SetofElts* type to confirm the cost estimates given in the text.

20.39 Perform a worst-case analysis for the cost of *FindElt*. That is, determine the tree organization that will cause the slowest execution time for *FindElt* and determine what that time is.

20.40 Perform a best-case analysis for the cost of *FindElt*. That is, determine the tree organization that will cause the fastest execution time for *FindElt* and determine what that time is.

Chapter 21

Generators: Writing Loops that Operate on Sets

Previous chapters have paved the way for the subject of this chapter: specifications for finite homogeneous sets were given in Section 9.2.3, and two distinct implementations to fit these specifications were given in Section 10.5; Chapter 20 developed a richer representation based on trees and lists. We now address a new issue about the separation of specifications from implementations—namely, how to write loops that operate on all the elements stored in a data structure—and we investigate a way to write loops that operate on sets without commiting the user's program to a particular choice of implementation.

21.1 Loops that Operate on Set Elements

It is quite common to need to write a loop that operates on each element of a data structure. It is clear how to do this for vectors:

$\quad$ **for** $i := 1$ **to** N **do begin** $\cdots$ $V[i]$ $\cdots$ **end**;

You also know how to write loops that operate on lists:

$\quad$ $p := L$; **while** $p \neq nil$ **do begin** $\cdots$ $p\uparrow.Data$ $\cdots$; $p := p\uparrow.Next$; **end**;

However, we have taken great pains to separate specification of the properties of a type from the actual data structures used for implementing the type. Thus we need to find a way to write loops that sequence through abstract data structures without explicitly referring to the structure.

There are two reasons for wanting to do this. First, your programs will be easier to write and to understand if they use only the operations provided by the type. Second, you may need to change the data representation some day. If all operations are carried out through the established routines, this is reasonably easy to do. If not, it can be quite hard.

The strategy we will describe for sets works well for many kinds of structures. We declare a new data type, *SetGen*, to represent the current status of a loop that is operating on a set. (Intuitively, we know this plays the role of the integer counter in a **for** loop.) Suppose *G* is of type *SetGen*. We define three procedures, *StartSetLoop*, *NextSetElt*, and *StopSetLoop*, intended to be used on a set *S* in the style

> *StartSetLoop*(*G, S*);
> **while** *NextSetElt*(*G, x*) **do**
>   **begin** · · · *operations on x* · · · **end**;
> *StopSetLoop*(*G*);

StartSetLoop initializes *G* in preparation for a loop that will operate on each element of *S*. *NextSetElt* indicates through its *boolean* return value whether the loop is completed. If the loop has not completed, *NextSetElt* returns a new value from set *S* in parameter *x* and updates *G*'s information about the status of the loop. *StopSetLoop* resets *G* to indicate that no loop is currently in progress.

We want the loops to have the following properties, and the specifications for sets must account for them:

- Every element of the set is delivered to the loop body exactly once, provided that the loop runs to completion and makes no changes to the set.
- The loop body is allowed to perform at least the operations on the set that do not modify it.
- It is safe to exit from the loop without waiting for the supply of elements to be exhausted.

A set of procedures such as *StartSetLoop*, *NextSetElt*, and *StopSetLoop*, which together serve to systematically supply values to a loop body, is called a *generator*.

21.1.1 SPECIFICATIONS FOR GENERATOR TYPE

We specify the properties of a *SetGen*, *G*, in terms of G_S, the set whose elements must be delivered to the loop, and G_R, the set of elements that have already been delivered.

Domain: $\{<G_R, G_S> \mid G_R \text{ and } G_S \text{ are sets}\}$

procedure *StartSetLoop*(**var** *G*: *SetGen*; *T*: *Set*);
 post $G = <\{\}, T>$

function *NextSetElt*(**var** *G*: *SetGen*; **var** *x*: *integer*): *boolean*;
 pre *StartSetLoop* must have been called for *G*
 post $RESULT \supset (x \text{ is in } G'_S - G'_R \text{ and } G_R = G'_R \cup \{x\})$, and
  $\sim RESULT \supset G_R = G'_R = G'_S$, and
  $G'_S = G_S$

procedure *StopSetLoop*(**var** *G*: *SetGen*);
 post $G_R = G'_S = G_S$

In other words, *StartSetLoop* initializes the generator to a state in which none of the set elements has been delivered to the loop. Each call on *NextSetElt* delivers an element that hasn't yet been delivered; after all elements have been delivered, the functional return is *false*. *StopSetLoop* terminates the loop by indicating that all elements have been delivered.

To facilitate the implementation, we will later have to impose certain additional restrictions on the use of *SetGen*s. These restrictions will not change the postconditions above, but will have the effect of adding some preconditions.

21.2 A Solution for the Implementation with Threaded Trees

An efficient implementation of the generator for a type will usually need to exploit information about the representation of the type. In the implementation of sets in Chapter 20, each set contains a linear thread; if the *ListThread* links are followed from the header of the set to the end of the thread, each node of the data structure will be visited exactly once. We take advantage of this in the implementaton of set generators.

We maintain information about the status of a loop using a pair of data elements. One of these indicates which set is associated with the generator; the other is a cursor, or pointer, into the linear thread that separates the elements that have been generated from those that have not. The implementations of the operations, then, must do the following things:

- *StartSetLoop*: Record which set is to be processed and set the cursor to indicate that no elements have been generated.

- *NextSetElt*: Scan for the next valid element. If there is one, get its value and return *true*. Otherwise, return *false*.

- *StopSetLoop*: Advance cursor past the last element.

Because the *SetGen* retains a pointer to the representation of the *SetofElts* originally handed to *StartSetLoop*, this choice of strategy will not work if the representation of this set changes during the loop. For example, execution of an *Insert* during a loop over a set might or might not cause the inserted element to come up later in the loop (see Exercises 21.6 and 21.7). Therefore, we'll add an extra precondition to *NextSetElt*:

> **function** *NextSetElt*(**var** *G*: *SetGen*; **var** *x*: *integer*): *boolean*;
> **pre** *StartSetLoop* has been called for *G* and
>   there has been no call on *Insert*, *Remove*, *Union*, or
>   *Intersect* since the last call on *StartSetLoop*
> **post** $RESULT \supset (x$ is in $G'_S - G'_R$ and $G_R = G'_R \cup \{x\})$, and
> $\sim RESULT \supset G_R = G'_R = G'_S$, and
> $G'_S = G_S$

Figure 21.1 depicts the situation when the following program is in the third execution of the inner body and the second execution of the outer loop:

> **var** *V*, *W*: *SetGen*;
> *StartSetLoop*(*V*, *S*);
> **while** *NextSetElt*(*V*, *x*) **do**
>   **begin**
>   *StartSetLoop*(*W*, *S*);
>   **while** *NextSetElt*(*W*, *y*) **do** · · · *LOOP BODY* · · ·
>   *StopSetLoop*(*W*);
>   **end**;
> *StopSetLoop*(*V*);

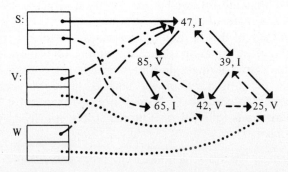

Fig. 21.1 Snapshot of double loop on a set

21.3 Programs for the Solution

21.3.1 THE DATA STRUCTURE

Our data structure consists simply of a record with a pointer, *Beg*, to the first element on the linear thread and a pointer, *Cur*, to the next element to be examined by *NextSetElt*.

> *SetGen* = **record**
> *Cur*: *RefSetElt*;
> *Beg*: *RefSetElt*;
> **end**;

21.3.2 LOOP CONTROL OPERATIONS

The purpose of *StartSetLoop* is to initialize the loop by indicating that no elements have been generated yet. We do this by setting both the *Beg* and *Cur* fields to the beginning of the set's *ListThread*. The code is:

> **procedure** *StartSetLoop*(**var** *G*: *SetGen*; *S*: *SetofElts*);
> {Initialize *ForAllInSet* Loop}
>   **begin** *G.Beg* := *S.ListThread*; *G.Cur* := *G.Beg*; **end**;

StopSetLoop terminates the loop, whether it has run to completion or not. It does this by setting the cursor to *nil*, which is the value it has after the threaded list has been completely traversed.

> **procedure** *StopSetLoop*(**var** *G*: *SetGen*);
> {Terminate *ForAllInSet* loop}
>   **begin** *G.Cur* := *nil*; **end**;

NextSetElt, shown in Fig. 21.2, scans the list thread beginning with the node currently pointed to by the *Cur* field of the generator record. If this is a valid node, its value is returned and the cursor is advanced. If it is not a valid node, the routine advances the cursor and tries again. If the cursor is *nil* initially or if the list is exhausted before a valid element is found, the routine returns the value *false* and does not update the data parameter.

For example, the standard loop

> *StartSetLoop*(*S*);
> **while** *NextSetElt*(*S*, *x*) **do** · · · *LOOP BODY* · · ·
> *StopSetLoop*(*S*);

executed with the example of Fig. 20.7 scans elements in the following order. Only the indicated elements are delivered to the loop body.

function *NextSetElt*(**var** *G*: *SetGen*; **var** *x*: *EltType*): *boolean*;
{Put next valid data element in *x*, leaving *G.Cur* pointing
beyond the element found}
var *R*: *RefSetElt*;
  begin
  R := *G.Cur*;
  NextSetElt := *false*;
  while *R* < > *nil* **do**
    if *R↑.Valid* **then**
      begin
      x := *R↑.Data*;
      NextSetElt := *true*;
      R := *R↑.Thread*;
      goto 1
      end
    else *R* := *R↑.Thread*;
1: *G.Cur* := *R*;
  end; {NextSetElt}

Fig. 21.2 *NextSetElt* implementation

StartSetLoop

  sets *Cur* to head of list

NextSetElt—first call

  [65, *I*] is invalid; skip it

  [85, *V*] is valid; return to loop body with *true*

NextSetElt—second call

  [42, *V*] is valid; return to loop body with *true*

NextSetElt—third call

  [25, *V*] is valid; return to loop body with *true*

NextSetElt—fourth call

  [39, *I*] is invalid; skip it

  [47, *I*] is invalid; skip it

  that was last element on thread; return to loop body with *false*

StopSetLoop

  makes certain that *Cur* = *nil*

21.4 Reimplementation of Set operations

Some of the set operations that were implemented by directly scanning the linear thread could have been written to use generators instead. In this section, we rewrite some of these functions to illustrate the use of generators.

21.4.1 THE OPERATION IsSubset

IsSubset returns the value *true* exactly when every element of its first parameter is a member of its second parameter. The code says just that:

```
function IsSubset(S, T: SetofElts): boolean;
  label 1;
  var x: EltType; G: SetGen;
    begin
    IsSubset := false;
    StartSetLoop(G, S);
    while NextSetElt(G, x) do if not IsMember(T, x) then goto 1;
    IsSubset := true;
 1: StopSetLoop(G);
    end;   {IsSubset}
```

The functional result is initially set to *false.* It can be set to *true* only if the **while** loop exits normally—that is, if every element of *S* is a member of *T.*

21.4.2 THE OPERATION Union

Union is implemented just as you might expect: every element of the second parameter is inserted into the first.

```
procedure Union(var S: SetofElts; T: SetofElts);
  var x: EltType; G: SetGen;
    begin
    StartSetLoop(G, T);
    while NextSetElt(G, x) do Insert(S, x);
    StopSetLoop(G);
    end;
```

Recall that the restrictions prohibit this routine from being called while a loop on *S* is in progress. The reason is that *Insert* is called from this routine, and elements inserted into *S* might or might not be delivered to the loop operating on *S.* That sort of unpredictability is highly undesirable.

Exercises

21.1 Write an implementation of the set-generator type that (a) works and (b) is as faithful as possible to the design given in this chapter. You may use this program in the following exercises.

21.2 Use the implementation of set generators given in this chapter to solve the following problem. Compare the resulting program to the similar program assigned in Chapter 20.

a) Let a class record consist of a class number, a set of prerequisites, and the set of students requesting the course for next semester.

b) Let a student record consist of a student number, the set of courses completed, and the set of courses requested for next semester.

c) Read values for class records and for the student numbers and completed-course parts of student records.

d) Then fill in the courses-requested part of the student records. As you do this, check to see that each student has completed the prerequisites for all courses that he or she has requested.

e) Keep track of the students who have made illegal requests (in a set, of course—you want to see each bad student once only). When processing is complete, print the results.

21.3 Write a print routine that uses a generator to produce elements. As usual, print only the values actually in the set. Note: You do not need to modify the *SetofElts* type; you only need to use a generator.

21.4 Write the loop control procedures for the bit-vector representation of sets given in Section 10.5.

21.5 Write the loop control procedures for the vector–pointer representation of sets given in Section 10.5.

21.6 Change the definition of sets to allow values to be inserted into a set on which a loop is active—that is, for which there has been a *StartSetLoop* with no intervening *StopSetLoop* and for which *NextSetElt* has not yet returned *false*. The problem requires you to be sure that each newly inserted value will be produced later in the loop. You will find that the values passed to procedure *Insert* fall into three classes:

1. *Value was already in the set.* Nothing special needs to be done— this value was already scheduled to be produced sometime during the loop.

2. *Value was never in the set.* In this case, you can be sure the value will be generated if you link it at the tail of the loop thread.

3. *Value was previously in the set, but is now invalid:* In this case, the value may have been passed over already as the loop searched down the thread. However, *Insert* has just switched the *Valid* bit from *false* to *true*, so you know this value has not been generated yet. Change the links in the loop thread so the new value is at the end of the thread (do not change its position in the tree, of course).

In two of these cases you need to operate on the thread in a fashion that is inefficient for singly linked lists. Change the data structure so the thread is *doubly linked* instead of singly linked, and make the necessary changes to allow insertions while a loop is running.

Changing the specification of *SetGen* to reflect this new behavior is rather tricky. What is the problem? How could you fix it?

21.7 Change the loop control procedures to delete elements as they process the set and to remove the nodes from the tree. Explain why it is safe to delete nodes in this way (that is, explain why the data structure remains well-formed). You can now eliminate one of the fields from the set header. Which one? Why? Rewrite the programs to do this.

The following exercises generalize the generator discussion of this chapter and apply it to other types.

21.8 Define a type represented by a linked list of integers. Implement a generator for the integer values in head-to-tail order.

21.9 Define a type represented by a binary tree of elements of type *T*. Implement generators for the values in the tree according to

    a) An LNR traversal.

    b) An NLR traversal.

You are free to design the tree representation to facilitate the generator construction.

21.10 Let two binary trees be defined as equal iff they have identical LNR traversals. Write a function that tests this equality using the generator of Exercise 21.9(b).

21.11 Change the representation of the binary tree of Exercise 21.9 so as to store the nodes consecutively in an array. Redo the LNR generator for the new representation.

21.12 Design a space-efficient representation for sparse matrices, in addition to routines for evaluating and assigning to array elements. Write a generator that delivers the elements of the matrix in row-major order, including zero elements. Write another generator that delivers them in column-major order.

21.13 Use the generators of Exercise 21.12 in the construction of a sparse matrix multiplication routine.

Chapter 22

Formula Manipulation: An Exercise in Defining a Data Type

In Part Two we argued that the use of abstract data types can help solve problems. In this chapter we illustrate such a use of abstract data types in a problem that involves formula manipulation. Our objective here is to write a program that computes the derivative* of a symbolic algebraic expression. We base the program organization on a data type *Expression*.

We attack the problem of formula manipulation by separating specification concerns from implementation concerns. Section 22.1 defines essential terms and states the precise problems to be solved; Section 22.2 then gives an initial specification for the abstract type *Expression*, which then allows us, in Section 22.3, to solve the problems in a natural way; Section 22.4 gives one possible implementation of the type *Expression*; and some considerations of performance, in Section 22.5, lead us to a revised set of specifications and a revised implementation.

22.1 Some Definitions

Intuitively, of course, an algebraic expression is a collection of constants and variables connected with operators such as + or *. For our purposes, we define a simple class of expressions as follows:

---

*Although differentiation is one of the operations of calculus, an understanding of calculus is not necessary for the understanding of this chapter.

Definition 22.1. An *expression* is either

- Undefined,
- A constant (that is, a number),
- A variable (in the mathematical sense, not as in programming languages), or
- A composite, consisting of two expressions (called the left and right operands)—neither of which is the undefined expression—and an operator, which can be addition $(+)$, subtraction $(-)$, multiplication $(*)$, or exponentiation $(\uparrow)$. (For convenience, we are using operator symbols typical of programming languages—or more importantly, of common input and output devices—rather than the standard ones of mathematics.) □

The only unfamiliarity about Definition 22.1 is the inclusion of an undefined expression. We included this to give us a convenient way of describing, for example, the value of expressions such as $0 \uparrow -1$.

We carefully cast Definition 22.1 to be unspecific about how composite expressions are to be written on paper. There are several standard ways to write the composite consisting of the operator α and the two operands u and v:

- *Infix* notation: $u \, \alpha \, v$.
- *Prefix* notation: $\alpha \, u \, v$
- *Postfix* notation: $u \, v \, \alpha$

Although a number of commercial pocket calculators use postfix notation, infix notation is the most familiar. It suffers from problems of ambiguity, however. For example, the expression written

$$a + b * c$$

is either a composite whose first operand is a or a composite whose first operand is $a + b$. We must sometimes resort to parentheses or conventions about order of association (for instance, "associate terms from left to right," or "do multiplication before addition") to resolve such ambiguities. In prefix or postfix notation, no ambiguities occur and parentheses are unnecessary. For example, "$a + (b * c)$" is "$+ a * bc$" in prefix, or "$abc* +$" in postfix, and "$(a + b) * c$" is "$* + abc$" in prefix, or "$ab+ c*$" in postfix.

An expression denotes a function. For each set of constants substituted for its variables, the function denotes a value (or is undefined). Several expressions (an infinite number, in fact) can represent a given function; we say that these expressions are *equivalent*. For example, the expressions "$X + Y$" and "$Y + X + 0$" are equivalent.

For any expression u and variable x, there is an expression called the *derivative* of u with respect to x, and written

$$\frac{du}{dx}.$$

Figure 22.1 gives several rules from the calculus, which give the derivatives for one class of expressions.

1. $\dfrac{dc}{dx} \;=\; 0$

2. $\dfrac{dx}{dx} \;=\; 1$

3. $\dfrac{dy}{dx} \;=\; 0$

4. $\dfrac{d(u+v)}{dx} = \dfrac{du}{dx} + \dfrac{dv}{dx}$

5. $\dfrac{d(u*v)}{dx} = v*\dfrac{du}{dx} + u*\dfrac{dv}{dx}$

6. $\dfrac{d(u\uparrow c)}{dx} = c*(u\uparrow(c-1))*\dfrac{du}{dx}$

where

c is any constant
u and v are any expressions
x and y are any distinct variables.

Fig. 22.1 Some simple derivatives

We now have defined enough terms to state the problems to be solved. We want a set of routines that

a) Read in expressions.
b) Print expressions.
c) Compute the derivatives of expressions.

We should keep in mind that in the future we may want to expand our set of routines to handle other problems having to do with algebraic expressions. In fact, we do just that in Chapter 23.

22.2 The Abstract Type Expression

It is natural to capture the notion of an expression in our programming language as a new data type. Actually, we need several types, since Definition 22.1 also mentions things called constants, variables, and operators. To simplify matters, let's just assume the following definitions:

> **type**
> | | | |
> |---|---|---|
> | *Variable* | = *char*; | {Any letter} |
> | *Constant* | = *integer*; | |
> | *Operator* | = *char*; | {In the set ['+','−','*','↑']} |

We need only the operations of assignment (:=) and comparison (= and ≠) on all three, together with the usual arithmetic operations on constants.

Definition 22.1 defines the domain of type *Expression* and indicates fairly clearly that the data type must meet the following requirements:

1. We should be able to determine whether any *Expression* denotes a constant, a variable, a composite expression, or an undefined expression.

2. We should be able to form a composite *Expression* from any two existing *Expression*s and any operator.

3. For any composite *Expression*, we must be able to determine its operands and operator.

4. There must be a way to denote the undefined expression.

To satisfy the first requirement, we can introduce a new enumerated type, whose values are the names of the possible types of *Expression*:

> *KindNames* = (*Undfnd, Vbl, Cnst, Cmpst*);

Now we can specify a function to determine the variety of any *Expression*:

> **function** *Kind*(*E*: *Expression*): *KindNames*;
> **post** *RESULT* = *Vbl* iff *E* is a variable,
>   *Cnst* iff *E* is a constant,
>   *Cmpst* iff *E* is *Operand*1 *Operator Operand*2,
>   and otherwise *Undfnd*;

The most obvious way to satisfy the requirement that we be able to form composite *Expression*s is to introduce a *constructor*:

> **function** *ConsExpr*(*Oper*: *Operator*; *Operand*1, *Operand*2: *Expression*): *Expression*;
> **post** *RESULT* = the expression "*Operand*1 *Oper Operand*2," if neither operand is undefined, and *Undefined* otherwise;

To satisfy the requirement that we be able to extract operators and operands, we can introduce three functions on *Expression*s:

> **function** *Oper*(*E*: *Expression*): *Operator*;
>   **pre** *Kind*(*E*) = *Cmpst*;
>   **axiom** *Oper*(*ConsExpr*(*a*, *b*, *c*)) = *a*;

> **function** *Operand*1(*E*: *Expression*): *Expression*;
>   **pre** *Kind*(*E*) = *Cmpst*;
>   **axiom** *Operand*1(*ConsExpr*(*a*, *b*, *c*)) = *b*;

> **function** *Operand*2(*E*: *Expression*): *Expression*;
>   **pre** *Kind*(*E*) = *Cmpst*;
>   **axiom** *Operand*2(*ConsExpr*(*a*, *b*, *c*)) = *c*;

Note that the use of axioms provides a convenient and concise description of the values of these functions.

Satisfying the requirement that we be able to denote the undefined expression is easy. We can simply define a constant called *Undefined*. In the same way, we could make *Undefined* a parameterless function.

Definition 22.1 says that variables and constants are expressions. In some programming languages, that definition would cause no problem. Unfortunately, the PASCAL language is strongly typed in such a way that it is not possible, for example, to treat an object both as a variable and as an *Expression*. Therefore, we define four *transfer functions*, which simply return the values of their arguments, but in the appropriate type.

> **function** *VblToExpr*(*V*: *Variable*): *Expression*;
>   **post** *RESULT* = the expression consisting of the variable *V*;

> **function** *ExprToVbl*(*E*: *Expression*): *Variable*;
>   **pre** *KindNames*(*E*) = *Vbl*;
>   **post** *RESULT* = the variable comprising the expression *E*;

> **function** *CnstToExpr*(*C*: *Constant*): *Expression*;
>   **post** *RESULT* = the expression consisting of the single constant *C*;

> **function** *ExprToCnst*(*E*: *Expression*): *Constant*;
>   **pre** *KindNames*(*E*) = *Cnst*;
>   **post** *RESULT* = the constant comprising the expression *E*;

Finally, we allow assignment of *Expression*s. We will not, however, allow tests for equality (=) among expressions, at least at present. Of course, in PASCAL there is nothing in particular to prevent a programmer from writing *X* = *Y* for two *Expression*s *X* and *Y*. We merely mean that the result of that test is not guaranteed to indicate the equality or inequality of X and Y. See Exercise 22.6 for more discussion of equality.

The preceding specifications define, for now, the data type *Expression* and provide a notation for writing them in our dialect of PASCAL. Figure 22.2 gives several examples of how to write expressions in this notation. We have given only specifications in this section, in keeping with the principle of separating abstract and concrete definitions. The abstract definitions alone give us sufficient information to write the routines required in Section 22.1, as we do in the next section. Section 22.4 gives one possible implementation for the type *Expression*.

Infix form	Equivalent Program Expression
x	*VblToExpr*('x')
5	*CnstToExpr*(5)
$3 + (x \uparrow 2)$	*ConsExpr*(' + ', *CnstToExpr*(3),
	ConsExpr('↑', *VblToExpr*('x'), *CnstToExpr*(2)))

Fig. 22.2 Some simple expressions in infix notation and as *Expression*s

22.3 Solutions to the Original Problems

Even without an implementation for the type *Expression*, we can proceed to solutions of the original problems, since we have transcribed all the necessary terms into our programming language.

22.3.1 READING EXPRESSIONS

The problem statement was not specific about the form in which a user is to write expressions for input. As we are mostly concerned with the internal manipulation of *Expression*s in this chapter, let's choose an input format that is convenient for processing. Prefix notation (see Section 22.1) is especially easy for a program to read. Thus, we define

> **procedure** *ReadExpn*(**var** *Where*: *text*; **var** *E*: *Expression*);
>   **post** An expression in prefix notation has been read from the file
>   *Where* into *E*. *E* is set to *Undefined* if the next datum on
>   *Where* was not a valid expression.
>   The file *Where* is advanced past the expression read;

Note that the file containing the expression to be read is a parameter to *ReadExpn*. This has the advantage of making the things affected by *ReadExpn*, including the input file, explicitly visible.

To facilitate the implementation of *ReadExpn*, we first define some simple auxiliary routines, whose implementations are left to the reader.

function *IsCapLet*(*ch*: *char*): *boolean*;
  post *RESULT* ≡ *true* iff *ch* is a capital letter;

function *IsDigit*(*ch*: *char*): *boolean*;
  post *RESULT* ≡ *true* iff *ch* is a digit;

function *IsOperator*(*ch*: *char*): *boolean*;
  post *RESULT* ≡ *true* iff *ch* is an operator symbol;

procedure *ScanPast*(**var** *Where*: *text*; *ch*: *char*);
  post File *Where* positioned to first position such that
     Where↑ ≠ *ch* or *eof*(*Where*);

Now we need only refer to the definition of prefix notation to see that reading a prefix expression involves either reading a simple variable or constant, or reading an operator and then two more prefix expressions. This transcribes almost immediately into a program, as follows:

```
var ch: char; lhs, rhs: Expression; n: Constant;
  begin
  E := Undefined;
  ScanPast(Where,' ');
  if IsOperator(Where↑) then
      begin
      read(Where, ch);
      ReadExpn(Where, lhs);  ReadExpn(Where, rhs);
      E := ConsExpr(ch, lhs, rhs)
      end
  else if IsDigit(Where↑) then
      begin  read(Where, n);  E := CnstToExpr(n)  end
  else if IsCapLet(Where↑) then
      begin  read(Where, ch);  E := VblToExpr(ch)  end
  end;  { ReadExpn }
```

22.3.2 PRINTING EXPRESSIONS

Just for variety, let's print expressions in a different format: infix notation. To ensure that there is no ambiguity, we place parentheses around all composite subexpressions, giving a "fully parenthesized" expression. Printing a composite amounts to printing a left parenthesis, followed by the left operand, the operator, the right operand, and a closing parenthesis. It is, in other words, essentially an NLNRN tree traversal, since we print three things for each composite—a left parenthesis, an operator, and a right parenthesis:

```
procedure PrintExpn(var Where: text; S: Expression);
  post S in parenthesized infix form is appended to file Where;
    begin
    if Kind(S) = Undfnd then write(Where,'undefined') else
    case Kind(S) of
        Vbl:    write(Where, ExprToVbl(S));
        Cnst:   write(Where, ExprToCnst(S): 3);
        Cmpst:
            begin
            write(Where,'(');
            PrintExpn(Where, Operand1(S));
            write(Where, Oper(S));
            PrintExpn(Where, Operand2(S));
            write(Where,')');
            end;
        end;
    end; { PrintExpn }
```

22.3.3 DIFFERENTIATING EXPRESSIONS

The rules of Figure 22.1 tell us precisely how to differentiate a large class of expressions. They do not cover all cases, however—in particular, the case $u \uparrow v$, where v is not a constant. Let's treat the derivatives of expressions not covered by the rules as undefined for now. Elaborations are left as exercises. Thus, we specify

```
function Diff(S: Expression; V: Variable): Expression;
  post RESULT = if S contains no subexpressions of the form
                    E1 ↑ E2, where E2 is not a simple constant,
                  then derivative of S with respect to V
                  else Undefined;
```

The implementation is essentially a simple transcription of the rules given in Figure 22.1:

```
begin
if Kind(S) = Undfnd then Diff := Undefined
  else case Kind(S) of
    Vbl:
      if V = ExprToVbl(S)
        then Diff := CnstToExpr(1)
        else Diff := CnstToExpr(0);
```

```
Cnst:
    Diff := CnstToExpr(0);
Cmpst:
  case Oper(S) of
    '+','−':
      Diff := ConsExpr(Oper(S), Diff(Operand1(S), V),
                  Diff(Operand2(S), V));
    '*':
      Diff :=
        ConsExpr('+',
          ConsExpr('*', Operand1(S), Diff(Operand2(S), V)),
          ConsExpr('*', Operand2(S), Diff(Operand1(S), V)));
    '↑':
      if Kind(Operand2(S)) ≠ Cnst
        then Diff := Undefined
        else Diff :=
          ConsExpr('*', Operand2(S),
            ConsExpr('*',
              ConsExpr('↑', Operand1(S),
                CnstToExpr(ExprToCnst(Operand2(S)) − 1)),
              Diff(Operand1(S), V)))
    end
  end
end;  { Diff }
```

Indeed, the correctness of *Diff* follows precisely *because* it is a transcription of the known rules of calculus, assuming that we implement the type *Expression* correctly.

22.3.4 EXAMPLES

The following is the transcript of a short terminal session with a program that uses the routines *ReadExpn*, *PrintExpn*, and *Diff.* The lines beginning with the symbol > designate input from the user of the program. Each command references one or more expressions, designated by the letters *A*, *B*, and *C*. The expressions are in terms of the variables *X* and *Y*. The command interpreter itself is not shown.

```
> Read A    + * Y ↑ X2 * 14X
A read

> Type A
A = ((Y * (X ↑ 2)) + (14 * X))
```

> Differentiate B A X
B is derivative of A with respect to X

> Type B
$B = (((Y * (2 * ((X\uparrow 1) * {}^.1))) + ((X\uparrow 2) * 0)) + (14 * 1) + (X * 0)))$

> Differentiate C A Y
C is derivative of A with respect to Y

> Type C
$C = (((Y * (2 * ((X\uparrow 1) * 0))) + ((X\uparrow 2) * 1)) + ((14 * 0) + (X * 0)))$

22.4 Implementation of the Expression Data Type

Definition 22.1 strongly suggests a tree-like representation for the type *Expression*, as do many such recursive definitions. There are several kinds of *Expression*—variable, constant, composite, and undefined. This suggests that *Expression* is a *union* of the four. Our program is

> **type**
>   *Expression* = ↑*ExprStruct*;
>   *ExprStruct* =
>   **record**
>   **case** *Kind*: *KindNames* **of**
>   *Vbl*: (*v*: *Variable*);
>   *Cnst*: (*c*: *Constant*);
>   *Cmpst*: (*Op*: *Operator*; *Opnd*1, *Opnd*2: *Expression*)
>   **end**;

Notice that this is practically a transcription of Definition 22.1, the only difference being the introduction of pointers and *KindNames*.

The constant *Undefined* can be represented by a pointer to an *ExprStruct* whose *Kind* field is *Undfnd*. We could make *Undefined* a parameterless function (creating a new *ExprStruct* each time the programmer mentions *Undefined*). We could also implement *Undefined* as a variable, including a special initialization routine for the type *Expression*. A user of the *Expression* type would then call this routine at the beginning of his program. Either of these options would work, but instead let's take advantage of the fact that *Expression*s are represented as pointers, and define

> **const** *Undefined* = *nil*;

It is now a completely straightforward task to provide bodies for the routines of Section 22.2. For example, the body of *ConsExpr* might be:

```
var E: Expression;
  begin
  if (Operand1 = Undefined) or (Operand2 = Undefined)
    then  ConsExpr := Undefined
    else begin
      new(E,Cmpst);  ConsExpr := E;
      E↑.Op := Oper;
      E↑.Opnd1 := Operand1;  E↑.Opnd2 := Operand2;
    end
  end;  { ConsExpr }
```

Before proceeding further, however, let's analyze what we've done.

22.5 A Performance Problem and its Solution

It is fairly easy to calculate the space required by an *Expression* under the representation of Section 22.4. In a simple implementation of our language, for example, where all pointers, integers, and characters take up one word of storage,

- A variable occupies 3 words—one for the pointer, one for *Kind*, and one for *v*.
- A constant occupies 3 words—one for the pointer, one for *Kind*, and one for *c*.
- A composite occupies 3 words—one for the pointer, one for *Kind*, and one for *Op*—in addition to the space occupied by its two operands.

An expression has as many composite subexpressions as it has operators. Therefore, an expression occupies

$$3V + 3C + 3K$$

words, where V is the number of variables, C is the number of constants, and K is the number of operators. Actually, this expression is an upper bound. Since our representation uses pointers, there is a possibility that subexpressions may be *shared*. For example, the expression contained in the variable b after executing

$$a := VblToExpr('x'); b := ConsExpr('+', a, a)$$

which corresponds to the expression $x + x$, occupies only 7 words, rather than 9, because the *ExprStruct* part of a is shared between the operands of b. (Trace through the effect of *ConsExpr* at the end of the preceding section to see this.)

However, the analysis above does not tell everything. Consider the sequence

$$b := ConsExpr(\ \cdots\ );\quad b := ConsExpr(\ \cdots\ )$$

Looking at the implementation of *ConsExpr*, we see that two *ExprStruct*s get created, the first of which promptly becomes useless because it is unreachable. In some programming language implementations, such useless data structures, or garbage, get recycled automatically. Unfortunately, it is not easy to track down unreachable objects in general (the sequence above was not at all a subtle example) and many programming language implementations leave that task to the programmer. Thus, we have a procedure, *dispose*(x), which is essentially the opposite of *new*; it returns the object pointed to by x to the pool of storage available for allocation in future executions of *new* (and woe unto the programmer who uses a pointer after having disposed of the object it pointed to).

Serious applications of formula manipulation systems generate many very large expressions, so we may expect that any vigorous exercise of *Expression*'s routines will create large amounts of garbage. We must, therefore, make provisions for the disposal of useless *Expression*s in implementing the data type. It would be nice if we could entirely hide the existence of garbage from the user.* After all, in such simple types as *integer*, old values do not take up any space at all. Unfortunately, here we run into a limitation of our language; we do not have a way to keep track of which *Expression*s have become useless (see Exercise 22.16). There seems to be no alternative but to make the user aware that old instances of *Expression*s take up space, and to provide a procedure for cleaning them up, such as

> **procedure** *Release*(E: *Expression*);
>   **pre** The instance E must be part of no other existing
>   *Expression* instance;
>   **post** The instance E' no longer exists;

We say "instance" because the same expression can be generated many times, and we only destroy one such "generation" at a time.

When we *Release* an *Expression*, we naturally want to *Release* its

---

* The term "user" is notoriously ambiguous and vaguely condescending. It generally means those programs and programmers that will use the routines (or whatever) being implemented. Thus, in this section, it means whoever or whatever uses *Expression*s without caring about their implementation.

operands as well. This suggests the following simple implementation for
Release:

```
begin
if  E↑.Kind = Cmpst then
    begin  Release(E↑.Opnd1);  Release(E↑.Opnd2)  end;
dispose(E)
end;  { Release }
```

Unfortunately, we soon get into trouble. For example, trace the execution
of

$$b := ConsExpr('+', a, a); \quad Release(b)$$

using the implementation of *Release* above and the implementation of
ConsExpr from the last section. You will find that after the storage pointed
to by *a* gets disposed of, the program attempts to use the pointer again,
which, as we suggested, has dire consequences.

An obvious solution is to make copies, or new instances, of *a* in the
example above, to avoid the problem of sharing. That is, we define

```
function  Copy(E: Expression): Expression;
    post RESULT = a new instance of E;
```

in the specifications of *Expression*, and then insist that the user write

$$b := ConsExpr('+', a, Copy(a))$$

Rewriting the routine *Diff* to allow us to take advantage of the ability
to *Release* an *Expression* results in the program shown in Figure 22.3. Note
that we have carefully made sure that *Diff* includes only copies of existing
instances of *Expression*s and that it always returns a new instance of an
Expression. Thus, there is no sharing, and *Release*s are always safe.

It is now safe to *Release Expression*s returned by *Diff*, but we have
paid dearly for the privilege. Sharing *Expression*s did serve a purpose, after
all—it decreased the amount of space required. Now we are throwing away
that advantage. Not only that, but copying, at least in a naive
implementation where we actually recreate the entire *Expression*, takes
time proportional to the number of operators in the *Expression* to be
copied. That means that all of the *ConsExpr*s performed inside *Diff* now
take a good deal longer than they did. It seems unfortunate to have to pay
such a price in space and time just to keep garbage from accumulating.

Perhaps there is an alternative. Note that no operation ever *changes*
an *Expression* once it is formed; we only create new *Expression*s. That
means that an *Expression* is permanently indistinguishable from its copies.
As a result, we can use the *same* instance of an *Expression* to represent all

```
    begin
    if Kind(S) = Undfnd then Diff := Undefined
      else case Kind(S) of
        Vbl:
          if V = ExprToVbl(S)
            then Diff := CnstToExpr(1)
            else Diff := CnstToExpr(0);
        Cnst:
          Diff := CnstToExpr(0);
        Cmpst:
          case Oper(S) of
            '+','-':
              Diff := ConsExpr(Oper(S),
                              Diff(Operand1(S), V),
                              Diff(Operand2(S), V));
            '*':
              Diff := ConsExpr('+',
                              ConsExpr('*', Copy(Operand1(S)),
                                          Diff(Operand2(S), V)),
                              ConsExpr('*',Copy(Operand2(S)),
                                          Diff(Operand1(S), V)));
            '↑':
              if Kind(Operand2(S)) ≠ Cnst
                then Diff := Undefined
                else Diff :=
                    ConsExpr('*', Copy(Operand2(S)),
                      ConsExpr('*',
                        ConsExpr('↑', Copy(Operand1(S)),
                          CnstToExpr(ExprToCnst(Operand2(S))-1)),
                        Diff(Operand1(S), V)))
          end
      end
    end;  { Diff }
```

Fig. 22.3 Implementation of *Diff* using *Copy*

copies of it, just as long as we count how many copies there are supposed
to be and carefully retain the space for that instance until *all* copies have
been *Released*.

This rather clever technique is one form of what is known as *reference
counting*. To apply it to *Expressions*, we first change the definition of
ExprStruct so that each instance of an *Expression* contains a count of how

many copies of it have been made and not yet *Released*:

type
    ExprStruct =
        record
            CopyCount: *Integer*;
            case *Kind*: *KindNames* **of**

                .   .

                .   .

                .   .

Each newly created *ExprStruct* has a *CopyCount* of 0—it has not been copied. The *Copy* operation simply increases this count and the *Release* operation decreases the count. *Release* does not actually perform a *dispose*, or *Release* the operands of an *Expression* instance, unless there are no further copies of it. See Figure 22.4 for the final versions of *Release* and *Copy*.

   Of course, reference counts take up space. It is not immediately clear that we have saved any space with this fancy scheme, unless we write programs that do a lot of sharing. Indeed, it is also not clear that we have saved time, as we must take time to initialize reference counts each time we create a new *Expression*. Exercise 22.17 takes up this question in more detail. It is always a good idea to display a healthy skepticism toward such innovations as reference counts and to do some analyses before embracing them wholeheartedly.

   Finally, we should mention that we have used the name *CopyCount* rather than, say *RefCount*, because our use of reference counting is not quite standard. See Exercise 22.16 for more details.

22.6 Summary

This chapter illustrates several of the points we made in this textbook. We took a "domain of discourse"—algebraic expressions—and showed how the domain immediately suggested the definition of an abstract data type. This abstract type, in turn, allowed us to use something resembling the terminology of that domain of discourse to solve naturally certain problems within that domain. When it came time to implement the abstract type, we discovered performance problems that, together with limitations imposed by the programming language, led to a revision of the original specifications. The discussion of *Copy*, meanwhile, provided an example of the advantages of *separation of concerns*—in this case, separation of the implementation of a data type from its use. The routine *Diff* would have been identical whether we had chosen to use reference counting or had made actual physical copies.

```
function ConsExpr(Oper: Operator;  Operand1, Operand2: Expression):
                    Expression;
   var E: Expression;
   begin
   if (Operand1 = Undefined) or (Operand2 = Undefined)
      then ConsExpr := Undefined
      else begin
         new(E, Cmpst);  ConsExpr := E;  E↑.CopyCount := 0;
         E↑.Op := Oper;
         E↑.Opnd1 := Operand1;  E↑.Opnd2 := Operand2;
         end
   end; { ConsExpr }
function Copy(E: Expression): Expression;
   begin
   Copy := E;
   if E ≠ Undefined then E↑.CopyCount := E↑.CopyCount + 1
   end; { Copy }
procedure Release(E: Expression);
   begin
   if E ≠ Undefined then
      if E↑.CopyCount = 0 then
         begin
         if E↑.Kind = Cmpst then
            begin Release(E↑.Opnd1);  Release(E↑.Opnd2) end;
         dispose(E)
         end
      else E↑.CopyCount := E↑.CopyCount − 1;
   end; { Release }
function VblToExpr(V: Variable): Expression;
   var E: Expression;
   begin
   new(E, Vbl);  VblToExpr := E;  E↑.CopyCount := 0;
   E↑.v := V
   end; { VblToExpr }
function CnstToExpr(C: Constant): Expression;
   var E: Expression;
   begin
   new(E, Cnst);  CnstToExpr := E;  E↑.CopyCount := 0;
   E↑.c := C
   end; { CnstToExpr }
```

Fig. 22.4 Revised versions of procedures with *Copy* and *Release*

Exercises

22.1 Create a complete system from the routines of the chapter by implementing the routines *ScanPast*, *IsCapLet*, and so on, and writing a main program to serve as a driver, using *ReadExpn*, *PrintExpn*, and *Diff* to read expressions and compute their derivatives.

22.2 Write a function to substitute a given *Expression* for each occurrence of a given variable in an *Expression*. That is, implement

> **function** *Substitute*(*R*: *Expression*; *V*: *Variable*; *E*: *Expression*):
>   *Expression*;
>   **post** *RESULT* = the *Expression E* with all occurrences of the
>   variable *V* replaced with *R*;

22.3 Write a function that computes the value of an *Expression* that contains only constants (no variables), and returns *Undefined* for any nonconstant *Expression*.

22.4 In *Diff*, we required that in all subexpressions of the form $u \uparrow v$, v had to be constant. The rules would still work, however, if v were any *Expression* containing no variables. Rewrite *Diff* to take advantage of this fact (see Exercise 22.3).

22.5 Extend *Expression* and *Diff* to handle division.

22.6 Implement a predicate *Equal* with two arguments that is true iff its two arguments are "the same." What possible interpretations might one attach to this specification? Is it unambiguous? Explain. What is wrong with using PASCAL's (=) operator to implement *Equal* directly?

22.7 Rewrite *PrintExpn* so that it prints only as many parentheses as necessary to avoid ambiguity. To make things easier, assume that by convention, operators are always applied left-to-right in the absence of parentheses.

22.8 The *ReadExpn* routine as written always interprets the minus sign (−) as the operation of subtraction, making it difficult for us to enter negative numbers directly. Devise an alteration to *ReadExpn* to get around this difficulty.

22.9 Rewrite *PrintExpn* to produce prefix notation in a form acceptable to *ReadExpn*.

22.10 Rewrite *ReadExpn* to read fully parenthesized infix notation as produced by *PrintExpn*.

22.11 We allow only binary operators, and not such things as negation $(-x)$ or trigonometric functions (sin x), which are unary operators. Extend *Expression* to allow some set of unary operators in addition to the existing binary ones. Remember that the routines must be modified to take the existence of unary operators into account and that appropriate routines must be added to allow the construction of *Expression*s containing unary operators and also to distinguish such *Expression*s and extract their components. Rewrite *Diff*, *ReadExpn*, and *PrintExpn* to accommodate the extended type.

22.12 Write a routine which, given an *Expression*, returns an equivalent *Expression* in which all subexpressions of the form $0 * u$ and $u * 0$ have been replaced by 0.

22.13 Consider the problem of reimplementing *Expression* using lists or vectors of symbols (constants, variables, and operators) as the representation, rather than the recursive structure developed in this chapter. Under this new representation, an *Expression* would be represented in, for example, prefix notation. Rewrite the appropriate definitions and routine bodies to effect this change. The bodies of *Diff*, *ReadExpn*, and *PrintExpn* should not have to change at all to work under the new representation.

22.14 Section 22.5 gave an upper bound on the space required by an *Expression*. With sharing, the actual space used can be lower. How much lower? In other words, what is the lower bound on the amount of space required by an *Expression* with V variables, C constants, and K operators?

22.15 Another way to compute the space used by an *Expression* in the absence of reference counts and sharing is as follows:

- Each variable or constant accounts for two words (*Kind* and v, or c.)
- Each composite operator accounts for four words (*Kind*, *op*, and the pointers to the two operands.)
- Finally, there is one pointer to the entire *Expression*.

For an *Expression* with V variables, C constants, and K operators, this works out to

$$2V + 2C + 4K + 1$$

words. Doesn't this contradict the analysis of Section 22.5? Discuss. Can you come up with any other expressions for the space used? Explain or demonstrate.

22.16 This exercise is an attempt to see whether we can relieve users of the need to keep track of what *Expression*s should be *Release*d. Observe that every time we do an assignment statement, such as

$$E1 := E2$$

there is one less pointer to the instance of the *Expression* previously in $E1$, and one more pointer to the instance denoted $E2$. Also, whenever we compute

$$ConsExpr(\cdots , E1, E2)$$

there is one more pointer to the instances denoted $E1$ and $E2$. Suppose we were to define our own procedure for doing assignments between *Expression*s (in place of $:=$). Could we implement it in such a way, and perhaps reimplement the other routines of *Expression*, so that the user no longer had to explicitly *Copy* and *Release Expression*s? Explain. What problems remain, and how might one get around them? What additional language features would aid you, as the implementor of *Expression*, in this effort?

22.17 Analyze the reference count scheme to determine the space requirements of an *Expression*. Attempt to quantify the effects of sharing. Perform some empirical tests on the amount of sharing introduced by the action of *Diff* on various test data. Under exactly what circumstances, if any, is reference counting more advantageous than physical copying?

22.18 Try to find a smaller representation for *Expression*s. Using PASCAL's **packed** facility is one possibility; find out what advantage(s) this gains for your implementation; or, if you are not using PASCAL, find out how much space you can save by adapting some packing scheme. Do the answers to Exercise 22.17 change at all? In what way(s)?

Chapter 23

Production Systems and Simplification

This chapter continues the study of the type *Expression* begun in Chapter 22. There, we developed a sort of miniature language, embedded in our programming language as a data type, for talking about algebraic expressions. In this chapter, we confront a class of problems that leads us to develop a more elaborate and very powerful language for transforming expressions. The problem is to devise a simplifier for expressions—a program that transforms expressions into simpler equivalent forms. The language we develop is a form of what is often called a *production language*.

Consider a typical polynomial, such as

$$3X^3 + 2X^2 + X + 6,$$

which could be read by *ReadExpn* in prefix notation as

$$+ + + *3\uparrow X3 *2\uparrow X2\ X 6$$

into an *Expression* variable, say *A*. If after reading this polynomial, we were then to execute

PrintExpn(Diff(Diff(A, X), X))
{ Print the result of differentiating *A* twice. }

the present version of *Diff* would produce something like

```
(((((( 3*(( 3*(((X↑ 2)*0) + ( 1*( 2*((X↑ 1)* 1))
))) + (((X↑ 2)* 1)*0)))) + (( 3* ((X↑ 2)*1)) * 0)
) + (((X↑ 3)* 0) + ( 0*( 3*((X↑ 2)* 1))))) + (((
```

$$2*((\ 2*(((X\uparrow 1)* \ 0)+(\ 1*(\ 1*((X\uparrow 0)* \ 1)))))$$
$$+(((X\uparrow 1)* \ 1)* \ 0)))+((\ 2*((X\uparrow 1)* \ 1))* \ 0))$$
$$+(((X\uparrow 2)* \ 0)+(\ 0*(\ 2*((X\uparrow 1)* \ 1))))))+ \ 0)+ \ 0)$$

But this is ridiculous; a person doing this computation would get

$$18X + 4$$

or in *PrintExpn*'s notation,

$$((18 * X) + 4).$$

The difference, of course, results because reasonable people *simplify* as they go along, reducing each intermediate result to a simpler equivalent form. For example, they would automatically replace any expression of the form "0 * something" with 0, or "1 * something" with just the "something" part. Not only do unsimplified *Expression*s take up space, but they are almost useless to users of the program.

An obvious approach to the problem is to write a routine that looks for subexpressions that could be simplified (using the routines supplied in the data type *Expression*) and replaces them as appropriate. As long as the number of such simplifications is small, we can write such a routine that tests for each simplification with some **if** statements and that contains code to make the changes. But when the number of simplifications becomes large (as it would in a real formula-manipulating system), such a routine becomes rather clumsy, both to extend with further simplifications and to understand. Therefore, we might consider developing a notation specially tailored to the task of describing simplifications. Then we can build a set of routines that interpret the notation and cause the changes to be made as described. Together with a set of simplification rules written in the notation, these constitute a simplifier.

23.1 Production Systems

Let's take another look at a typical simplification:

"0 * something" simplifies to "0."

This rule fits a fairly general scheme:

<pattern> simplifies to <replacement>.

By <pattern> we mean "form" or "general structure." Most simplification rules can be written in just this way. For example,

"1 * something" simplifies to "something";

"0 + something" simplifies to "something";

"something ↑ 1" simplifies to "something."

A set of rules of this form is one example of a production system—any general scheme in which actions (in our case, replacements of subexpressions) are controlled by the recognition of patterns within the data. Each pattern/replacement pair is called a *production*.

Production systems provide a *language*, in effect, for describing transformations of formulas, as well as many other things. Let's try to be a little more rigorous about what this language is and how we are to use it. To do so, we must consider

1. What patterns we must be able to specify and what notation will be used to describe patterns.
2. What pattern replacements we must be able to specify and what notation will be used to describe replacements.
3. The order in which the set of productions will be applied.

23.1.1 PATTERNS

The long example near the beginning of this chapter serves as an excellent source of possible patterns. A little examination reveals the following patterns of expressions that could be simplified (u denotes any expression, and $c1$ and $c2$ any constants):

$u \uparrow 1, \quad u \uparrow 0,$

$1 * u, \quad u * 1, \quad 0 * u, \quad u * 0,$

$0 + u, \quad u + 0,$

$c1 * (c2 * u).$

All of these simplifiable expressions are composed of just a few kinds of components:

- A particular constant (for example, 0 or 1),
- A pattern matching any constant,
- A pattern matching any expression, or
- A composite consisting of two smaller patterns and an operator.

To give us some notation for defining patterns, let

- $\&A$ denote the pattern matching any expression, and

- &*C* denote the pattern matching any constant,

so that the patterns corresponding to the expressions above are written

$$\&A \uparrow 1, \quad \&A \uparrow 0$$

$$1 * \&A, \&A * 1, 0 * \&A, \&A * 0$$

$$0 + \&A, \&A + 0$$

$$\&C * (\&C * \&A)$$

23.1.2 REPLACEMENTS

Having matched an expression to a pattern, we can describe most of the replacements we might want either as

- A particular constant,
- Some subexpression of the expression matched by the pattern, or
- A composite consisting of an operator and two smaller replacements.

We can already specify constants and composites just as we do for patterns and ordinary expressions. All we need is some notation for designating the subexpressions of an expression. One possibility is to find a convention for numbering the subexpressions of any expression. For example, we can use a breadth-first numbering, defined as follows: $\&Si$ designates subexpression number i of the expression, E, that was matched, where

- E itself is $\&S1$.
- If $\&Si$ is a subexpression of E, then its left operand is $\&S(2i)$ and its right is $\&S(2i + 1)$.

For example, in the expression $(A + B) * (C + D)$, we have

$\&S1$ is $(A + B) * (C + D)$	$\&S5$ is B
$\&S2$ is $(A + B)$	$\&S6$ is C
$\&S3$ is $(C + D)$	$\&S7$ is D
$\&S4$ is A	

Note the resemblance to the way we stored full binary trees in vectors (see Section 10.4.3).

Now, for example, we can write the productions

$$1 * \&A \quad \text{simplifies to} \quad \&S3$$

$$\&A * 1 \quad \text{simplifies to} \quad \&S2$$

We would also like to perform another simplification, evaluation, and write something like

    $\&C * \&C$ simplifies to the value of $\&S2$ times $\&S3$, a single constant.

A simple extension that covers this case is the definition

- $\&Vi$ denotes the value of the subexpression numbered i, which must be of the form $\&C$ operator $\&C$.

For example, one application (see Section 23.1.1) could be

    $\&C * (\&C * \&A)$ simplifies to $K * \&S7$, where K is the product of the two constants.

This simplification is the result of *two* other productions:

    $\&C * (\&C * \&A)$ simplifies to $(\&C * \&C) * \&A$

and

    $\&C * \&C$ simplifies to $\&V1$.

The idea of using two productions in place of one can simplify both our "production language" and our set of rules.

23.1.3 ORDERING OF PRODUCTIONS

In the expression

    $(0 * 3) + (1 * X),$

several of the productions we've discussed are applicable, which raises the question of the order in which productions are to be applied. There are actually two questions involving ordering:

1. When several subexpressions can be processed by one or more productions, which gets processed?
2. When several productions apply to the same subexpression, which production gets applied?

It is not at all clear how these questions should be answered, and the answers can make a good deal of difference to the performance or correctness of a set of productions. For example, consider the following two rules:

    $\&C + \&C$ simplifies to $\&V1$

    $\&A + \&C$ simplifies to $\&S3 + \&S2$

The second production ensures that constants in an addition are moved to the front of the expression. As long as the first production is always applied before the second, this production system always works and eventually finishes (with no further transformations possible). However, if the second production were always applied when applicable, in favor of the first, the system would never terminate on expressions such as $2 + 3$.

Consider the efficiency of a set of rules such as

$\&A * (\&A + \&A)$ simplifies to $(\&S2 * \&S6) + (\&S2 * \&S7)$.

$\&C + \&C$ simplifies to $\&V1$.

$\&C * \&C$ simplifies to $\&V1$.

The first production is simply the distributive law of multiplication over addition. If in simplifying the expression

$$15 * (1 + 6)$$

we were to consider applying rules to the whole expression before looking at any of its subexpressions, we would have to apply four productions before getting the final result of 105:

$$15 * (1 + 6) \rightarrow (15 * 1) + (15 * 6) \rightarrow 15 + (15 * 6) \rightarrow$$

$$15 + 90 \rightarrow 105.$$

If we consider the subexpression $1 + 6$ first, on the other hand, we only need to apply two productions:

$$15 * (1 + 6) \rightarrow 15 * 7 \rightarrow 105.$$

For our purposes, we will select an algorithm for applying productions that is not very efficient, but that serves to illustrate the basic ideas. First, we choose to *order* the productions so that when two productions apply to the same expression, we apply the first. Second, we fix the order in which subexpressions are processed by defining the process of "applying all possible productions" as

repeat
    if the expression is a composite **then**
        begin
        Apply all possible productions to left operand;
        Apply all possible productions to right operand
        end;
    repeat
        Apply first applicable production to whole expression
    until No more productions apply
until No productions were applicable to the whole expression

In other words, we have an iterated LRN traversal, in which subexpressions are treated first, and then the entire expression. The entire process is repeated until no changes are made in a complete pass, because a change to the entire expression may change one of its operands, requiring that the operand be processed again. We've already seen an example where reprocessing is necessary; it involved the two rules:

$\&C * (\&C * \&A)$ simplifies to $(\&C * \&C) * \&A$

$\&C * \&C$ simplifies to $\&V1$

The algorithm just described guarantees that at termination (if in fact the algorithm does terminate), no more productions apply anywhere in the expression being processed. The following argument based on structural induction demonstrates this:

1. Processing of a noncomposite expression terminates only when no further productions can be applied. (This is the basis step.)
2. Assuming that the algorithm applies all possible productions to the operands of a composite (the inductive hypothesis), we conclude that processing of a composite expression does not terminate until no productions apply to the operands and no productions apply to the expression itself.

Proving the termination of this algorithm is, in general, not possible; the algorithm terminates for some sets of productions, and for others it does not.

23.2 Extending the Type Expression with Patterns and Replacements

Our purpose now is to add enough machinery to the programs of Chapter 22 so that we can write a set of routines to do simplifications in a convenient "production language." We could start by defining types *Pattern* and *Replacement*, but these seem so similar to the type *Expression* that it seems a shame to duplicate effort. Therefore, let's simply *extend* the type *Expression* so that it encompasses patterns and replacements. Then we can define a routine we call *ApplyProductions* that implements the algorithm described in Section 23.1.3.

23.2.1 EXTENSIONS TO THE SPECIFICATIONS OF *EXPRESSION*

To extend the domain of *Expression* to include patterns (generalizations of $\&A$ and $\&C$) and replacements (generalizations of $\&Si$ and $\&Vi$), we must introduce constants and constructors for these objects.   The following, together with the existing *ConsExpr*, should suffice:

> **function** *MatchAny: Expression*;
>   **post** *RESULT* = the pattern matching any *Expression*;

> **function** *MatchAnyCnst: Expression*;
>   **post** *RESULT* = the pattern matching any constant;

> **function** *ReplSubExpr*(*n: integer*): *Expression*;
>   **post** *RESULT* = the replacement specifying subexpression *n* (&*Sn*);

> **function** *ReplValSubExpr*(*n: integer*): *Expression*;
>   **post** *RESULT* = the replacement specifying the value of
>   subexpression *n* (& *Vn*);

The primitive operations must allow us to determine if a given pattern matches a given expression and to form the expression specified by a given replacement pattern:

> **function** *Match*(*E*, *P*: *Expression*): *boolean*;
>   **pre** *E* must contain no pattern or replacement parts (&*A*, &*C*, &*S*,
>   or & *V*);
>   **post** *RESULT* = *true* iff pattern *P* matches expression *E*,
>   *false* if *P* is not a proper pattern;

> **function** *Replace*(*E*, *R*: *Expression*): *Expression*;
>
>   **pre** *E* must not contain pattern or replacement parts.
>   *R* must contain no pattern parts (&*A* or &*C*);
>   **post** *RESULT* = the replacement for *E* specified by *R*;

23.2.2 THE ROUTINE *APPLYPRODUCTIONS*

The data type *Expression* has now been extended enough that we can use it to implement the production system scheme discussed in Section 23.1.3. First, we must define a production to be two patterns, or, *Expression*s:

> **type**
>   *Production* = **record** *Patn, Repl: Expression* **end**;

In Section 9.1.4.1 we saw how to create the type *listofProduction*, meaning an ordered list of *Production*s, given the type *Production*.   The type *listofProduction* provides appropriate routines for manipulating lists (although unfortunately, limitations of PASCAL have forced us to modify

FirstItem to be a procedure rather than a function.)   We also add the
definition

  type *RefListofProduction* = ↑ *listofProduction*;

The algorithm of Section 23.1.3 now is:

```
function ApplyProductions(E: Expression; L: RefListofProduction):
                              Expression;
    pre  E contains no &A, &C, &S, or &V subexpressions.  L is a list
         of valid productions (that is, valid patterns
         and valid replacements);
    post RESULT = result of applying ordered productions L to E,
                  iteratively in LRN order until no further
                  productions apply;
    var Changed: boolean;  NewExpr, PrevExpr: Expression;
        RemainingProds: RefListofProduction;  P: Production;
    begin
    NewExpr := Copy(E);
    repeat
        Changed := false;
        if Kind(NewExpr) = Cmpst
          then begin
            PrevExpr := NewExpr;
            NewExpr := ConsExpr(Oper(NewExpr),
                          ApplyProductions(Operand1(NewExpr), L),
                          ApplyProductions(Operand2(NewExpr), L));
            Release(PrevExpr)
          end;
        RemainingProds := L;
        while not IsEmpty(RemainingProds) do
          begin
          FirstItem(RemainingProds, P);
          RemainingProds := Tail(RemainingProds);
          if Match(NewExpr, P.Patn) then
              begin
              PrevExpr := NewExpr;
              NewExpr := Replace(NewExpr, P.Repl);
              Release(PrevExpr);
              Changed := true;  RemainingProds := L
              end
          end
    until not Changed;
    ApplyProductions := NewExpr
    end;  { ApplyProductions }
```

Note that, as in the final version of *Diff*, we are careful to make sure that *ApplyProductions* returns a totally new *Expression* with judicious use of *Copy*. Also, we are careful to eliminate partial results with *Release*.

23.2.3 IMPLEMENTING THE EXTENDED *EXPRESSION* TYPE

It only remains to implement the extensions made to *Expression* in Section 23.2.1. First, it's pretty clear what extensions are needed to the data structure:

```
type
    Expression = ↑ExprStruct;
    KindNames  = (Undfnd, Vbl, Cnst, Cmpst,
                        Any, AnyCnst, SubExpr, ValSubExpr);
    ExprStruct =
       record
           CopyCount: integer;
           case Kind: KindNames of
               Vbl:   (v: Variable);
               Cnst:  (c: Constant);
               Cmpst: (Op: Operator;  Opnd1,Opnd2: Expression);
               SubExpr, ValSubExpr: (n: integer)
       end;
```

The new *KindNames* (*Any*, *AnyCnst*, *SubExpr*, and *ValSubExpr*) correspond directly to the new constructors. The new field named *n* in *ExprStruct* corresponds to the subexpression numbers accompanying &S and &V replacements. We leave the implementations of the new constructors to the reader, as they are immediate.

A simple case analysis of the possible types of patterns leads almost directly to an implementation of *Match*.

```
function Match(E, P: Expression): boolean;
    begin
    if E = Undefined then Match := false
    else case Kind(P) of
        Vbl, SubExpr, ValSubExpr:
                Match := false;
        Cnst:   Match := (E↑.c = P↑.c);
        Any:    Match := true;
        AnyCnst: Match := (Kind(E) = Cnst);
        Cmpst:  if Kind(E) ≠ Cmpst
                    then Match := false
                    else Match :=
                            (Oper(E) = Oper(P))
```

> **and** *Match(Operand1(E), Operand1(P))*
> **and** *Match(Operand2(E), Operand2(P))*
>   **end**;
>   **end**; { Match }

To implement *Replace*, it is useful to have some utility routines to evaluate constant expressions and to fetch subexpressions by number. These are shown in Fig. 23.1. With them, the body of *Replace* is

```
function Replace(E, R: Expression): Expression;
  begin
    case Kind(R) of
    Cnst:  Replace := Copy(R);
    Cmpst: Replace :=
            ConsExpr(Oper(R),
                        Replace(E, Operand1(R)),
                        Replace(E, Operand2(R)));
    ValSubExpr:
          Replace :=
              CnstToExpr(Eval(ExtractExpr(E, R↑.n)));
    SubExpr:
          Replace := Copy(ExtractExpr(E, R↑.n))
    end
  end;  { Replace }
```

23.3 A Simplification Routine

Our original task was the simplification of *Expression*s. We can do that now with a simple routine such as

```
function Simplify(E: Expression): Expression;
  pre  E must contain no &A, &C, &S, or &V subexpressions;
  post RESULT = simplification of E;
  begin  Simplify := ApplyProductions(E, SimpList)  end;
```

where the list *SimpList* has been appropriately initialized beforehand. One way to initialize *SimpList* is to extend *ReadExpn* (see Section 22.3.1) so it reads the extended *Expression*s discussed in this chapter (an easy task). Then, we can place the desired simplification rules in an *initialization file* to be read by an appropriate routine at the beginning of the entire program. (Note that the design decision to make the input file an argument to *ReadExpn* makes it easy to use *ReadExpn* for both initialization and normal use.)

```
function Exponentiate(a, b: Constant): Constant;
  pre a to the b must be integral;
  post RESULT = a to the b;
  { Body left to the reader. }

function Eval(E: Expression): Constant;
  pre E is a Cmpst whose operands are constants;
  post RESULT = value of E;
  begin
  case Oper(E) of
    '+':
      Eval := ExprToCnst(Operand1(E))
              + ExprToCnst(Operand2(E));
    '−':
      Eval := ExprToCnst(Operand1(E))
              − ExprToCnst(Operand2(E));
    '*':
      Eval := ExprToCnst(Operand1(E))
              * ExprToCnst(Operand2(E));
    '↑':
      Eval := Exponentiate(ExprToCnst(Operand1(E)),
              ExprToCnst(Operand2(E)))
    end
  end;  { Eval }

function ExtractExpr(E: Expression; n: integer): Expression;
  post RESULT = subexpression of E indexed by n,
                or Undefined if none;
  var parent: Expression;
  begin
  ExtractExpr := Undefined;
  if n = 1 then ExtractExpr := E
  else begin
    parent := ExtractExpr(E, n div 2);
    if parent ≠ Undefined then
      if Kind(parent) = Cmpst then
        if odd(n)
          then ExtractExpr := Operand2(parent)
          else ExtractExpr := Operand1(parent)
    end
  end;  { ExtractExpr }
```

Fig. 23.1 Auxiliary functions for Replace

The following initialization file covers the simplifications discussed in this chapter and serves to reduce the horrendous example at the beginning to its simplest form.

a)	$+\&C\&C$	$\&V1$	{ Do constant additions }
b)	$*\&C\&C$	$\&V1$	{ Do constant multiplications }
c)	$-\&C\&C$	$\&V1$	{ Do constant subtractions }
d)	$\uparrow\&C\&C$	$\&V1$	{ Do constant exponentiations }
e)	$*\&A\&C$	$*\&S3\&S2$	{ Constants in front in products }
f)	$+\&A\&C$	$+\&S3\&S2$	{ Constants in front in sums }
g)	$\uparrow\&A0$	1	{ Exponentiation by 0 }
h)	$\uparrow\&A1$	$\&S2$	{ Exponentiation by 1 }
i)	$*1\&A$	$\&S3$	{ Multiplication by 1 }
j)	$*0\&A$	0	{ Multiplication by 0 }
k)	$+0\&A$	$\&S3$	{ Addition of 0 }
l)	$*\&C*\&C\&A$	$**\&S2\&S6\&S7$	{ Reassociate constant products }

The productions are in prefix form, as used by *ReadExpn*. Each line contains a pattern followed by the corresponding replacement. The letters to the left of each production serve only for identification. Descriptive comments follow each production. Note the use of commutativity to reduce the number of rules (the constants 0 and 1 can be assumed to be the left operands.) Also, note that we depend on the order of application, as discussed above.

To see some of these productions in action, consider their application to the expression

$$(((3 * (2 * ((X\uparrow 1) * 1))) + ((X\uparrow 2) * 0)) + 1),$$

(as printed by *PrintExpn*). Simplification by our rules would result in the following transformations:

$(((3*(2*(X* 1))) + ((X\uparrow 2)* 0)) + 1$	
$(((3*(2*(1*X))) + ((X\uparrow 2)* 0)) + 1$	by (e)
$(((3*(2*X)) + ((X\uparrow 2)* 0)) + 1$	by (i)
$((((3* 2)*X) + ((X\uparrow 2)* 0)) + 1$	by (l)
$(((6*X) + ((X\uparrow 2)* 0)) + 1$	by (b)
$(((6*X) + (0*(X\uparrow 2))) + 1$	by (e)
$(((6*X) + 0) + 1$	by (j)
$((0+(6*X)) + 1$	by (f)
$((6*X) + 1$	by (k)
$(1+(6*X))$	by (f)

We have put the productions applied to the right of the resulting *Expressions*.

The production system above terminates for all possible *Expressions.* To show this, we show that all productions eventually lead to smaller *Expressions.* Since there is a limit to how small an *Expression* can get, the process must terminate. We proceed by cases:

1. Productions (a), (b), (c), (d), (g), (h), (i), (j), and (k) immediately make an *Expression* smaller.

2. Production (*l*) does not increase the size of an *Expression.* After applying production (*l*), the next transformation will be to apply (*b*) to the left operand, reducing the *Expression*'s size, unless the *Expression* was of the form $\&C * (\&C * \&C)$. However, this last expression (which would cause an infinite loop) is impossible, since there is no way the subexpression $(\&C * \&C)$ could have escaped being reduced to a single constant. Note that if we did not simplify operands first, we would have introduced a potentially infinite loop.

3. Productions (e) and (f) do not increase the size of an *Expression.* After either is applied, one of the previously covered productions must also apply. (Note that the cases $(\&C * \&C)$ and $(\&C * \&C)$ cannot occur here because of the way in which the productions are ordered.)

23.4 Summary

In this chapter, we again invented a specialized language—the language of productions—to deal with a problem. You might reasonably argue that we took the long way around, and that for the *particular* problem of simplification we might just as well have written a single function with the simplifications "wired in." Of course, part of our purpose here was simply to introduce production systems, which have much broader applicability than the single application we've shown here. Beyond that, however, we can also argue that our solution lends itself readily to expansions, more simplifications, and also to understanding. Our production language is especially tailored to describing simplifications, whereas our general-purpose programming language is not.

Exercises

23.1 Fill in the missing details of the routines described in this chapter and provide a main routine so that you can test the result. Include the routines *Diff*, *PrintExpn*, and *ReadExpn* from the last chapter.

23.2 Extend the routines and rules to handle division.

23.3 Write productions that

a) Change subtractions to additions by multiplying the subtrahend by -1.

b) Use the equality

$$(x \uparrow y) \uparrow z = x \uparrow (y * z)$$

to simplify some repeated exponentiations.

c) Associate all additions to the left, so that all additions of three or more terms are of the form

$$\cdots (((a + b) + c) + \cdots$$

23.4 Suppose you want a "collect common terms" simplification, so that for example,

$$12x + 5x \text{ simplifies to } 17x.$$

To do so, you need to extend patterns so that *Match* can perform matches that depend on the two parts of an *Expression*'s being equal. Let's define a new pattern, $\&Ei$, for i any integer, which is defined

- $\&Ei$ matches any subexpression that is the same as expression number i of the expression being matched.

Thus, to say that $x - x$ simplifies to 0, we could write

$$\&A - \&E2 \text{ simplifies to } 0.$$

Implement the $\&E$ pattern. Write some simplifications to "collect common terms," as shown in this exercise.

23.5 Differentiation is, in effect, a binary operator taking an expression and a variable as its two operands. Extend the set of operators for *Expression* to include one for differentiation. Write simplification rules that do the same thing as *Diff*. You may assume that the only variable is X, or you may use the equality pattern of Exercise 23.4 to implement the entire set of rules. Hint: Take advantage of the ordering on production rules.

23.6 The routine *ApplyProductions* is quite inefficient. It tries to do simplifications on subexpressions that have not changed. Optimize it to avoid re-matching whenever possible.

23.7 As a function of the number of pattern matches applied and the size of an *Expression*, what upper bounds can you place on the time required by *ApplyProductions*?

23.8 Suppose we trade arithmetic operations for *logical* operations, such as PASCAL's **and** and **or**. Specifically, let + represent **or**; let * represent **and**; let 0 represent *false*; let 1 represent *true*; and let - represent **nand** ("not and": *a* **nand** *b* iff **not** (*a* **and** *b*)). Devise a set of simplification rules for logic and use them with *ApplyProductions*.

23.9 In what ways do the productions in Section 23.3 depend on the order in which they are applied? Devise a set of simplification rules that will work *regardless* of the order in which they are applied.

23.10 Suppose that we wanted to make *ApplyProductions* into a *noniterative* routine—a pure LRN traversal. In other words, having finished simplifying a subexpression, we never want to return to it. How should our simplification productions (Section 23.3) be modified to work?

23.11 There is a severe problem involved in finding a set of simplification rules that will reduce expressions such as

$$((((3 * X + 4 * Y) + 2 * Z) + Q) + X).$$

The two terms involving the variable X can be arbitrarily "far apart." In the example above, a very large pattern would be required to see, match, and combine the two X terms. Further, that pattern won't work if there are more intervening terms.

   In this exercise, you are to try to devise a way around the problem. To begin with, note that if all sums involving single variables are rewritten so their terms are in alphabetical order by variable name, it is easy to combine terms. See if you can develop this idea until it works for a larger class of expressions. Extend the set of possible patterns to include $\&Li$, which matches any expression that comes before subexpression $\&Si$ in an ordering that you define. Try to find productions that combine all terms in the expressions of the form

$$E1 + E2 + E3 + \cdots + Ei$$

where each Ei is a product of variables or variables raised to constant powers. A large number of possible orderings will work. Most polynomials, for example, are written in decreasing order of exponents. Your ordering, of course, must apply to terms containing several variables.

References

Aho, Alfred V., John E. Hopcroft, and Jeffrey D. Ullman (1974). *The Design and Analysis of Computer Algorithms.* Addison-Wesley, Reading, MA.

Aho, A. V. and J. D. Ullman (1977). *Principles of Compiler Design.* Addison-Wesley, Reading, MA.

Bentley, J. L., D. Haken, and J. B. Saxe (1978). *A General Method for Solving Divide-and-Conquer Recurrences.* Carnegie-Mellon University Technical Report, Pittsburgh.

Boehm, C. and G. Jacopini (1966). Flow-Diagrams, Turing Machines, and Languages With Only Two Formation Rules, *Communications of the ACM*, vol. 9, no. 5, pp. 366–371.

Borodin, A., and I. Munro (1975). *The Computational Complexity of Algebraic and Numeric Problems.* American Elsevier, New York.

Carroll, Lewis (1963). *The Annotated Alice.* World Publishing, New York.

Chomsky, N. (1956). Three Models for the Description of Language, *IRE Transactions on Information Theory* vol. 2, no. 3.

Chomsky, N. (1959).   On Certain Formal Properties of Grammars, *Information and Control*, vol. 2, no. 2.

Dahl, O. J., E. W. Dijkstra, and C. A. R. Hoare (1972).   *Structured Programming.* Academic Press, New York.

Dijkstra, E. W. (1976). *A Discipline of Programming.* Prentice-Hall, Englewood Cliffs, NJ.

Elson, Mark (1973).   *Concepts of Programming Languages.*   Science Research Associates, Chicago.

Elson, Mark (1975). *Data Structures.* Science Research Associates, Chicago.

Floyd, R. W. (1967).   Assigning Meanings to Programs.   *Proceedings of the Symposium on Applied Mathematics,* vol. 19, J. T. Schwartz (ed.).   American Mathematical Society, Providence, RI, pp. 19–32.

Gries, D. (1971). *Compiler Construction for Digital Computers.*   John Wiley and Sons, New York.

Grogono, Peter (1978). *Programming in PASCAL.*   Addison-Wesley, Reading, MA.

Guttag, J. V. (1975).   The Specificaton and Application to Programming of Abstract Data Types, Ph.D. dissertation, University of Toronto.

Guttag, J. V. (1977). Abstract Data Types and the Development of Data Structures, *Communications of the ACM*, vol. 20, no. 6.

Hoare, C. A. R. (1969).   An Axiomatic Basis for Computer Programming, *Communications of the ACM*, vol. 12, no. 10.

Hoare, C. A. R. (1972).   Proof of Correctness of Data Representations, *Acta Informatica,* vol. 1, no. 4, pp. 271–281.

Hopcroft, John E., and Jeffrey D. Ullman (1969).   *Formal Languages and their Relation to Automata.* Addison-Wesley, Reading, MA.

Horner, W. G. (1819).   A new method of solving numerical equations of all orders by continuous approximation, *Philosophical Transactions of the Royal Society of London,* vol. 109, pp. 308–335.

Horowitz, Ellis, and Sartaj Sahni (1976). *Fundamentals of Data Structures.* Computer Science Press, Woodland Hills, CA.

Horowitz, Ellis, and Sartai Sahni (1978). *Fundamentals of Computer Algorithms.* Computer Science Press, Woodland Hills, CA.

Jensen, Kathleen, and Niklaus Wirth (1974). *PASCAL User Manual and Report.* Springer-Verlag, New York.

Kernighan, Brian W., and P. J. Plauger (1974).   *The Elements of Programming Style.* McGraw-Hill, New York.

Knuth, Donald E. (1973). *Fundamental Algorithms*, vol. 1, Addison-Wesley, Reading, MA.

Knuth, Donald E. (1969).   *Seminumerical Algorithms*, vol. 2, Addison-Wesley, Reading, MA.

Knuth, Donald E. (1973). *Sorting and Searching*, vol. 3, Addison-Wesley, Reading, MA.

Knuth, Donald E. (1974).   Structured Programming with Go To Statements. *Computing Surveys*, vol. 6, no. 4, pp. 261–301.

Knuth, Donald E. and Jack Merner (1961), Algol 60 Confidential, *Communications of the ACM*, vol. 4, no. 6, pp. 268–273.

Liskov, B. H. and Zilles, S. N. (1975).   Specification Techniques for Data Abstractions, *IEEE Transactions on Software Engineering*, vol. 1, no. 1.

Liskov, B., A. Snyder, R. Atkinson, and C. Schaffert (1977).   Abstraction Mechanisms in CLU, *Communications of the ACM*, vol. 20, no. 8, pp. 564–576.

Liu, Chung L. (1977). *Elements of Discrete Mathematics*, McGraw-Hill, New York.

Manna, Zohar (1974).   *Mathematical Theory of Computation*, McGraw-Hill, New York.

Manna, Z. and R. Waldinger (1978).   The Logic of Computer Programming, *IEEE Transactions on Software Engineering*, vol. 4, no. 3.

Martin, William A. (1971). Sorting, *Computing Surveys*, vol. 3, no. 4.

Minsky, Marvin L. (1967). *Computation: Finite and Infinite Machines*, Prentice-Hall, Englewood Cliffs, NJ.

Naur, P. (ed.) (1963). Revised Report on the Algorithmic Language Algol 60, *Communications of the ACM*, vol. 6, no. 1.

Newton, I. (1711). *Analysis per quantitatem series, fluxiones ac differentias*, edited by William Jones, London, vol. 10.

Parnas, David (1972). A Technique for Software Module Specifications with Examples, *Communications of the ACM*, vol. 15, no. 5, pp. 330–336.

Pfaltz, John L. (1977). *Computer Data Structures*, McGraw-Hill, New York.

Polya, G. (1973). *How to Solve It. A New Aspect of Mathematical Method.* Princeton University Press, Princeton.

Pratt, T. W. (1975). *Programming Languages: Design and Implementation.* Prentice Hall, Englewood Cliffs, NJ.

Preparata, Franco P., and Raymond T. Yeh (1973). *Introduction to Discrete Structures*, Addison-Wesley, Reading, MA.

Randell, B. and D. J. Russell (1964). *Algol 60 Implementation.* Academic Press, London.

Salomaa, A. (1973). *Formal Languages.* Academic Press, New York.

Stanat, Donald F., and David F. McAllister (1977). *Discrete Mathematics in Computer Science.* Prentice-Hall, Englewood Cliffs, NJ.

Stone, Harold S., and Daniel P. Siewiorek (1975). *Introduction to Computer Organization and Data Structures: PSP-11 Edition.* McGraw-Hill, New York.

Strunk, W. and E. B. White. *The Elements of Style.* Macmillan, New York.

Tremblay, J. P., and R. Manohar (1975). *Discrete Mathematical Structures with Applications to Computer Science.* McGraw-Hill, New York.

Tremblay, J. P., and P. G. Sorenson (1976). *An Introduction to Data Structures with Applications.* McGraw-Hill, New York.

van Wijngaarden, A. et al (1975).   Revised Report on the Algorithmic
Language Algol 68, *Acta Informatica*, vol. 5, nos. 1–3, pp. 1–236.

Wickelgren, Wayne A. (1974). *How to Solve Problems*, W. H. Freeman, San
Francisco.

Wirth, N. (1971).   The Programming Language Pascal, *Acta Informatica*,
vol. 1, no. 1, pp. 35–63.

Wulf, W. A., D. B. Russell, and A. N. Habermann (1971). BLISS: A
Language for Systems Programming, *Communications of the ACM*, vol. 1,
no. 12, pp. 780–790.

Wulf, W. A., R. L. London, and M. Shaw (1976).   An Introduction to the
Construction and Verification of Alphard Programs, *IEEE Transactions on
Software Engineering*, vol. 2, no. 4.

Appendix A

Glossary and Notation

A.1 Standard Mathematical and Logical Notation

Notation	Interpretation
$a \lor b$	a *or* b: At least one of the propositions a and b is true.
$a \land b$	a *and* b: Both of the propositions a and b are true.
$\sim a$	*Not* a: The logical proposition a is not true.
$a \equiv b$ (also a iff b)	
	A *logically equivalent to* b, or a *iff* b: a and b are either both true or both false.
$a \supset b$	a *implies* b: Either a is false, or both a and b are true.
$\forall x \cdots$	*Universal quantification*: For all values x, $\cdots$
$\forall L \leq i \leq U \cdots$	For all values of i between L and U, $\cdots$
$\forall x$ in $A \cdots$	For all values x in the set $A \cdots$
$A \subset B$	The set A is a (not necessarily proper) subset of the set B; (that is, every member of A is a member of B).
$A \cup B$	A *union* B: The set containing exactly the elements in either of the sets A and B.
$A \cap B$	A *intersect* B: The set containing exactly the elements that appear in both the sets A and B.

571

$A - B$ The *set difference* of A and B: The set of all elements in set A, but not in set B.

2^A The *powerset of* A: The set of all subsets of A.

$A \times B$ *Cartesion product of* A *and* B: the set of all ordered pairs (a, b), with a in A and b in B.

$\{a, b, c, \cdots\}$ The set containing the elements $a, b, c, \cdots$

$\{x \mid Q\}$ The set of all values x satisfying the predicate Q.

$\{x \text{ in } A \mid Q\}$ The set of all values x in the set A satisfying the proposition Q.

$f: A \rightarrow B$ f is a function mapping elements of A into the set B.

$\lceil x \rceil$ The smallest integer not exceeded by x.

$|x|$ The absolute value of x.

$\sum_{i=m}^{n} E$ The sum of all values of the expression E for the variable i going from m to n. The value is 0 if $m > n$.

A.2 Standard Definitions from Discrete Mathematics

Notation	Interpretation
DAG	Directed Acyclic Graph. See Directed Graph.
Domain	(of a function) See *Function*.
Directed graph	A *directed graph* is a set of objects called *nodes* together with a relation on that set. Alternatively, a directed graph is a set of nodes and a set of *arcs* each leading from one node to another. A directed graph is *acyclic* if there is no sequence of arcs from any node leading back to the same node.
Function	A function f from a set A (the *domain*) to the set B (the *codomain*) is an assignment of a unique value in B to each value in A. Alternatively, a function from A to B is a relation on A and B in which for each element, a in A, there is a unique pair (a, b) in the relation.
Graph	See *Directed Graph*.
Linear Ordering	A *linear ordering* on a set A is a partial ordering in

which every pair of elements is comparable (that is, if $\leq$ is the linear ordering, then for all a and b in A, either $a \leq b$ or $b \leq a$).

Partial Ordering   A *partial ordering* on a set A is a relation, $\leq$, on A which is *reflexive* ($x \leq x$ for all x in A), *transitive* ($x \leq y$ and $y \leq z$ implies $x \leq z$), and *antisymmetric* ($x \leq y$ and $y \leq x$ is impossible unless $x = y$).

Undirected Graph   An *undirected graph* is a set of nodes and a set of unordered pairs of those nodes.

A.3 Notation Introduced in the Text

Section references are given in brackets following each definition.

Notation	Interpretation
$P\{S\}Q$	If the predicate P is true, then execution of the program segment S will successfully terminate with Q being true. [5.1]
E'	When used in program assertions, E' means the value of the expression E before execution of a given program segment. [5.1]
$C1$ **cand** $C2$	*Conditional and*: In programs, **cand** is the same as $C1$ **and** $C2$, but is assumed not to require evaluation of $C2$ if $C1$ is false. [Appendix C]
$C1$ **cor** $C2$	*Conditional or*: In programs, **cor** is the same as $C1$ **or** $C2$, but is assumed not to require evaluation of $C2$ if $C1$ is true. [Appendix C]
$R_1 + R_2$	The regular expression defined by the union of RE's R_1 and R_2. [1.3.2]
$R_1 R_2$	The regular expression defined by the concatenation of RE's R_1 and R_2. [1.3.2]
R^*	The regular expression defined by the concatenation of an arbitrary number, including 0, occurrences of the RE R. [1.3.2]
$x \rightarrow y$	In a state transition diagram, the label of an arc indicating the input symbol x and the output symbol y. [1.1.2]
	In the context of a grammar or production system, $x \rightarrow y$ means "y can be derived from x through

	application of some production."	[13.3.4]
$\rightarrow y$	The label of an FSG arc.	[1.1.3]
$x \rightarrow$	The label of an FSR arc.	[1.1.3]
ϵ	The null string.	[1.1.3]
	Also the empty list.	[9.1.4.1]
$< >$	The null sequence.	[7.4.2.2]
$x \sim y$	The concatenation of sequences x and y.	[7.4.2.2]
$\#L$	The memory location corresponding to label L, as in $x := \#L$.	[Chapter 4]
$@x$	The indirect value of x, as in **goto** $@x$.	[Chapter 4]
$\alpha(A, i, e)$	The array A with its ith element replaced by expression E.	[7.2.2]
$<d_1, d_2, \cdots D_n>$	Same as the sequence $< > \sim <d_1> \sim <d_2> \sim \cdots <d_n>$.	[7.4.2.2]
$\uparrow T$	The type whose values are references to elements of type T.	[8.2]
$x\uparrow$	For x an element of type $\uparrow T$, $x\uparrow$ is the element of type T currently referenced by x.	[8.2]
βx	For x, an element of type T, βx is an element of type $\uparrow T$ that references x.	[8.2]
AR	Activation record.	[16.1]
$cp(S, Q)$	Cost precondition. Weakest precondition extended to account for execution time.	[6.5.2]
DPDA	Deterministic push-down automaton.	[13.1.1]
ERROR	A distinguished value of all types representing an undefined or error result.	[7.4.2.2]
FCL/1	The flow chart language containing the control statements "**goto** L" and "**if** C **then goto** L."	[2.2.1]
FCL/2	The flow chart language containing the control statements "**begin** S_1; S_2; $\cdots S_n$ **end**," "**if** C **then** S_1 **else** S_2," and "**while** C **do** S."	[2.2.2]
FCL/3	FCL/2 extended with nonrecursive routines.	[3.3.5]
first(x)	The first element of sequence x.	[7.4.2.2]
FSA	Finite state automaton.	[1.1]
FSG	Finite state generator.	[1.1.3]
FSM	Finite state machine.	[1.1]

FSR	Finite state recognizer.	[1.1.3]
last(x)	The last element of sequence *x*.	[7.4.2.2]
leader(x)	The sequence *x* without its last element.	[7.4.2.2]
length(x)	The number of elements in sequence *x*.	[7.4.2.2]
NDFSR	Nondeterministic finite state recognizer.	[1.2]
NDPDA	Nondeterministic push-down automaton.	[13.1.1]

$O(f(n))$ Order notation: Stands for some function, g, of n such that there is a constant M satisfying

$$|g(n)| \leqq M f(n)$$

for all positive integers n. Multiple instances of a term $O(f(n))$ in an expression may not be assumed to refer to the same function, g.   [6.2]

PDA	Push-down automaton.	[13.1.1]
RE	Regular expression.	[1.3.2]

rep(S)   The representation of the statement, phrase, or value *S*.   [4.1]

sc   Space cost. Gives space required for a data structure as a function of type and other information.   [12.1]

SMAL   A primitive programming language with control statements very close to those of assembly language.
[Chapter 4]

TM	Turing machine.	[13.2]
trailer(x)	The sequence *x* without its first element.	[7.4.2.2]

wp(S, Q)   The weakest precondition (that is, the necessary and sufficient condition, that assures the proper termination of statement *S* in a state satisfying predicate *Q*).
[5.2]

Appendix B

Standards for Evaluating Programs

A perennial question from students in programming courses is "Why did I get a *C* on this program? ··· but, *it WORKS!*" This appendix explains what we believe is an acceptable, better-than-*C* solution to any programming problem.

B.1 Introduction

Remember our fundamental premise: programming is, or ought to be, an engineering discipline. The criteria for evaluating a program are similar to those for evaluating *any* engineering design. First, if a program does not perform its intended function, it is useless. To be *minimally* acceptable, a program must be correct. However, getting the program right is only one facet of building a well-engineered program. Let's consider an analogy to emphasize the point.

Suppose you are a civil engineer assigned the task of designing a bridge to cross the Allegheny river. No matter what other nice properties your design might have, it would be useless if it didn't quite reach all the way across the river, or if it were too narrow for even a bicycle, or if it would not bear the weight of the expected peak traffic load. You certainly wouldn't expect any design that failed to meet these functional specifications to be accepted. Among the bridge designs that *do* meet the minimal functional specifications, however, you recognize that some are superior to others: some may cost less, or be easier to build, or are prettier, or are easier to widen when the traffic load gets worse. Some bridge designs are, in fact, outstanding.

Now back to programming. Given several computer programs that all work, we recognize that some are better than others. Some of the characteristics that make one program better than another are discussed in the following sections.

B.2 Programs are Read by People

Programs are read by people as well as by machines. There is a limit to the amount of complexity that the human mind can deal with at one time. Most interesting programs exceed this limit. Therefore, it is futile to try to keep track of an entire program in your head. You are going to have to read your own programs in the process of writing them, and you cannot expect to be able to remember all of the details. You will save yourself untold grief as you move on to bigger and bigger programs if you learn to keep them tidy as you build them.

Real programs are likely to hang around for a long time. They get changed, both because errors are discovered and because someone decides that they should do something different. Changes, moreover, are often made by programmers other than the author of the program; if by chance the author of a program does make a change in it, he has probably forgotten many of the details and assumptions that were originally an active part of the thinking that went into it. Writing programs so you can live with them is a habit you should develop now. If you don't do it now, you will have even more trouble learning how later.

B.3 What is Involved in Presenting a Program Well?

When you wonder about how to present or document a program, or when you wonder whether you've done a good job on one, you should ask yourself if your program can be read *and understood* by a colleague. If it takes someone more than ten or fifteen minutes to understand a three-page program that solves the same problem as fifty other programs he has seen, even written, something must be *wrong*.

A useful principle to apply to your programs is the Law of Least Astonishment: a program should clearly describe the algorithm that is being used and the data to which it is applied. After reading the program, you should not be astonished by what it does.

Many things contribute to a good presentation of a program. The most important is good taste, but that cannot be taught in a few written words. Some aspects that *can* be described are

- Program organization
- Comments and other documentation
- Programming style
- Clarity and efficiency
- Reasonable decisions about unspecified matters
- Selection of test data

Further, you should certainly get a copy of *The Elements of Programming Style* by Kernighan and Plauger. You should also read it. This book is written in the manner of the famous *The Elements of Style* by Strunk and White, a book about writing in English that every student should also own. *Programming Style* gives many simple, and direct suggestions about several important aspects of program style, supporting them with lots of examples.

B.3.1 PROGRAM ORGANIZATION

You were probably taught in a previous course that you should write a general outline for your program and, in successive steps, make it more precise and specific until it falls directly into whatever language you are using. This is good advice. Follow it. Your program organization should reflect your outline. If you want to review this technique of developing programs or learn more about it, read *Structured Programming* by Dahl, Dijkstra, and Hoare.

Procedures are not merely intended for making it possible to invoke a piece of code usable from two or more different locations in the program. They are also intended for isolating the code for a task so that you can debug and develop it without worrying about other parts of the program.

Design your procedures so that each procedure will perform one well-defined operation, not several unrelated ones. Sets of related operations that are performed several places in the program should be turned into single procedures, even if the problem statement did not explicitly make that requirement. If you do this, you make it possible to change one section of text rather than many when you later discover you want to do things differently. In addition, group your procedures so that logically related procedures are physically close together.

Avoid writing any procedure that alters variables that are not parameters to that procedure. According to the Law of Least Astonishment, the person reading the code that calls a procedure is entitled to know what values it may change. For example,

$C := 3;$

$f(X);$

print $C;$

should print a 3. There are a few exceptions to this (even in this text); a group of procedures might share an array and expect no other code to touch it. This fact, then, should be the subject of a comment. Make sure that the documentation for a group of procedures like this clearly states its data-sharing assumptions.

B.3.2 COMMENTS AND OTHER DOCUMENTATION

Comments are intended to help people figure out what a program is doing. The phrase

{ The next statement adds 1 to K }

is not a useful comment. There should be a block of comments at the beginning of the program that indicates

1. Who wrote the program, and when.
2. What the program is supposed to do.
3. Information needed to use the program: input and output formats, restrictions on input, etc.
4. A sketch of the overall plan of the program and the major program variables used to carry out the plan.
5. Anything special about the way variables are used, particularly the meanings of special values. (for example, -1 means no input yet; 0 means input was ok; 2 means end of input; etc.)

Comments like these are often most legible if they are presented in a tabular format. It should always be immediately clear what the comment is about, even when a large block of commentary is unavoidable.

There should be a comment on each routine indicating its effects. Our preconditions and postconditions are examples of this sort of comment. Groups of related routines (all those that implement primitive operations on a list type, for example) should have comments indicating the relation.

It is sometimes reasonable to include separate documentation, that is, documentation separate from the listing. It is reasonable to do this when the information is not likely to change and when it applies to the program as a whole rather than to individual sections. This type of documentation might include general instructions for use (minimal information on input/output requirements still belongs in the program), explanation of the overall algorithm or implementation strategy, reasons for certain decisions, and details about the design of data structures.

A well-selected variable or routine name can serve as a comment each place it is used. If you regularly choose meaningful names, it may reduce the number of comments you need. Be very careful, however, not to misname variables. For example, using a variable named *Count* to accumulate an inner product would be most confusing.

B.3.3 PROGRAMMING STYLE

By "programming style" we mean the way things are said, and the way they are laid out on the page.

B.3.3.1 Indentation
By using the same indentation for related portions of your program, you can substantially increase the ease with which it can be read. Indentation should reflect the logical structure of the program; text that is part of a statement should be indented with respect to that statement. It is important, furthermore, that such indentation be used consistently. Inconsistent indentation can be worse than none at all, since it may lead the reader into a false sense of understanding.

B.3.3.2 Grouping
Related parts of a program should be close together. Blank lines should be used to get white space between two blocks of text. When you are stuck with an all-upper-case printer, you need to help the reader with judicious white space. (NOTE HOW MUCH HARDER IT IS TO READ TEXT WHEN IT IS ALL IN UPPER CASE. LOWER CASE PROVIDES MORE VISUAL CUES TO THE READER BY LESS MONOTONOUS LETTER SIZES AND A GREATER VARIETY OF SHAPES.) This is another reason for tabulating comments rather than packing them onto as few lines as possible.

B.3.3.3 Mystery Constants
If you use constants that encode some special meaning, such as the numerical representation of "?" or the current Restaurant Tax rate in North Carolina, declare a constant to hold this value. Only 3.1415926535 is familiar enough that you can use it without explanation, but it's much easier to say once only,

   PI = 3.1415926535,

and then type two characters instead of 12.

B.3.3.4 Control Structures

You need not overwork yourself getting rid of **goto** statements; that is not the point. In fact, the **goto** is often the best way to exit a **for** or **while** loop before its natural termination. Do try to use **if–then–else** and **while** instead of **goto** where such constructs are more natural, as they are almost all the time.

B.3.3.5 Global Variables

If a variable is accessed and altered from many places in your program, someone is likely to do something wrong to it. When that happens, you will have to read the *entire* program to figure out who is doing what to the variable, and why. A better idea is to have access to any particular variable limited to as few sections of the program as possible.

B.3.4 EFFICIENCY AND CLARITY

It is important for you to choose an algorithm that will run with reasonable efficiency.   However, it is counterproductive to write an obscure, convoluted program in an attempt to trick the compiler into doing a little bit better than it would if you simply wrote out the algorithm directly. If the program will be used a lot, you will often be able to use a smarter, or optimizing, compiler that will produce more efficient code without interference from you.

B.3.5 MAKING REASONABLE DECISIONS ABOUT PERIPHERAL MATTERS

Most "real-world" problem specifications state the major objective of the program, but leave to the programmer decisions about auxiliary matters such as input and output formats, error handling, and treatment of certain special cases. You must make and implement *reasonable* decisions about those matters, bearing in mind that programs are read and used by humans. For example, when you write your input and output statements, you usually have a choice of formats. You should print the value of your inputs as well as your results. You need not make elaborate formats, but your listings should be laid out neatly and everything should be labeled so the reader does not need to consult the program listing to understand the output.

B.3.6 SELECTION OF TEST DATA

You should realize by now that a program cannot be assumed to be correct just because it prints output in the right format. You cannot assume that a program is correct when it executes one test case properly, or even when it has been "proven correct" (proofs and specifications can be erroneous, too.) A very important part of writing a program is designing test data that will demonstrate to you that the program is, indeed, correct. Then you need to test at least all of the common correct cases. You should also test the "boundary conditions"—those situations that are at the fringe of allowed input conditions. Typical examples of boundary conditions are an empty input stream, cases that cause loops to execute zero times, and the largest allowed input values. Finally, you should test some illegal input data just to be sure that your program behaves gracefully when data are incorrect.

Appendix C

The Example Programming Language

In order to present programming examples it is necessary to use some particular notation, or programming language, to write them. In this textbook we use a dialect of PASCAL for descriptive purposes; we have made minimal revisions in order to make some material easier to present. (The programs in Part Four are written, for the most part, in standard PASCAL.) Readers totally unfamiliar with PASCAL should read one of the PASCAL texts listed in the references.

Since this is not a textbook on programming languages, and since we expect that all our readers will have some programming experience, we do not present a full definition of the language we use. Rather, we point out its major deviations from PASCAL, and then rely on the reader's intuition and good sense to fill in the details.

The following are the major deviations from PASCAL:

1. Statement labels can also be identifiers as in ALGOL. The definition of <label> is extended with

   <label> ::= <identifier>

   The <label declaration part> is optional.

2. We adopt ALGOL block structure, changing the placement of declarations, where appropriate. That is, we freely place variable declarations inside compound statements, as in

```
begin
var x: integer;
x := 1;
end
```

3. Formal types in procedure headers have been liberalized. Thus,

   procedure $F(x: \uparrow Q;\ y:$ **array** $[1..n]$ **of** $T)$;

   is legal.

4. The conditional operators **cand** and **cor** are defined as follows:

   A **cand** B is equivalent to **if** A **then** B **else false**

   A **cor** B is equivalent to **if** A **then true else** B

5. Phrases beginning **pre** or **post** following a routine header are comments.

Appendix D

A Simple Computer

Many students first learn to program in a higher-level language and hence know very little about the nature of the programs that computers actually execute. The machine language of a computer, or the set of instructions it can actually execute, is quite different from FORTRAN, ALGOL, COBOL, or any of the other higher-level languages.

There is really very little reason for most programmers to know the intimate, often obscure, details of the machine language for a *particular* computer. On the other hand, without some knowledge of the general structure of machines, it is difficult to appreciate why some programming language constructs are designed the way they are, why some programs are more efficient than others, and so on. This appendix is, therefore, devoted to a description of a very simple computer. Although this hypothetical computer does not closely resemble any existing machine, it has enough in common with all of them to suffice for our purposes. An understanding of it will help you understand a real machine, should you need to.

Essentially all computers in use today are what we call von Neumann machines (after the famous twentieth century mathematician, John von Neumann). As shown in Figure D.1, there are generally three major, interconnected components of such a computer: a memory, a central processor, and an input-output unit.

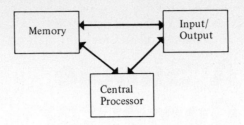

Fig. D.1 The major components of a computer

D.1 The Memory Unit

The memory is divided into a number of storage cells, each having several important properties:

- Each storage cell has a name, usually called an *address* or *location*. For reasons that become clear later, the nonnegative integers are generally used as the addresses, or names, of storage cells. Thus, we speak of location 0 or the storage cell whose address is 0, and location 1 or the storage cell whose address is 1, and so on.

- Each storage cell is capable of holding a relatively small amount of information. Generally this information is encoded as a fixed-length sequence of *bits*. (A bit, which is short for "binary digit," is an entity that may assume only one of two distinguished values—usually denoted 0 and 1.) The information in a storage cell is generally called a *word*; machines have been built with words containing anywhere from eight to sixty or more bits. Since we are not concerned with details, we usually do not specify the size of the word held in a storage cell. You may simply assume that the storage cell is "big enough" unless told otherwise.

- The contents of a storage cell may be both read and written. Reading the contents of a storage cell is a *nondestructive* operation; that is, the contents remain the same after the operation is complete. Writing, on the other hand, is *destructive*: a new bit pattern is inserted into the cell and the previous contents of the cell are lost.

It is extremely important to observe that a storage cell contains merely a sequence of bits, a pattern of ones and zeros, that have no intrinsic meaning. The storage cell is one place where we run into the issue of representation: a pattern of bits may be used to represent a number, an instruction to the computer, one or more alphabetic characters, or any of a large number of other things.

The memory unit is a passive device in a computer system. It may receive commands from either the central processor or input-output units, but it does not initiate actions of its own. The commands it receives are either "read the contents of the cell whose address is A," or "write the word W into the cell whose address is A."

D.2 The Input-Output Unit

Input-output, or I/O as it is usually called, varies widely from computer to computer. In addition, I/O is generally rather complex at the machine language level. We don't have need for information about the I/O unit in this textbook and shall, therefore, only note that the I/O unit (or units, since there are often several of them) generally receive commands from the central processor to move information either from memory to an output device (such as a printer), or to move information from an input device (such as a card reader) to memory. One such command to the I/O unit will cause a series of operations involving the memory.

D.3 The Central Processor

The central processor is the component of the computer that actually executes programs. The program itself is stored in the memory unit, typically as one instruction per storage cell. The central processor reads one word from memory and then performs the operation specified by this instruction. It then reads the next instruction, performs that operation, and so on until it encounters an instruction that tells the computer to stop, or halt. The series of actions involved in reading and executing an instruction is called the *fetch-execute cycle.*

Again, a word stored in a memory cell has *no* intrinsic meaning; it is simply a pattern of bits. Only when the contents of the storage cell is *used* in a given context is an interpretation placed upon it. The same bit pattern used in two different contexts can have two different interpretations. For example, when a word is fetched and used as an instruction, it is interpreted as an instruction. If the same bit pattern, indeed the same storage location, were used in a context that needs an integer, it would be interpreted as an integer.

The instructions that the central processor can perform, or execute, fall into two broad categories: *data manipulating* operations and *control* operations.

Data manipulating operations are those that perform arithmetic, logical, or other data manipulation operations on information stored in the memory unit. Generally these operations read one or more words (bit patterns) from the memory unit; then, placing an interpretation on the word(s), perform the desired manipulation, and leave the result in a specified place.

Control operations alter the order of instruction evaluation; they control what is meant by the "next" instruction. By convention the next instruction is usually the one contained in the storage cell whose address is one greater than that of the current instruction. Thus, typically, the consecutive steps (or instructions) of a machine language program are found in storage cells with consecutive, ascending addresses. However, *all* computers provide one or more instructions that override this precept — that is, they force the next instruction to be taken from a specified storage cell and for the conventional sequencing of instructions to resume from this new location. Such instructions are called *branch*, *transfer*, or *jump* instructions and are analogous to the **goto** statement in most higher level programming languages.

The data and control operations interact through a subset of the control instructions called *conditional branches* (or *conditional jumps*). These instructions typically examine a word, place an interpretation on it ("This is an integer.") and then branch *only* if the word satisfies some property ("This integer is equal to zero."). These conditional control instructions are extremely important since they are the building blocks out of which such higher-level language constructs as the **if** and **for** (or **DO**) statements are constructed.

D.4 An Instruction Set

To repeat, the central processor can execute many kinds of instructions stored in memory. Here we'll describe only some of the kinds of instructions that may be available in a typical computer because we don't want to get bogged down in unnecessary detail. We'll describe the machine language as though it were a higher-level language and simply place restrictions on the language so that each statement corresponds roughly to what can be done in one machine-language instruction. For this purpose we use the restricted language SMAL (pronounced "small" and standing for "Similar to MAchine Language"). SMAL is similar to the language we have used throughout the text with the following exceptions.

1. There is only one scalar data type, *word*. Literals of type *word* are written as integers. Both *integer* arithmetic operations and *boolean*

(logical) operations are defined on the words. Thus, though it may seem strange, the following is legal:

begin var x, y: word; $\cdots$ $x := x + y$; $x := x$ **and** 7; $\cdots$

Boolean operations are performed bit-by-bit.

2. There is only one structured type, **array** [<integer>..<integer>] **of** *word*.

3. The only statements are assignment, **goto**, and conditional **goto**.

   a) The assignment statement must have a very simple expression on the right-hand side; it may have at most one operator.

   b) The **goto** statement is of the form **goto** <place>, where the definition of <place> is either <label> or @ <variable> as is explained in (4).

   c) The conditional **goto** is of the form **if** <relation> **then goto** <place>, where <relation> must be a simple comparison between variables (or a variable and a constant).

4. Statements may be labeled. In addition, the memory address of a labeled statement can be stored in a variable and a **goto** may branch to the statement whose label is stored in a variable. Thus,

   L: $x := x + 1$; $\cdots$

   $y := \#L$; $\cdots$

   goto @y

   is a legal program in which

   a) The statement "$x := x + 1$" is labeled L.

   b) The notation $\#L$ denotes "the memory address in which is stored the statement labeled L." Thus, "$y := \#L$" stores this address in the variable y.

   c) The notation @y means "the statement whose address is stored in the variable y."

Thus, assuming no other assignments to y, the statement **goto** @y will cause control to branch back to the statement labeled L.

It is worth noting some of the things that are not in SMAL. Such convenient facilities as compound statements, procedures and functions, more complex expressions, and **for**, **while**, and **case** statements are all missing.

Appendix E

Collected programs from Part Four

This appendix contains the final implementations of the various programs developed in Part Four. These programs are ordered as required by standard PASCAL (constant declarations first, then type declarations, etc.) As in the text, several of the bodies have been left to the reader.

E.1 Lexical Analyzer

```
type
    State =          (NEW, ID, ASGN, COM, DOT, STR, STRQ, NUM,
                      REAL, QUIT);
    Action =         (S, AS, E, AES, FAIL);
    CharClass =      (Let, Dig, LBrace, RBrace, Col, Eql, Point, Sem,
                      Plus, Minus, Mult, Divd, Comma, LPar, RPar,
                      EndFl, Quote, Space, Othr);
    Word = array [1..4] of char;
    LexType = char;

var
{ The three components of the state transition matrix }
    F:  array [State] of array [CharClass] of State;
    G:  array [State] of array [CharClass] of Action;
    LT: array [State] of array [CharClass] of LexType;
```

593

```
{ Current state and character of the FSM }
    CurState, NextState: State;
    CurCharClass: CharClass;
    CurCharItself: char;

{ The Buffer }
    Buffer: array [1..100] of char;
    BufPtr: integer;

{ Used by SetUpFSM }
    CharsDone: array [CharClass] of Boolean;


{ Utilities left to reader }

function GetFlag:char;
    { Read 4 characters from input, leaving last char in ch }
function GetState: State;
    { Read 4-char state literal and convert to type State }
function GetAction: Action;
    { Read 4-char action literal and convert to type Action }
function GetCharClass: CharClass;
    { Read 1-char name for character and convert to type CharClass }
function GetLexType: LexType;
    { Read 1-char LexType for labeling lexeme. }
procedure SetUpLiteralConversion;
    { Sets up character equivalents for enumerated types. }


{ Reading input tape }

function FindClass(Ch: char): CharClass;
    { Return character class corresponding to input character }
    var ChCl: CharClass;
    begin
    FindClass := Othr;
    if Ch is a letter then FindClass := Let else
    if Ch is a digit then FindClass := Dig else
    for ChCl := succ(Dig) to pred(Othr) do
        if CharClassLit[ChCl] = Ch then FindClass := ChCl;
    end;

procedure Scan;
    { Sets CurCharItself to next input character and CurCharClass
      to the class of that character. }
    begin
```

```
    read(CurCharItself);
    if eof(input) then CurCharItself:= CharClassLit[EndFl];
    CurCharClass := FindClass(CurCharItself);
    end;
```

{ Actions }

```
procedure AppendChar;
  { Appends current input character to buffer }
  begin BufPtr := BufPtr + 1; Buffer[BufPtr] := CurCharItself; end;

procedure EmitLex(Tg: LexType);
  { Outputs lexeme now in buffer (tagged Tg) and clears buffer }
  var i: integer;
  begin
  write('!', Tg,'! ');
  for i := 1 to BufPtr do write(Buffer[i]);
  BufPtr := 0;
  writeln(' ');
  end;

procedure FailMsg(CurSt: state; CurCh: char);
  { Error message }
  begin
  writeln(' Failed scanning ', CurCh,
          ' in state', StateLit[CurState]);
  end;
```

{ Prime the machine }

```
procedure StartFSM;
  { Set up FSM in starting state }
  begin  CurState := NEW; Scan; BufPtr := 0;  end;
```

{ Table initialization routines }

```
procedure ClearCharsDoneList;
  { Clear list of characters processed for current state }
  var ThisCh: CharClass;
  begin
  for ThisCh := Let to Othr do CharsDone[ThisCh] := false;
  end;

procedure RecordCharDone(ThisCh: CharClass);
```

```
{ Record current character as having been processed }
begin
CharsDone[ThisCh] := true;
end;

procedure ProcessExplicitTransition( ThisSt: State);
{ Get CharClass, record it as defined, move values to tables }
var ThisCh: CharClass; ch: char;
begin
ThisCh                 := GetCharClass; read(ch);
G[ThisSt, ThisCh]   := GetAction;      read(ch);
LT[ThisSt, ThisCh]  := GetLexType;     read(ch);
F[ThisSt, ThisCh]   := GetState;       read(ch);
RecordCharDone( ThisCh);
end;

procedure ProcessDefaultTransition( ThisSt: State);
{ Read default transition and use for all chars not yet defined }
var TSt: State; TAct: Action; TCh: CharClass; TType: LexType;
    ch: char;
begin
read(ch); read(ch);
TAct   := GetAction;    read(ch);
TType  := GetLexType;  read(ch);
TSt    := GetState;     read(ch);
for TCh := Let to Othr do
   if not CharsDone[TCh] then
   begin
   G[ThisSt, TCh]    := TAct;
   LT[ThisSt, TCh]   := TType;
   F[ThisSt, TCh]    := TSt;
   end;
end;

procedure SetUpTransitionTable;
{ Fill in State Transition Table }
var ThisSt: State; ThisFlag: char;
begin
readln; ThisFlag := GetFlag;
while ThisFlag = '>' do
   begin
   ThisSt := GetState;
   ClearCharsDoneList;
   readln; ThisFlag := GetFlag;
   while ThisFlag = ' ' do
      begin
      ProcessExplicitTransition( ThisSt);
```

```
            readln; ThisFlag := GetFlag;
          end;
       ProcessDefaultTransition( ThisSt);
       readln; ThisFlag := GetFlag;
       end;
    end;

procedure SetUpFSM;
  { Load state transition tables }
  begin
  SetUpLiteralConversion;
  SetUpTransitionTable;
  end;

begin  { Main program }
SetUpFSM;
StartFSM;
while CurState ≠ Quit do
   begin
   NextState := F[CurState, CurCharClass];
   case G[CurState, CurCharClass] of
       S:   Scan;
       E:   EmitLex(LT[CurState, CurCharClass]);
       AS:  begin AppendChar; Scan; end;
       AES: begin
               AppendChar;
               EmitLex(LT[CurState, CurCharClass]);
               Scan;
               end;
       Fail: FailMsg(CurState, CurCharItself);
       end;
   CurState := NextState;
   end;
end.
```

E.2 The Types SetofElts and SetGen

```
type
     EltType = integer;

     RefSetElt = ↑SetElt;
     SetElt = record
          Data: EltType;
```

```
            Valid: boolean;
            LeftLink, RightLink, Thread: RefSetElt;
            end;

        SetofElts = record
            TreeHead, ListThread: RefSetElt;
            end;

        DiscrimClass = (SameClass, LeftClass, RightClass);

        SetGen = record
            Cur: RefSetElt;
            Beg: RefSetElt;
            end;
```

{ Utilities }

```
function Discrim(x, y: EltType): DiscrimClass;
   post (RESULT = SameClass ≡ x = y)
      ∧ (RESULT = LeftClass ≡ x < y)
      ∧ (RESULT = RightClass ≡ x > y);
   begin
   if x = y
      then Discrim := SameClass
      else if x < y
         then Discrim := LeftClass
         else Discrim := RightClass
   end; {Discrim}


function FindElt(var S: RefSetElt; x: EltType): boolean;
   post (RESULT ⊃ S↑.Data = x) ∧
      (~ RESULT ⊃
         x not in S'↑ and if it is to be added, it should
         be a son of S↑; S = nil denotes the set header);
   label 1;
   var Next: RefSetElt;
   begin
   Next := S; FindElt := false;
   while Next ≠ nil do
      begin
      S := Next;
      case Discrim(Next↑.Data, x) of
         SameClass:  begin FindElt := true; goto 1; end;
         LeftClass:   Next := Next↑.LeftLink;
         RightClass:  Next := Next↑.RightLink
         end;
      end;
```

1: **end**; {FindElt}

procedure *MakeCell*(**var** *S*: *RefSetElt*; *Val*: *boolean*; *Dat*: *EltType*);
  post *S*↑*Valid* = *Val* ∧ *S*↑.*Data* = *Dat* ∧ *S*↑ has no sons;
    begin
    new(*S*); *S*↑.*Valid* := *Val*; *S*↑.*Data* := *Dat*;
    S↑.*LeftLink* := *nil*; *S*↑.*RightLink* := *nil*; *S*↑.*Thread* := *nil*;
    end; {MakeCell}

{ Routines for type *SetofElts* }

procedure *Clear*(**var** *S*: *SetofElts*);
  post S = { };
    begin
    S.TreeHead := *nil*; *S.ListThread* := *nil*;
    end;   {*Clear*}

procedure *Remove*(**var** *S*: *SetofElts*; *x*: *EltType*);
  post S = *S'* ∼ {*x*};
  var *R*: *RefSetElt*;
    begin
    R := *S.TreeHead*;
    if *FindElt*(*R*, *x*) **then** *R*↑.*Valid* := *false*;
    end; {Remove}

procedure *Insert*(**var** *S*: *SetofElts*; *x*: *EltType*);
  post *S* = *S'* ∪ {*x*};
  var *R*, *T*: *RefSetElt*;
    begin
    R := *S.TreeHead*;
    if *FindElt*(*R*, *x*)
        then *R*↑.*Valid* := *true*
        else
            begin
            MakeCell(*T*, *true*, *x*);
            T↑.*Thread* := *S.ListThread*; *S.ListThread* := *T*;
            if *R* = *nil*
                then *S.TreeHead* := *T*
                else if *Discrim*(*R*↑.*Data*, *x*) = *LeftClass*
                    then *R*↑.*Leftlink* := *T*
                    else *R*↑.*RightLink* := *T*;
            end;
    end; {Insert}

function *IsMember*(*S*: *SetofElts*; *x*: *EltType*): *boolean*;
  post *RESULT* ≡ (*x* is in *S*);

```
var R: RefSetElt;
  begin
  R := S.TreeHead;
  IsMember := FindElt(R, x) cand R↑.Valid;
  end;   {IsMember}

function IsEmpty (S: SetofElts): boolean;
  post RESULT ≡ (S = { });
  label 1;
  var R: RefSetElt;
  begin
  IsEmpty := true;
  R := S.ListThread;
  while R ≠ nil do
      begin
      if R↑.Valid then begin IsEmpty := false; goto 1; end;
      R := R↑.Thread;
      end;
1: end;  {IsEmpty}

function IsSubset(S, T: SetofElts): boolean;
  post RESULT ≡ (S ⊂ T);
  label 1;
  var x: EltType; R: RefSetElt;
  begin
  IsSubset := false;
  R := S.ListThread;
  while R ≠ nil do
      begin
      if R↑.Valid cand not IsMember(T, R↑.Data) then goto 1;
      R := R↑.Thread;
      end;
  IsSubset := true;
1: end;  {IsSubset}

procedure Union(var S: SetofElts; T: SetofElts);
  post = S' ∪ T;
  var x: EltType; R: RefSetElt;
  begin
  R := T.ListThread;
      while R ≠ nil do
          begin
          if R↑.Valid then Insert(S, R↑.Data);
          R := R↑.Thread;
          end;
  end; {Union}
```

```
procedure Intersect(var S: SetofElts; T: SetofElts);

  post S = S' ∩ T;
  var RS, RT: RefSetElt;
  begin
  RS := S.ListThread;
  while RS ≠ nil do
      begin
      RT := T.TreeHead;
      RS↑.Valid := RS↑.Valid cand FindElt(RT, RS↑.Data)
                    ˜ cand RT↑.Valid;
      RS:= RS↑.Thread;
      end;
  end; {Intersect}

procedure Compact(var S: SetofElts);
  var B: SetofElts; x: EltType; R: RefSetElt;
  begin
  Clear(B);
  R := S.ListThread;
  while R ≠ nil do
      begin
      if R↑.Valid then Insert(B, R↑.Data);
      R := R↑.Thread;
      end;
  S := B;
  end; {Compact}

procedure Copy(var S: SetofElts; T: SetofElts);
  var R: RefSetElt;

  function Icopy(Q: RefSetElt): RefSetElt;
  {Returns pointer to root of copy of tree rooted at Q}
  var c: RefSetElt;
      begin
      if Q = nil
                  then Icopy := nil
                  else
                    begin
                    MakeCell(c, Q↑.Valid, Q↑.Data);
                    c↑.LeftLink := Icopy(Q↑.LeftLink);
                    c↑.RightLink := Icopy(Q↑.RightLink);
                    c↑.Thread := R; R := c; Icopy := c;
                    end;
      end; {Recursive inner Icopy}

  begin {Copy body}
```

```
R := nil;
S. TreeHead := Icopy( T. TreeHead);
S. ListThread := R;
end; {Copy}
```

{ Routines for type *SetGen* }

```
procedure StartSetLoop(var G: SetGen; S: SetofElts);
  {Initialize ForAllInSet Loop}
    begin  G.Beg := S.ListThread;  G.Cur := G.Beg; end;
```

```
procedure StopSetLoop(var G: SetGen);
  {Terminate ForAllInSet loop}
    begin  G.Cur := nil; end;
```

```
function NextSetElt(var G: SetGen; var x: EltType): boolean;
  {Put next valid data element in x, leaving G.Cur pointing
  beyond the element found}
  var R: RefSetElt;
    begin
    R := G.Cur;
    NextSetElt := false;
    while R <> nil do
      if R↑. Valid then
          begin
          x := R↑. Data;
          NextSetElt := true;
          R := R↑. Thread;
          goto 1
          end
        else R := R↑. Thread;
1:  G.Cur := R;
    end; {NextSetElt}
```

E.3 Differentiator and Simplifier

```
const  Undefined = nil;
```

```
type
    Variable   = char;     {Any letter}
    Constant   = integer;
```

```
Expression = ↑ExprStruct;
KindNames  = (Undfnd, Vbl, Cnst, Cmpst,
                    Any, AnyCnst, SubExpr, ValSubExpr);

ExprStruct =
  record
     CopyCount: integer;
     case Kind: KindNames of
        Vbl:   (v: Variable);
        Cnst:  (c: Constant);
        Cmpst: (Op: Operator;  Opnd1,Opnd2: Expression);
        SubExpr, ValSubExpr: (n: integer)
  end;

Production = record  Patn, Repl: Expression end;
RefListofProduction = ↑listofProduction;
```

{ List of productions to be used by *Simplify*.
  Initialization not shown. }

```
var
    SimpList: RefListofProduction;
```

{ Various utilities (also need routines for *listofProduction*) }

```
function IsCapLet(ch: char): boolean;
  post RESULT ≡ true iff ch is a capital letter;

function IsDigit(ch: char): boolean;
  post RESULT ≡ true iff ch is a digit;

function IsOperator(ch: char): boolean;
  post RESULT ≡ true iff ch is an operator symbol;

procedure ScanPast(var Where: text; ch: char);
  post File Where positioned to first position such that
        Where↑ ≠ ch or eof(Where);
```

{ Routines for type *Expression* }

```
function ConsExpr(Oper: Operator;  Operand1, Operand2: Expression):
                    Expression;
  var E: Expression;
  begin
  if (Operand1 = Undefined) or (Operand2 = Undefined)
```

```
      then  ConsExpr := Undefined
      else begin
        new(E, Cmpst);  ConsExpr := E;  E↑.CopyCount := 0;
        E↑.Op := Oper;
        E↑.Opnd1 := Operand1;  E↑.Opnd2 := Operand2;
        end
   end;  { ConsExpr }
function Copy(E: Expression): Expression;
   begin
   Copy := E;
   if E ≠ Undefined then E↑.CopyCount := E↑.CopyCount + 1
   end;  { Copy }
procedure Release(E: Expression);
   begin
   if E ≠ Undefined then
      if E↑.CopyCount = 0 then
         begin
         if E↑.Kind = Cmpst then
            begin  Release(E↑.Opnd1);  Release(E↑.Opnd2)  end;
            dispose(E)
         end
      else  E↑.CopyCount := E↑.CopyCount − 1;
   end;  { Release }
function VblToExpr(V: Variable): Expression;
  var E: Expression;
   begin
   new(E, Vbl);  VblToExpr := E;  E↑.CopyCount := 0;
   E↑.v := V
   end;  { VblToExpr }
function CnstToExpr(C: Constant): Expression;
  var E: Expression;
   begin
   new(E, Cnst);  CnstToExpr := E;  E↑.CopyCount := 0;
   E↑.c := C
   end;  { CnstToExpr }


{ Applications of simple Expressions }

procedure ReadExpn(var Where: text; var E: Expression);
  var ch: char;  lhs, rhs: Expression;  n: Constant;
   begin
   E := Undefined;
   ScanPast(Where,' ');
   if IsOperator(Where↑) then
      begin
      read(Where, ch);
```

```
          ReadExpn(Where, lhs);  ReadExpn(Where, rhs);
          E := ConsExpr(ch, lhs, rhs)
        end
      else if IsDigit(Where↑) then
        begin  read(Where, n);  E := CnstToExpr(n)  end
      else if IsCapLet(Where↑) then
        begin  read(Where, ch);  E := VblToExpr(ch)  end
      end;  { ReadExpn }


procedure PrintExpn(var Where: text; S: Expression);
  post S in parenthesized infix form is appended to file Where;
    begin
    if Kind(S) = Undfnd then write(Where,'undefined') else
    case Kind(S) of
        Vbl:    write(Where, ExprToVbl(S));
        Cnst:   write(Where, ExprToCnst(S): 3);
        Cmpst:
          begin
          write(Where,'(');
          PrintExpn(Where, Operand1(S));
          write(Where, Oper(S));
          PrintExpn(Where, Operand2(S));
          write(Where,')');
          end;
        end;
      end;  { PrintExpn }


function Diff(S: Expression; V: Variable): Expression;
begin
if Kind(S) = Undfnd then Diff := Undefined
  else case Kind(S) of
    Vbl:
      if V = ExprToVbl(S)
        then Diff := CnstToExpr(1)
        else Diff := CnstToExpr(0);
    Cnst:
        Diff := CnstToExpr(0);
    Cmpst:
      case Oper(S) of
          '+','-':
            Diff := ConsExpr(Oper(S),
                             Diff(Operand1(S), V),
                             Diff(Operand2(S), V));
          '*':
            Diff := ConsExpr('+',
                             ConsExpr('*', Copy(Operand1(S)),
                                 Diff(Operand2(S), V)),
```

```
                              ConsExpr('*',Copy(Operand2(S)),
                                  Diff(Operand1(S), V)));
        '↑':
        if Kind(Operand2(S)) ≠ Cnst
          then Diff := Undefined
          else Diff :=
              ConsExpr('*', Copy(Operand2(S)),
                ConsExpr('*',
                  ConsExpr('↑', Copy(Operand1(S)),
                    CnstToExpr(ExprToCnst(Operand2(S)) − 1)),
                  Diff(Operand1(S), V)))
      end
    end
end;  { Diff }
```

{ Utilities for pattern matching and replacement }

```
function Exponentiate(a, b: Constant): Constant;
  pre a to the b must be integral;
  post RESULT = a to the b;
{ Body left to the reader. }
```

```
function Eval(E: Expression): Constant;
  pre E is a Cmpst whose operands are constants;
  post RESULT = value of E;
  begin
  case Oper(E) of
    '+':
    Eval := ExprToCnst(Operand1(E))
          + ExprToCnst(Operand2(E));
    '−':
    Eval := ExprToCnst(Operand1(E))
          − ExprToCnst(Operand2(E));
    '*':
    Eval := ExprToCnst(Operand1(E))
          * ExprToCnst(Operand2(E));
    '↑':
    Eval := Exponentiate(ExprToCnst(Operand1(E)),
          ExprToCnst(Operand2(E)))
    end
  end;  { Eval }
```

```
function ExtractExpr(E: Expression; n: integer): Expression;
  post RESULT = subexpression of E indexed by n,
              or Undefined if none;
  var parent: Expression;
```

```
begin
ExtractExpr := Undefined;
if n = 1 then ExtractExpr := E
else begin
    parent := ExtractExpr(E, n div 2);
    if parent ≠ Undefined then
      if Kind(parent) = Cmpst then
        if odd(n)
            then ExtractExpr := Operand2(parent)
            else ExtractExpr := Operand1(parent)
    end
end; { ExtractExpr }


{ Primitives on patterns }

function Match(E, P: Expression): boolean;
begin
if E = Undefined then Match := false
else case Kind(P) of
    Vbl, SubExpr, ValSubExpr:
            Match := false;
    Cnst:    Match := (E↑.c = P↑.c);
    Any:     Match := true;
    AnyCnst: Match := (Kind(E) = Cnst);
    Cmpst:   if Kind(E) ≠ Cmpst
                then Match := false
                else Match := (Oper(E) = Oper(P))
                            and Match(Operand1(E), Operand1(P))
                            and Match(Operand2(E), Operand2(P))
  end;
end; { Match }

function Replace(E, R: Expression): Expression;
begin
case Kind(R) of
    Cnst:   Replace := Copy(R);
    Cmpst:  Replace :=
              ConsExpr(Oper(R),
                        Replace(E, Operand1(R)),
                        Replace(E, Operand2(R)));
    ValSubExpr:
        Replace :=
              CnstToExpr(Eval(ExtractExpr(E, R↑.n)));
    SubExpr:
        Replace := Copy(ExtractExpr(E, R↑.n))
  end
```

end; { Replace }

{ Application of pattern routines }

```
function ApplyProductions(E: Expression; L: RefListofProduction):
                          Expression;
  pre  E contains no &A, &C, &S, or &V subexpressions.  L is a list
       of valid productions (that is, valid patterns
       and valid replacements);
  post RESULT = result of applying ordered productions L to E,
                iteratively in LRN order until no further
                productions apply;
  var Changed: boolean;  NewExpr, PrevExpr: Expression;
    RemainingProds: RefListofProduction;  P: Production;
  begin
  NewExpr := Copy(E);
  repeat
      Changed := false;
      if Kind(NewExpr) = Cmpst
        then begin
          PrevExpr := NewExpr;
          NewExpr := ConsExpr(Oper(NewExpr),
                      ApplyProductions(Operand1(NewExpr), L),
                      ApplyProductions(Operand2(NewExpr), L));
          Release(PrevExpr)
        end;
      RemainingProds := L;
      while not IsEmpty(RemainingProds) do
        begin
        FirstItem(RemainingProds, P);
        RemainingProds := Tail(RemainingProds);
        if Match(NewExpr, P.Patn) then
            begin
            PrevExpr := NewExpr;
            NewExpr := Replace(NewExpr, P.Repl);
            Release(PrevExpr);
            Changed := true;  RemainingProds := L
            end
        end
    until not Changed;
    ApplyProductions := NewExpr
  end; { ApplyProductions }

function Simplify(E: Expression): Expression;
  pre  E must contain no &A, &C, &S, or &V subexpressions;
  post RESULT = simplification of E;
  begin  Simplify := ApplyProductions(E, SimpList)  end;
```

INDEX

α (array assignment), 185
β (reference operator), 208, 239
$\sim$ (sequence concatenation), 192
$\uparrow$ (dereference operator), 208